Alberto Guzman

Continuous Functions of Vector Variables

Birkhäuser
Boston • Basel • Berlin

Alberto Guzman
The City College of New York, CUNY
Department of Mathematics
New York, NY 10031
U.S.A.

Library of Congress Cataloging-in-Publication Data

Guzman, Alberto, 1947-
Continuous functions of vector variables / Alberto Guzman.
p. cm.
Includes bibliographical references and index.
ISBN 0-8176-4273-0 (acid-free paper) — ISBN 3-7643-4273-0 (acid-free paper)
1. Functions, Continuous. 2. Normed linear spaces. 3. Vector spaces. I. Title.

QA331.5.G89 2002
515'.73–dc21

2002018232
CIP

AMS Subject Classifications: 26-01

Printed on acid-free paper

Birkhäuser

ISBN 0-8176-4273-0 SPIN 10851398
ISBN 3-7643-4273-0

Typeset by TEXniques, Inc., Cambridge, MA.
Printed and bound by Hamilton Printing Company, Rensselaer, NY.
Printed in the United States of America.

9 8 7 6 5 4 3 2 1

A member of BertelsmannSpringer Science+Business Media GmbH

To "Berto grande" and Ramona,
whose work and sacrifice,
made my life possible;

and to Susan and Harold,
whose love has made my life complete

Contents

Preface

This text is appropriate for a one-semester course in what is usually called advanced calculus of several variables. The focus is on expanding the concept of continuity; specifically, we establish theorems related to extreme and intermediate values, generalizing the important results regarding continuous functions of one real variable.

We begin by considering the function $f(x, y, \ldots)$ of multiple variables as a function of the single vector variable $(x, y, \ldots)$. It turns out that most of the treatment does not need to be limited to the finite-dimensional spaces $\mathbf{R}^n$, so we will often place ourselves in an arbitrary vector space equipped with the right tools of measurement. We then proceed much as one does with functions on $\mathbf{R}$. First we give an algebraic and metric structure to the set of vectors. We then define limits, leading to the concept of continuity and to properties of continuous functions. Finally, we enlarge upon some topological concepts that surface along the way.

A thorough understanding of single-variable calculus is a fundamental requirement. The student should be familiar with the axioms of the real number system and be able to use them to develop elementary calculus, that is, to define *continuous function, derivative*, and *integral*, and to prove their most important elementary properties. Familiarity with these properties is a must. To help the reader, we provide references for the needed theorems.

For our material, a course in linear algebra is also prerequisite. The student needs to be familiar with elementary principles of vector spaces, linear maps, and matrices, for which we also give references. For vector spaces, the list of areas in which this knowledge is essential includes spaces and subspaces; linear combinations, linear span, spanning; linear independence; basis and dimension. From the study of linear maps, only the definition and the most basic properties are needed. For matrices, their use to represent linear maps should be understood, and it is helpful to understand their use (via row reduction, determinants, and the like) to solve systems of linear equations.

For a course that is shorter than a full semester, the most natural omission is the last section in each chapter. In these five sections the material is closer to functional analysis than to calculus. Moreover, the sections form a tight sequence, but nothing in the remainder of the book depends on them.

I wish to express my appreciation to the many people who helped bring the book to life. I must single out the three who had to endure the most. David Kramer was the copyeditor, and added clarity and better flow to almost every paragraph. Elizabeth Loew managed the production of the project turning a series of lecture notes into printable material, particularly the faithful rendering of figures. Finally, I will always be grateful to Ann Kostant, Executive Editor at Birkhäuser Boston, for finding merit in a quirky book by an unknown author.

Alberto Guzman
April, 2002

1
Euclidean Space

1.1 Multiple Variables

We start by defining the variables we will use.

Definition. If $x_1, \ldots, x_n$ are (not necessarily distinct) real numbers, we call the expression $(x_1, \ldots, x_n)$ **an (ordered) n-tuple of real numbers**. We call $x_1, \ldots, x_n$ the **coordinates** of the n-tuple. The set of all n-tuples of real numbers is denoted by $\mathbf{R}^n$.

An n-tuple is a list of names of real numbers, and the order in which the names are listed is significant. Therefore, we could define **n-tuple** the same way we define **finite sequence**, as a mapping from a segment $\{1, 2, \ldots, n\}$ to $\mathbf{R}$. With sequences, though, we are typically interested in looking for a pattern from term to term, especially with regard to behavior as n increases. Here we look at $(x_1, \ldots, x_n)$ as a whole, with no focus on the separate coordinates.

When $n = 1$, an n-tuple is a disguised form of a real number. Hence we will make no distinction between members of $\mathbf{R}$ and $\mathbf{R}^1$.

When $n = 2$, an n-tuple is called an **ordered pair**. We have such a strong association between ordered pairs and our picture of the Cartesian plane that we will almost invariably use (x, y) to denote an ordered pair rather than (x_1, x_2), and similarly with $n = 3$, **ordered triple**, Cartesian space, and coordinates (x, y, z). Almost everything we write can be easily understood in terms of pictures in the plane or in 3-space.

Example 1. A subset of $\mathbf{R}^2$ is variously called "a relation on $\mathbf{R}$" or "the graph of a relation on $\mathbf{R}$." The corresponding set of points in the Cartesian plane is always "the graph of the relation." If the relation is given by an equation in x and y, the set of pairs or of points is the "graph of the equation." Finally, if the equation is equivalent to $y = f(x)$ for some function f, then we speak of the "graph of the function."

(a) For the relation ("greater than") described by $x > y$, the graph consists of those points in the plane whose x-coordinate exceeds the y-coordinate. We picture the graph in Figure 1.1; the line $y = x$ is shown dashed to indicate that it is not part of the graph.

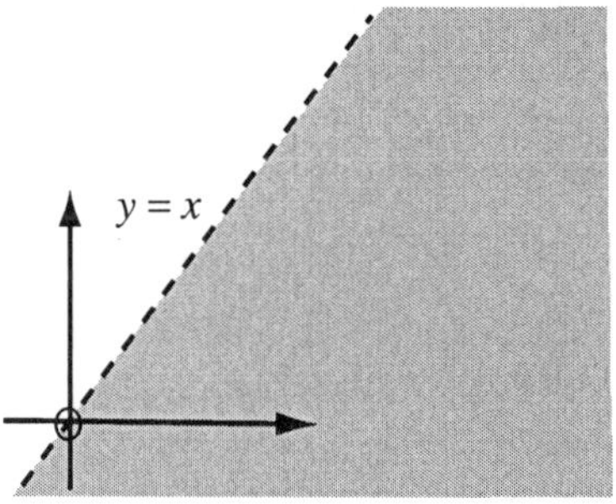

Figure 1.1.

(b) The equation $ax + by + c = 0$ always has a line for its graph, provided that the constants a and b are not both zero. If $b = 0$, then we have the vertical line described by $x = -c/a$. If $b \neq 0$, then we may solve for y:

$$y = \frac{-a}{b}x + \frac{-c}{b},$$

giving us the familiar slope–intercept form of the equation of a line that has a slope. In this case, we may also refer to the line as the graph of the function

$$f(x) := -\frac{a}{b}x - \frac{c}{b}.$$

For $n \geq 4$ we will use not "quadruple," "quintuple," and so on, but the generic "n-tuple." At that level, our ability to visualize becomes limited, but not eliminated. For example, any three members of $\mathbf{R}^n$ can be pictured as three points in a plane, and relations among them explored accordingly.

Exercises

Throughout the exercises, an asterisk (*) indicates an unusually difficult exercise or part.

1. (a) Sketch the graph of the equation $x^2 - y^2 = 1$.
   (b) Is there a function g such that $x^2 - y^2 = 1$ iff $y = g(x)$?

2. (a) Geometrically, given a set of points in the plane, under what condition is it the graph of some function?

(b) What is the corresponding algebraic condition for a subset of $\mathbf{R}^2$ to be the graph of some function?

3. In $\mathbf{R}^3$, the set of triples (x, y, z) that satisfy

$$x + 2y + 3z = 4$$

is called the **solution set** of the equation, and the points with those coordinates make up the **graph of the equation**.

(a) Sketch the graph of the equation.

(b) Knowing what the graph in (a) looks like, what do you think is the shape of the graph of the solution set (the set of simultaneous solutions) of the system

$$\begin{aligned} x + 2y + 3z &= 4, \\ 5x + 6y + 7z &= 8? \end{aligned}$$

(c) Solve the system in (b), and then sketch the graph of the solution set.

4. In $\mathbf{R}^2$, what is the solution set of the equation $ax + by = c$? Consider all possible combinations of values of the real numbers a, b, and c.

5. In $\mathbf{R}^3$, considering the possible values of the coefficients, what are the possible solution sets of:

(a) The equation $ax + by + cz = d$?

(b) The system $ax + by + cz = d$, $ex + fy + gz = h$?

1.2 Points and Lines in a Vector Space

In $\mathbf{R}^n$ the usual operations are **addition** and **scalar multiplication**, described by

$$\begin{aligned} (x_1, \dots, x_n) + (y_1, \dots, y_n) &:= (x_1 + y_1, \dots, x_n + y_n), \\ a(x_1, \dots, x_n) &:= (\alpha x_1, \alpha x_2, \dots, \alpha x_n). \end{aligned}$$

The algebraic structure in $\mathbf{R}^n$ can be reduced to ten fundamental properties of these two operations:

(a) closure	if $\mathbf{x}$ and $\mathbf{y}$ are in $\mathbf{R}^n$, then so is $\mathbf{x} + \mathbf{y}$;
(b) associativity	if $\mathbf{x}, \mathbf{y}, \mathbf{z}$ are in $\mathbf{R}^n$, then $\mathbf{x} + (\mathbf{y} + \mathbf{z}) = (\mathbf{x} + \mathbf{y}) + \mathbf{z}$;
(c) commutativity	if $\mathbf{x}$ and $\mathbf{y}$ are in $\mathbf{R}^n$, then $\mathbf{x} + \mathbf{y} = \mathbf{y} + \mathbf{x}$;
(d) identity	there is an element $\mathbf{O}$ in $\mathbf{R}^n$ such that $\mathbf{x} + \mathbf{O} = \mathbf{O} + \mathbf{x} = \mathbf{x}$ for every $\mathbf{x}$ in $\mathbf{R}^n$;
(e) inverse	given $\mathbf{x}$ in $\mathbf{R}^n$, there exists $-\mathbf{x}$ in $\mathbf{R}^n$ with $\mathbf{x} + (-\mathbf{x}) = -\mathbf{x} + \mathbf{x} = \mathbf{O}$;
(f) closure	if α is real and $\mathbf{x}$ is in $\mathbf{R}^n$, then $\alpha\mathbf{x}$ is in $\mathbf{R}^n$;

(g) associativity	if α, β are real and $\mathbf{x}$ is in $\mathbf{R}^n$, then $\alpha(\beta\mathbf{x}) = (\alpha\beta)\mathbf{x}$;
(h) distributivity	if α, β are real and $\mathbf{x}$ is in $\mathbf{R}^n$, then $(\alpha + \beta)\mathbf{x} = \alpha\mathbf{x} + \beta\mathbf{x}$;
(i) distributivity	if α is real and $\mathbf{x}, \mathbf{y}$ are in $\mathbf{R}^n$, then $\alpha(\mathbf{x} + \mathbf{y}) = \alpha\mathbf{x} + \alpha\mathbf{y}$;
(j) unity	if 1 is the unity in $\mathbf{R}$ and $\mathbf{x}$ is in $\mathbf{R}^n$, then $1\mathbf{x} = \mathbf{x}$.

Statements (a)–(j) are called the vector space axioms. The combination of a set and two operations satisfying these axioms is called a **vector space** or **linear space**. From these statements we can prove fairly directly many other properties of the operations; for example, $\alpha\mathbf{x} = \mathbf{O}$ iff either $\alpha = 0$ or $\mathbf{x} = \mathbf{O}$. [See Lay, Section 4.1, for material in this section.] We will refer to this class of results by the collective name "algebra of vector spaces."

Since $\mathbf{R}^n$ is a vector space, we henceforth call its members **vectors**. However, $\mathbf{R}^n$ is also a model of Euclidean geometry, a connection that allows us to use geometric intuition to understand analysis. For that reason, we also refer to members of $\mathbf{R}^n$ as **points**.

To begin along the geometric path, we define certain sets associated with subspaces of $\mathbf{R}^n$ that have familiar geometric attributes.

Let S be a subspace and $\mathbf{b}$ a member of $\mathbf{R}^n$. The set $\mathbf{b} + S := \{\mathbf{b} + \mathbf{x} : \mathbf{x} \in S\}$ is called a **translate** of S. The most important translates for us are those of one-dimensional subspaces. The set $\langle\langle\mathbf{c}\rangle\rangle := \{\alpha\mathbf{c} \colon \alpha \in \mathbf{R}\}$ of multiples of the nonzero vector $\mathbf{c}$ is a one-dimensional subspace of $\mathbf{R}^n$. We always think of $\langle\langle\mathbf{c}\rangle\rangle$ as a line through the origin, and we will call it the **line of c**. A translate $\mathbf{b} + \langle\langle\mathbf{c}\rangle\rangle$ is a **line through b**.

Example 1. Consider $\mathbf{c} := (3, 4)$ in $\mathbf{R}^2$.

(a) The elements of $\langle\langle\mathbf{c}\rangle\rangle$ have the form $\alpha(3, 4) = (3\alpha, 4\alpha)$. Because the x-coordinate 3 is not 0, the multiples all satisfy

$$y = 4\alpha = 4\frac{x}{3} = \frac{4}{3}x,$$

which we recognize as the equation of a line through the origin with slope determined by $\mathbf{c}$.

(b) Given a second point $\mathbf{b} := (5, 6)$, the translate $\mathbf{b} + \langle\langle\mathbf{c}\rangle\rangle = \{\mathbf{b} + \alpha\mathbf{c} \colon \alpha \text{ real}\}$ consists of vectors of the form

$$(5, 6) + \alpha(3, 4) = (5 + 3\alpha, 6 + 4\alpha).$$

These all satisfy the equation

$$4(x - 5) = 3(y - 6), \qquad \text{(Justify!)}$$

which we know represents a line through (5, 6) with direction specified by $\mathbf{c}$.

More commonly, we determine a line by giving two of its points. Suppose $\mathbf{a}$ and $\mathbf{b}$ are unequal. The difference $\mathbf{b} - \mathbf{a}$ specifies the direction of the line and either $\mathbf{a}$ or $\mathbf{b}$ determines the location of the line connecting the two points. In Figure 1.2, any vector $\mathbf{a} + \alpha(\mathbf{b} - \mathbf{a})$ is above $\mathbf{b}$ along the line through $\mathbf{a}$ and $\mathbf{b}$ if $\alpha > 1$, is between $\mathbf{a}$ and $\mathbf{b}$ (inclusive) if $0 \leq \alpha \leq 1$, and is below $\mathbf{a}$ on the line if $\alpha < 0$. Therefore, we call $\mathbf{a} + \langle\!\langle \mathbf{b} - \mathbf{a} \rangle\!\rangle$ **the line determined by** (synonyms: **joining, connecting**) $\mathbf{a}$ and $\mathbf{b}$.

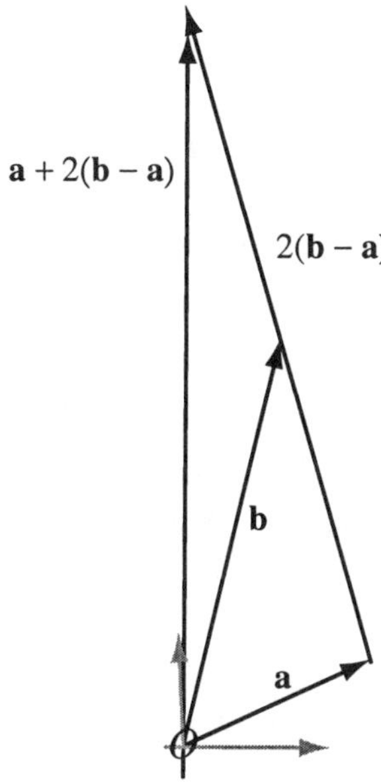

Figure 1.2.

Example 2. In $\mathbf{R}^2$, $\mathbf{a} := (1, 2)$ and $\mathbf{b} := (8, 5)$ are distinct. The line joining them consists of vectors of the form

$$(x, y) = \mathbf{a} + \alpha(\mathbf{b} - \mathbf{a}) = (1 + 7\alpha, 2 + 3\alpha).$$

These are precisely the pairs determined by the following familiar forms:

(a) **Point–Slope**: Since $8 - 1 \neq 0$, we have

$$y - 2 = 3\alpha = \left(\frac{3}{7}\right) 7\alpha = \frac{3}{7}(x - 1).$$

(b) **Parametric**: Clearly, we have

$$\begin{aligned} x &= 1 + 7t, \\ y &= 2 + 3t, \qquad \text{as } t \text{ varies over } \mathbf{R}. \end{aligned}$$

(c) **General Linear**: We can see that $3(x - 1) = 7(y - 2)$; since both sides of this equation equal 21α. Therefore,

$$3x + (-7)y + 11 = 0.$$

The line is of the form by $ax + by + c = 0$, with not both a and b zero.

A vector on the line joining **a** and **b** is of the form

$$\mathbf{a} + \alpha(\mathbf{b} - \mathbf{a}) = (1 - \alpha)\mathbf{a} + \alpha\mathbf{b}.$$

We conclude that the line joining **a** and **b** consists of exactly those linear combinations $\beta\mathbf{a} + \alpha\mathbf{b}$ for which $\alpha + \beta = 1$.

Now consider in Figure 1.2 the set of points $\mathbf{a} + \alpha(\mathbf{b} - \mathbf{a})$ for which $0 \le \alpha \le 1$. We observed that these are between **a** and **b**. We refer to this set as the **line segment from a to b**. Notice that any such point is of the form

$$\mathbf{a} + \alpha(\mathbf{b} - \mathbf{a}) = (1 - \alpha)\mathbf{a} + \alpha\mathbf{b} = \beta\mathbf{a} + \alpha\mathbf{b}$$

with $\alpha + \beta = 1$ and each of α and β between 0 and 1.

Example 3. Let $\mathbf{a} := (1, 2, 6)$, $\mathbf{b} := (4, 10, -5)$ in $\mathbf{R}^3$. The line joining them consists of the linear combinations

$$(1 - \alpha)(1, 2, 6) + \alpha(4, 10, -5) = (1 + 3\alpha, 2 + 8\alpha, 6 - 11\alpha).$$

Thus $(31, 82, -104)$, corresponding to $\alpha = 10$, is on the line.

The easiest way to represent the line is parametrically:

$$x = 1 + 3t, \quad y = 2 + 8t, \quad z = 6 - 11t.$$

The other standard way is to use the paired equations

$$\frac{x - 1}{3} = \frac{y - 2}{8} = \frac{z - 6}{-11}.$$

Notice that $1\mathbf{a} + 0\mathbf{b}$ and $0\mathbf{a} + 1\mathbf{b}$ are the endpoints of the line segment. As α increases from 0 to 1, the combination $(1 - \alpha)\mathbf{a} + \alpha\mathbf{b}$ travels from **a** to **b**. Thus,

$$\mathbf{c} := \frac{2}{3}\mathbf{a} + \frac{1}{3}\mathbf{b} = \left(2, \frac{14}{3}, \frac{7}{3}\right)$$

is one-third of the way from **a** to **b**. In particular, we have the **midpoint formula**: If $\mathbf{c} = (x_1, \ldots, x_n)$ and $\mathbf{d} = (y_1, \ldots, y_n)$, then the midpoint of the segment joining them is

$$\frac{1}{2}\mathbf{c} + \frac{1}{2}\mathbf{d} = \left(\frac{x_1 + y_1}{2}, \ldots, \frac{x_n + y_n}{2}\right).$$

Exercises

1. Find vectors **b** and **c** in $\mathbf{R}^2$ such that the indicated line is the translate $\mathbf{b} + \langle\langle \mathbf{c} \rangle\rangle$:

   (a) the line through $(3, 5)$ with slope 4.

(b) the line through $(-2,-1)$ and $(4,2)$.

(c) the line with equation $y = mx + b$.

2. Find the equations of the following lines:

(a) $(-2,1) + \langle\langle(2,-4)\rangle\rangle$ (b) $(-2,1) + \langle\langle(4,-2)\rangle\rangle$.

3. (a) Is $(5,6)$ on the line joining $(1,2)$ and $(3,4)$? Is $(1,2)$ on the line joining $(3,4)$ and $(5,6)$?

   (b) Given distinct points $\mathbf{a}$, $\mathbf{b}$, $\mathbf{c}$, show that $\mathbf{c}$ is on the line joining $\mathbf{a}$ and $\mathbf{b}$ iff $\mathbf{a}$ is on the line joining $\mathbf{b}$ and $\mathbf{c}$.

4. Let $\mathbf{a}$, $\mathbf{b}$, and $\mathbf{c}$ be distinct points. Prove that $\mathbf{a}$, $\mathbf{b}$, and $\mathbf{c}$ are collinear (all on one line $\mathbf{d} + \langle\langle\mathbf{e}\rangle\rangle$) iff $\mathbf{b} - \mathbf{a}$ and $\mathbf{c} - \mathbf{a}$ are dependent vectors.

5. Are $(1,2)$, $(3,4)$, and $(6,5)$ dependent? Are they collinear?

6. Are $(1,2,3,4,5)$, $(6,7,8,9,10)$, and $(11,12,13,14,15)$ dependent? Are they collinear?

7. Show that three or more collinear vectors are necessarily dependent, but not conversely.

8. Show that if a and b are not both zero, then the solutions of $ax+by = c$ form a line in $\mathbf{R}^2$; that is, they can be put in the form $(x, y) = (a_1, a_2) + \alpha(b_1, b_2)$. (Hint: The solutions of a linear system can be written as the sum of one solution and the general solution of the associated homogeneous system.)

9. *Show that the expression "the line joining $\mathbf{a}$ and $\mathbf{b}$" is justified by showing that if $\mathbf{c} + \langle\langle\mathbf{d}\rangle\rangle$ is another line that contains both $\mathbf{a}$ and $\mathbf{b}$, then the two lines are the same (set).

1.3 Inner Products and the Geometry of $\mathbf{R}^n$

Our description of lines gives $\mathbf{R}^n$ some elements of geometry. In this section we add some of the geometry of triangles.

Picture two vectors $\mathbf{a}$ and $\mathbf{b}$ as arrows with the same initial point O (see Figure 1.3). If $\mathbf{c}$ goes from the end of $\mathbf{b}$ to the end of $\mathbf{a}$, then $\mathbf{b}+\mathbf{c} = \mathbf{a}$, and so $\mathbf{c} = \mathbf{a} - \mathbf{b}$. We will make this identification in every space; that is, we will always think of $\mathbf{a}$, $\mathbf{b}$, and $\mathbf{a} - \mathbf{b}$ as forming a triangle (albeit possibly one with a side of zero length or a straight angle).

Now make the picture part of $\mathbf{R}^2$, so that we attach Cartesian coordinates to all three vectors. We have

$$\mathbf{a} = (a_1, a_2), \quad \mathbf{b} = (b_1, b_2), \quad \mathbf{c} = \mathbf{a} - \mathbf{b} = (a_1 - b_1, a_2 - b_2).$$

Let a, b, and c represent the lengths of the corresponding vectors, and θ the angle between $\mathbf{a}$ and $\mathbf{b}$. Applying the law of cosines to the triangle, we write

$$c^2 = a^2 + b^2 - 2ab\cos\theta.$$

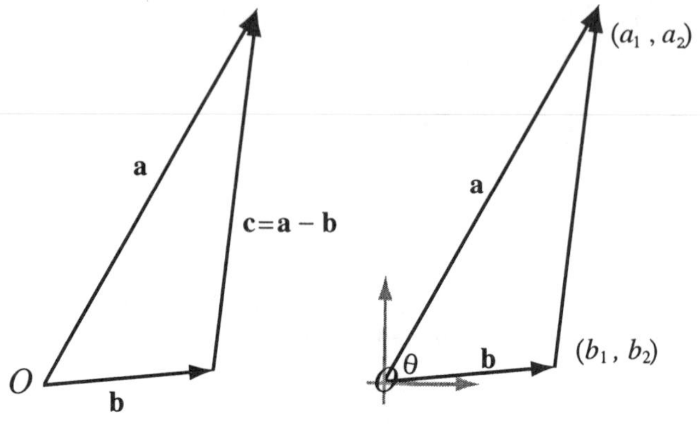

Figure 1.3.

Putting the lengths in terms of coordinates, we substitute $(a_1 - b_1)^2 + (a_2 - b_2)^2 = (a_1{}^2 + a_2{}^2) + (b_1{}^2 + b_2{}^2) - 2ab\cos\theta$. Therefore,

$$\begin{aligned} ab\cos\theta &= \left(\left[a_1{}^2 + a_2{}^2\right] + \left[b_1{}^2 + b_2{}^2\right] - \left[a_1 - b_1\right]^2 - \left[a_2 - b_2\right]^2\right)/2 \\ &= a_1 b_1 + a_2 b_2. \end{aligned}$$

This last relationship is the basis for the geometric ideas we want to export to $\mathbf{R}^n$.

Definition. Given $\mathbf{x} = (x_1, \ldots, x_n)$ and $\mathbf{y} = (y_1, \ldots, y_n)$ in $\mathbf{R}^n$, the real number

$$\mathbf{x} \bullet \mathbf{y} := x_1 y_1 + \cdots + x_n y_n$$

is called the **dot product** (or **scalar product** or **inner product**) of $\mathbf{x}$ and $\mathbf{y}$.

The notation $(\mathbf{x}, \mathbf{y})$ and the name "inner product" are more standard, but we will stick to the "dot" name, because it is familiar, and to the dot notation, because our parentheses are already overworked.

We will call the dot product an "operation on $\mathbf{R}^n$" even though it combines two vectors to produce a nonvector. We previously took a similar liberty with "operation" when we applied it to scalar multiplication, which combines a nonvector with a vector.

Theorem 1.1. *The dot product in* $\mathbf{R}^n$ *has the following properties*:

(a) *Symmetry* $\quad \mathbf{x} \bullet \mathbf{y} = \mathbf{y} \bullet \mathbf{x}.$

(b) *Distributivity* $\quad (\mathbf{x} + \mathbf{y}) \bullet \mathbf{z} = (\mathbf{x} \bullet \mathbf{z}) + (\mathbf{y} \bullet \mathbf{z})$ *and*
$\mathbf{a} \bullet (\mathbf{b} + \mathbf{c}) = (\mathbf{a} \bullet \mathbf{b}) + (\mathbf{a} \bullet \mathbf{c}).$

(c) *Homogeneity* $\quad (\alpha\mathbf{x}) \bullet \mathbf{y} = \alpha(\mathbf{x} \bullet \mathbf{y}) = \mathbf{x} \bullet (\alpha\mathbf{y}).$

(d) *Positive Definiteness* $\quad \mathbf{x} \bullet \mathbf{x} > 0$ *for all* $\mathbf{x}$,
with the lone exception that $\mathbf{O} \bullet \mathbf{O} = 0.$

Proof. We leave the proof to Exercise 7.

An operation that associates a real number to each pair of vectors in a space and for which Theorem 1.1 holds is called an **inner product**. A real vector space that admits such an operation is an **inner product space over the reals**. Just as all the properties of vector spaces follow from the vector space axioms, so the properties of inner product spaces follow from Theorem 1.1. [Consult Lay, Section 6.1.]

The most important of those properties is the Cauchy–Schwarz inequality, which says that $\mathbf{a} \bullet \mathbf{b}$ is no greater in absolute value than the product ab of the two lengths. It is possible to give an entirely algebraic proof of the property. We, however, want to look at it geometrically. After all, we want to generalize $\mathbf{a} \bullet \mathbf{b} = ab \cos\theta$. Saying that $|\mathbf{a} \bullet \mathbf{b}|$ is no more than ab amounts to insisting that $|\cos\theta| \leq 1$. In terms of trigonometry, the last inequality says that in a right triangle, the side facing the right angle is at least as long as each of the other sides. We therefore look at properties of right angles.

Definition.

(a) We call $(\mathbf{a} \bullet \mathbf{a})^{1/2}$ the **length of a**, and denote it by $\|\mathbf{a}\|_2$.

(b) We say that **a** and **b** are **orthogonal** (or **perpendicular** or at **right angles**) if $\mathbf{a} \bullet \mathbf{b} = 0$.

(c) We call the multiple $([\mathbf{a} \bullet \mathbf{b}]/[\mathbf{b} \bullet \mathbf{b}])\mathbf{b}$ the **(vector) projection of a onto (the line of) b**, and denote it by $\mathbf{proj}(\mathbf{a}, \mathbf{b})$. If $\mathbf{b} = \mathbf{O}$, then we set $\mathbf{proj}(\mathbf{a}, \mathbf{b})$ equal to $\mathbf{O}$.

Example 1. Let $\mathbf{a} := (6, 8)$, $\mathbf{b} := (2, 1)$ in $\mathbf{R}^2$ (see Figure 1.4).

From

$$\|\mathbf{a}\|_2 := (\mathbf{a} \bullet \mathbf{a})^{1/2} = \left(6^2 + 8^2\right)^{1/2}$$

we see that our definition of length generalizes the formula for distance from the origin to a vector, which is in turn a manifestation of the Pythagorean theorem.

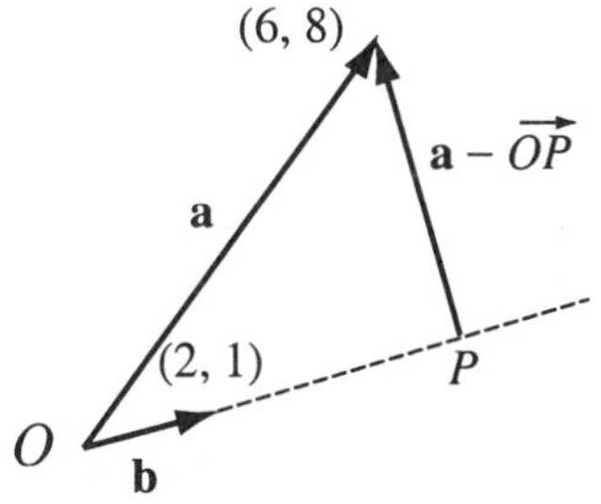

Figure 1.4.

Draw the line through **a** perpendicular to the line of **b**, as in Figure 1.4. The line of **b** has equation $y = x/2$; the perpendicular has equation $y - 8 = -2(x - 6)$.

They intersect at P, which has coordinates $(8, 4)$. Vector $\overrightarrow{OP}$ is what we think of as "the (vector) component of **a** in the direction of **b**." "Vector projection" has the same meaning:

$$\mathbf{proj}(\mathbf{a}, \mathbf{b}) := ([\mathbf{a} \bullet \mathbf{b}]/[\mathbf{b} \bullet \mathbf{b}])\mathbf{b} = (20/5)\mathbf{b} = 4(2, 1) = (8, 4).$$

Because P is the foot of the perpendicular from **a** to the line of **b**, we expect two geometric consequences. First, $\mathbf{a} - \overrightarrow{OP}$ should be orthogonal to **OP**. We find that

$$\overrightarrow{OP} \bullet (\mathbf{a} - \mathbf{OP}) = (8, 4) \bullet (-2, 4) = 0.$$

Thus in $\mathbf{R}^2$, our definition of orthogonality generalizes the principle that slopes of perpendicular lines are negative reciprocals. Second, P should be closer to **a** than any other point on the line of **b**. We codify these in the next theorem.

Theorem 1.2. *In any inner product space the following statements hold*:

(a) *For every* **x** *and* **y**, $\mathbf{x} - \mathbf{proj}(\mathbf{x}, \mathbf{y})$ *is orthogonal to* **y** *(and therefore to* $\mathbf{proj}(\mathbf{x}, \mathbf{y})$*).*

(b) (The Pythagorean Theorem) *In a right triangle, the square of the (length of the) hypotenuse equals the sum of the squares of the other two sides. In symbols, if* **x** *is orthogonal to* **y**, *then (see Figure 1.5)*

$$\|\mathbf{x} - \mathbf{y}\|_2^2 = \|\mathbf{x}\|_2^2 + \|\mathbf{y}\|_2^2.$$

(c) *The shortest length from* **x** *to a multiple of* **y** *is from* **x** *to* $\mathbf{proj}(\mathbf{x}, \mathbf{y})$.

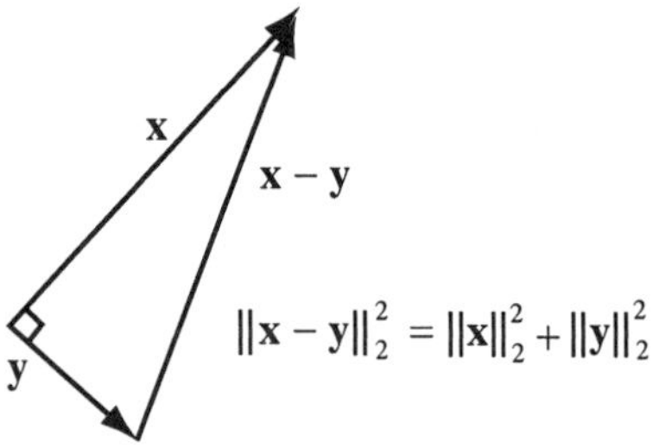

Figure 1.5.

The proofs are left as Exercise 8. However, more important than the proofs is the geometric interpretation. Example 1 illustrates (a). It also states (c), whose illustration should be done in connection with Exercise 1 as well as 8. Figure 1.5 shows (b). There, **x** and **y** are orthogonal. Therefore, the triangle is a right triangle, with $\mathbf{x} - \mathbf{y}$ opposite the right angle. Theorem 1.2(b) says that the Pythagorean theorem applies to a right triangle *in any inner product space.*

Theorem 1.3. (Cauchy–Schwarz Inequality) *For vectors* **x** *and* **y** *in an inner product space*:

(a) $|\mathbf{x} \bullet \mathbf{y}| \leq \|\mathbf{x}\|_2 \|\mathbf{y}\|_2$.

(b) *Equality holds iff one of* $\mathbf{x}$ *and* $\mathbf{y}$ *is a multiple of the other.*

Proof. If $\mathbf{y} = \mathbf{O}$, then the left side is zero by Exercise 3a, the right side is also zero by positive definiteness, and $\mathbf{y} = 0\mathbf{x}$; this takes care of both (a) and (b).

Assume instead that $\mathbf{y}$ is not $\mathbf{O}$. We always have

$$\mathbf{x} = [\mathbf{x} - \mathbf{proj}(\mathbf{x}, \mathbf{y})] + \mathbf{proj}(\mathbf{x}, \mathbf{y}).$$

By Theorem 1.2(a), the summands on the right are orthogonal. By (the Pythagorean) Theorem 1.2(b),

$$\|\mathbf{x}\|_2^2 = \|\mathbf{x} - \mathbf{proj}(\mathbf{x}, \mathbf{y})\|_2^2 + \|\mathbf{proj}(\mathbf{x}, \mathbf{y})\|_2^2.$$

Therefore,

$$\|\mathbf{x}\|_2^2 \geq \|\mathbf{proj}(\mathbf{x}, \mathbf{y})\|_2^2 = \left(\frac{\mathbf{x} \bullet \mathbf{y}}{\mathbf{y} \bullet \mathbf{y}}\mathbf{y}\right) \bullet \left(\frac{\mathbf{x} \bullet \mathbf{y}}{\mathbf{y} \bullet \mathbf{y}}\mathbf{y}\right) = (\mathbf{x} \bullet \mathbf{y})^2 / \|\mathbf{y}\|_2^2.$$

(Explain all three!) Consequently, $\|\mathbf{x}\|_2^2 \|\mathbf{y}\|_2^2 \geq (\mathbf{x} \bullet \mathbf{y})^2$, and the inequality (a) follows.

Indeed, the inequality is strict, unless $\|\mathbf{x} - \mathbf{proj}(\mathbf{x}, \mathbf{y})\|_2^2 = 0$, that is, unless $\mathbf{x} = \mathbf{proj}(\mathbf{x}, \mathbf{y})$. What will make $\mathbf{x}$ match its projection? If $\mathbf{x}$ equals its projection, then $\mathbf{x} = ([\mathbf{x} \bullet \mathbf{y}]/[\mathbf{y} \bullet \mathbf{y}])\, \mathbf{y}$ is a multiple of $\mathbf{y}$. Conversely, if $\mathbf{x}$ is a multiple $\mathbf{x} = \alpha\mathbf{y}$ of $\mathbf{y}$, then

$$\mathbf{proj}(\mathbf{x}, \mathbf{y}) = ([\alpha\mathbf{y} \bullet \mathbf{y}]/[\mathbf{y} \bullet \mathbf{y}])\, \mathbf{y} = \alpha([\mathbf{y} \bullet \mathbf{y}]/[\mathbf{y} \bullet \mathbf{y}])\, \mathbf{y} = \mathbf{x}.$$

We conclude that the inequality is strict, except if $\mathbf{x}$ is a multiple of $\mathbf{y}$. □

Cauchy's inequality (as we will call it) allows us to define angle measure in $\mathbf{R}^n$, or any other inner product space. If $\mathbf{x}$ and $\mathbf{y}$ are nonzero, then

$$-1 \leq \frac{\mathbf{x} \bullet \mathbf{y}}{\|\mathbf{x}\|_2 \|\mathbf{y}\|_2} \leq 1.$$

Hence the number $(\mathbf{x} \bullet \mathbf{y})/(\|\mathbf{x}\|_2 \|\mathbf{y}\|_2)$ is in the domain of the inverse cosine function.

Definition. If $\mathbf{x}$ and $\mathbf{y}$ are nonzero vectors in $\mathbf{R}^n$, the **angle between x and y** means $\cos^{-1}(\mathbf{x} \bullet \mathbf{y} / \|\mathbf{x}\|_2 \|\mathbf{y}\|_2)$. If either vector is $\mathbf{O}$, we arbitrarily set the angle to be $\pi/2$.

We will write as a theorem a string of results that summarize angle information.

Theorem 1.4. *In an inner product space, let* θ *be the angle between vectors* $\mathbf{x}$ *and* $\mathbf{y}$*. Then:*

(a) $\mathbf{x} \bullet \mathbf{y} = \|\mathbf{x}\|_2 \|\mathbf{y}\|_2 \cos\theta$.

(b) $\theta = \pi/2$ *iff* $\mathbf{x}$ *and* $\mathbf{y}$ *are orthogonal.*

(c) (Law of Cosines) $\|\mathbf{x} - \mathbf{y}\|_2^2 = \|\mathbf{x}\|_2^2 + \|\mathbf{y}\|_2^2 - 2\|\mathbf{x}\|_2 \|\mathbf{y}\|_2 \cos\theta$.

(d) $\theta = 0$ *iff* $\mathbf{x} \neq \mathbf{O}$ *and* $\mathbf{y}$ *is a positive multiple of* $\mathbf{x}$.

(e) $\theta = \pi$ *iff* $\mathbf{x} \neq \mathbf{O}$ *and* $\mathbf{y}$ *is a negative multiple of* $\mathbf{x}$.

Proof. (a) and (b) are immediate from the definitions.

(c) Exercise 9.

(d) If $\mathbf{x} \neq \mathbf{O}$ and $\mathbf{y} = \alpha\mathbf{x}$ with $\alpha > 0$, then

$$\cos\theta := (\mathbf{x} \bullet \alpha\mathbf{x})/(\|\mathbf{x}\|_2 \|\alpha\mathbf{x}\|_2) = \alpha(\mathbf{x} \bullet \mathbf{x})/(\alpha\|\mathbf{x}\|_2^2) = 1,$$

and $\theta = 0$.

Conversely, if $\theta = 0$, then by definition neither $\mathbf{x}$ nor $\mathbf{y}$ can be zero. Also, by part (a), $\mathbf{x} \bullet \mathbf{y} = \|\mathbf{x}\|_2 \|\mathbf{y}\|_2$. By the second provision of Cauchy's inequality, one of $\mathbf{x}$ and $\mathbf{y}$ is a multiple of the other. Since both are nonzero, write $\mathbf{y} = \alpha\mathbf{x}$. From

$$\alpha = \frac{\alpha(\mathbf{x} \bullet \mathbf{x})}{(\mathbf{x} \bullet \mathbf{x})} = \frac{\mathbf{x} \bullet \mathbf{y}}{\|\mathbf{x}\|_2^2} = \frac{\|\mathbf{y}\|_2}{\|\mathbf{x}\|_2},$$

we conclude that $\alpha > 0$. □

(e) Exercise 4.

Exercises

1. In the situation of Example 1, show (using either algebra or calculus) that on the line with equation $y = x/2$, the point closest to $(6, 8)$ is $(8, 4)$.

2. Let $\mathbf{a} = \left(\sqrt{3}, 1, 0\right)$, $\mathbf{b} = \left(\sqrt{3}, -1, 0\right)$.

   (a) Use the definition to calculate the angle between $\mathbf{a}$ and $\mathbf{b}$.

   (b) Use a sketch of $\mathbf{a}$ and $\mathbf{b}$ to predict the angle between $\mathbf{a}$ and $-\mathbf{b}$. Does the prediction match the definition?

   (c) Calculate the angle between $\mathbf{a}$ and $\mathbf{a} + \mathbf{b}$. Was it predictable?

3. (a) Using just Theorem 1.1, prove that $\mathbf{x} \bullet \mathbf{O} = 0$ for all $\mathbf{x}$.

   (b) Can $\mathbf{x} \bullet \mathbf{y} = 0$ with neither $\mathbf{x}$ nor $\mathbf{y}$ equal to $\mathbf{O}$?

   (c) Is there any vector other than $\mathbf{O}$ that is orthogonal to all vectors?

   (d) Is there any vector $\mathbf{a}$ other than $\mathbf{O}$ that is orthogonal to all vectors that are not scalar multiples of $\mathbf{a}$?

4. In an inner product space, show that the angle between nonzero vectors $\mathbf{x}$ and $\mathbf{y}$ is π iff $\mathbf{x} = \alpha\mathbf{y}$ with $\alpha < 0$.

5. (a) Prove that if $\mathbf{x}$ and $\mathbf{y}$ are orthogonal, then $\|\mathbf{x}+\mathbf{y}\|_2 = \|\mathbf{x}-\mathbf{y}\|_2$.
   (b) Draw a picture of the situation in (a). What is the geometric meaning of the equation?

6. (a) Prove that if $\|\mathbf{x}\|_2 = \|\mathbf{y}\|_2$, then $\mathbf{x}+\mathbf{y}$ is orthogonal to $\mathbf{x}-\mathbf{y}$.
   (b) Draw a picture of the situation. What is the geometric meaning of the equation?

7. Prove Theorem 1.1.

8. Prove Theorem 1.2. (Hint: (a) and (b) are straightforward applications of the properties of dot product, that is, of Theorem 1.1. For (c), consider that if $\alpha\mathbf{y}$ is any multiple of $\mathbf{y}$, then $\alpha\mathbf{y}$, $\mathbf{proj}(\mathbf{x}, \mathbf{y})$, and $\mathbf{x}$ are the vertices of a right triangle. Name the vector sides of that triangle; then apply part (b).)

9. Prove Theorem 1.4(c).

1.4 Norms and the Definition of Euclidean Space

We have introduced the quantity $\|\mathbf{x}\|_2 := (\mathbf{x} \bullet \mathbf{x})^{1/2}$ and named it "the length of $\mathbf{x}$." In this section we show that "length" is justified but "the" is not.

Theorem 1.5. *In any inner product space, the length function* $\|\mathbf{x}\|_2 := (\mathbf{x} \bullet \mathbf{x})^{1/2}$ *has the following properties*:

(a) Positive Definiteness	$\|\mathbf{x}\|_2 > 0$ *for every* $\mathbf{x}$, *with the usual exception that* $\|\mathbf{O}\|_2 = 0$.
(b) Radial Homogeneity	$\|\alpha\mathbf{x}\|_2 = \lvert\alpha\rvert\,\|\mathbf{x}\|_2$ *for real* α.
(c) Subadditivity	$\|\mathbf{x}+\mathbf{y}\|_2 \le \|\mathbf{x}\|_2 + \|\mathbf{y}\|_2$.

Proof. (a) and (b) are Exercise 5.
(c) (See Exercise 6 for an alternative proof.) We have

$$\|\mathbf{x}+\mathbf{y}\|_2^2 := (\mathbf{x}+\mathbf{y}) \bullet (\mathbf{x}+\mathbf{y}) = \mathbf{x} \bullet \mathbf{x} + 2\mathbf{x} \bullet \mathbf{y} + \mathbf{y} \bullet \mathbf{y} = \|\mathbf{x}\|_2^2 + 2\mathbf{x} \bullet \mathbf{y} + \|\mathbf{y}\|_2^2.$$

By Cauchy's inequality, $2\mathbf{x} \bullet \mathbf{y} \le 2\|\mathbf{x}\|_2\|\mathbf{y}\|_2$. Therefore,

$$\|\mathbf{x}+\mathbf{y}\|_2^2 \le \|\mathbf{x}\|_2^2 + 2\|\mathbf{x}\|_2\|\mathbf{y}\|_2 + \|\mathbf{y}\|_2^2 = (\|\mathbf{x}\|_2 + \|\mathbf{y}\|_2)^2.$$

Because all the terms are nonnegative, the conclusion follows. □

We discuss Theorem 1.5 along three lines.

First we look at names. A function that assigns a real number $\|\mathbf{x}\|$ to each member $\mathbf{x}$ of a vector space, in a way that satisfies Theorem 1.5, is called a **norm**. A vector space equipped with such a function is a **normed vector space** (or **normed linear space**). Of all the properties that our notion of length has, it turns out that the ones listed in Theorem 1.5 form a sufficient foundation for our work. In other words, for most of what we do, the fact that our Pythagorean length is a norm is all we need to know about the length. Consequently, we will use "length" and "norm" almost interchangeably.

The name "radial homogeneity" is not inspiring, but it does convey some information, which we will discuss in the next section. If we could factor out the α without the absolute value, we would call the property simply "homogeneity."

The name "subadditivity" is algebraic. Its geometric counterpart is much more familiar. In our usual picture of vectors (Figure 1.3) we think of **a**, **b**, and **a** − **b** as the sides of a triangle. The subadditivity property amounts to a statement that in a normed space, no side of a triangle exceeds the sum of the other two:

$$\begin{aligned} \|\mathbf{a}\| &\le \|\mathbf{b}\| + \|\mathbf{a}-\mathbf{b}\|; \\ \|\mathbf{b}\| &\le \|\mathbf{a}\| + \|\mathbf{b}-\mathbf{a}\| = \|\mathbf{a}\| + |-1|\,\|\mathbf{a}-\mathbf{b}\| = \|\mathbf{a}\| + \|\mathbf{a}-\mathbf{b}\|; \\ \|\mathbf{a}-\mathbf{b}\| &\le \|\mathbf{a}\| + \|-\mathbf{b}\| = \|\mathbf{a}\| + \|\mathbf{b}\|. \end{aligned}$$

For that reason, the relation

$$\|\mathbf{x}+\mathbf{y}\| \le \|\mathbf{x}\| + \|\mathbf{y}\|$$

has the geometric name the **triangle inequality**.

In fact, the name "triangle inequality" suggests more. In the triangles we know, no side can even *equal* the sum of the others. This turns out to be the case wherever there is an inner product.

Theorem 1.6. *In an inner product space, each side of a nondegenerate triangle is strictly smaller than the sum of the other sides. In symbols:*

(a) $\|\mathbf{a}-\mathbf{b}\|_2 < \|\mathbf{a}\|_2 + \|\mathbf{b}\|_2$, *unless either* $\mathbf{b} = \mathbf{O}$ *or* **a** *is a multiple* $\alpha\mathbf{b}$ *with* $\alpha \le 0$.

(b) *Equivalently,*

$$\|\mathbf{x}+\mathbf{y}\|_2 < \|\mathbf{x}\|_2 + \|\mathbf{y}\|_2,$$

unless one of **x** *and* **y** *is a nonnegative multiple of the other.*

Proof. (a) Let θ be the angle between **a** and **b**. By the law of cosines,

$$\|\mathbf{a}-\mathbf{b}\|_2^2 = \|\mathbf{a}\|_2^2 + \|\mathbf{b}\|_2^2 - 2\|\mathbf{a}\|_2\|\mathbf{b}\|_2\cos\theta < (\|\mathbf{a}\|_2 + \|\mathbf{b}\|_2)^2$$

unless $\|\mathbf{a}\|_2 = 0$ or $\|\mathbf{b}\|_2 = 0$ or $\cos\theta = -1$. In the last case, Theorem 1.4(e) tells us that **a** is a negative multiple of **b**.

We have shown that either $\mathbf{b} = \mathbf{O}$ or $\mathbf{a} = \alpha\mathbf{b}$ with $\alpha \le 0$.

(b) Exercise 7. □

Example 1. What are the longest chords within the unit disk?

Take two points **a** and **b** in $\mathbf{R}^2$ with $\|\mathbf{a}\|_2 = \|\mathbf{b}\|_2 = 1$ (see Figure 1.6). The chord from **b** to **a** is the vector $\mathbf{a} - \mathbf{b}$, which (Theorem 1.6) satisfies

$$\|\mathbf{a} - \mathbf{b}\|_2 < \|\mathbf{a}\|_2 + \|\mathbf{b}\|_2 = 2,$$

unless **a** is a nonpositive multiple $\alpha\mathbf{b}$. In the latter case,

$$1 = \|\mathbf{a}\|_2 = |\alpha|\ \|\mathbf{b}\|_2 = |\alpha|,$$

so that $\alpha = -1$ and $\|\mathbf{a} - \mathbf{b}\|_2 = \|-2\mathbf{b}\|_2 = 2$.

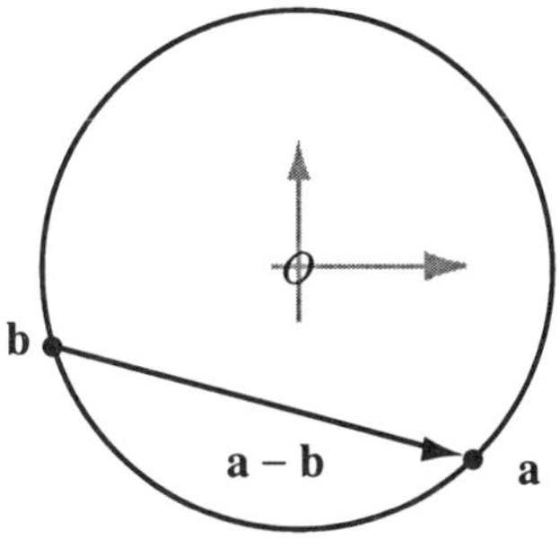

Figure 1.6.

The chord has length 2, which is as long as possible, iff **a** and **b** are diametrically opposite.

The second line of discussion is a declaration of the ground on which we will do battle. The main focus of our work will have a very specific setting. We will study functions on the set $\mathbf{R}^n$ with the vector space structure given by the "standard operations"

$$(x_1, \ldots, x_n) + (y_1, \ldots, y_n) := (x_1 + y_1, \ldots, x_n + y_n),$$
$$\alpha(x_1, \ldots, x_n) := (\alpha x_1, \ldots, \alpha x_n),$$

and the "standard norm"

$$\|\mathbf{x}\|_2 := (\mathbf{x} \bullet \mathbf{x})^{1/2}$$

derived from the "standard inner product"

$$\mathbf{x} \bullet \mathbf{y} := x_1 y_1 + \cdots + x_n y_n.$$

This combination of elements is called **Euclidean space**. Some books use a symbol like $\mathbf{E}^n$ to represent the whole package. We will use $\mathbf{R}^n$ even though this notation suggests just the underlying set of n-tuples.

At its most specific, our analysis will be based on $\mathbf{R}^n$ as a finite-dimensional real vector space with a norm derived from a dot product. To exhibit our results as instances of more general principles, we will also discuss them in infinite-dimensional spaces and in (finite-dimensional) spaces using other norms. The former setting is harder and is therefore set apart in the final sections of our chapters. The finite-dimensional locale using other norms is elementary enough for us to weave into the text.

That setting also brings us to the third avenue of pursuit related to Theorem 1.5. It is trivial that on any space, there are many ways to define a norm. (There are lots of inner products, too, but we state that one inner product is much like another.) For example, in $\mathbf{R}^n$, the function $\|\mathbf{x}\|_{12}$ defined by $\|\mathbf{x}\|_{12} := 12\|\mathbf{x}\|_2$ satisfies Theorem 1.5 (Verify!) and is therefore a norm. Observe that switching from $\|\mathbf{x}\|_2$ to $\|\mathbf{x}\|_{12}$ amounts to adopting inches in place of feet as the unit of measure. Any such rescaling produces a new but entirely similar norm. What we do next is to demonstrate a norm that is fundamentally different from the standard norm.

Definition. For $\mathbf{x} := (x_1, \ldots, x_n)$ in $\mathbf{R}^n$, **maxnorm** is the function defined by

$$\|\mathbf{x}\|_0 := \max\{|x_1|, \ldots, |x_n|\}.$$

Theorem 1.7. *In* $\mathbf{R}^n$:

(a) *Maxnorm satisfies Theorem 1.5; it is a norm.*

(b) *Maxnorm is not derived from an inner product.*

Proof. (a) Exercise 8.

(b) Notice the special character of this statement. It does not say merely that maxnorm is not the child of the standard dot product. It says that *no* dot product gives rise to maxnorm.

To prove a statement of this nature, we typically look for a property that every dot-product norm has but maxnorm lacks. Such a property is described by Theorem 1.6.

Thus, take $\mathbf{x} := (2, 0, 0, \ldots, 0)$ and $\mathbf{y} := (1, 1, 0, \ldots, 0)$. Then

$$\|\mathbf{x}+\mathbf{y}\|_0 := \max\{3, 1, 0, \ldots, 0\} = 3$$

and

$$\|\mathbf{x}\|_0 + \|\mathbf{y}\|_0 = 2 + 1 = 3.$$

We have $\|\mathbf{x}+\mathbf{y}\|_0 = \|\mathbf{x}\|_0 + \|\mathbf{y}\|_0$, even though neither of $\mathbf{x}$ and $\mathbf{y}$ is a multiple of the other. That would be impossible if maxnorm came from an inner product. □

We will give below one more illustration of how maxnorm is different in character from the Pythagorean norm. Before that, we give a theorem that shows an important property of norms. In this theorem and henceforward we will use phrases like "for any norm" and "in any normed linear space," together with the generic notation $\|\mathbf{x}\|$, to indicate results that depend only on the properties (Theorem 1.5) of norms.

Theorem 1.8. *The difference of the norms is never more than the norm of the difference. In symbols: For any vectors* $\mathbf{x}$ *and* $\mathbf{y}$ *in any normed space,*

$$|\|\mathbf{x}\| - \|\mathbf{y}\|| \le \|\mathbf{x} - \mathbf{y}\|.$$

Proof. From the triangle inequality, we have

$$\|\mathbf{x} + (\mathbf{y} - \mathbf{x})\| \le \|\mathbf{x}\| + \|\mathbf{y} - \mathbf{x}\|,$$

so

$$\|\mathbf{y}\| - \|\mathbf{x}\| \le \|\mathbf{y} - \mathbf{x}\|.$$

Similarly,

$$\|\mathbf{x}\| - \|\mathbf{y}\| \le \|\mathbf{x} - \mathbf{y}\|.$$

Since $\|\mathbf{y} - \mathbf{x}\| = \|\mathbf{x} - \mathbf{y}\|$, the two inequalities imply

$$-\|\mathbf{x} - \mathbf{y}\| \le \|\mathbf{x}\| - \|\mathbf{y}\| \le \|\mathbf{x} - \mathbf{y}\|.$$

The absolute value inequality follows. □

Example 2. In $\mathbf{R}^2$, what is the shortest segment from $\mathbf{b} := (2, 0)$ to the unit circle?

The answer depends on how we measure, that is, on which norm we employ.

(a) Let $\mathbf{x}$ belong to the unit circle. With the Pythagorean norm, Theorem 1.8 says that the length of the vector from $\mathbf{x}$ to $\mathbf{b}$ satisfies

$$\|\mathbf{b} - \mathbf{x}\|_2 \ge \|\mathbf{b}\|_2 - \|\mathbf{x}\|_2 = 1.$$

In fact, though, we know more. By Theorem 1.6(b), $\|\mathbf{b} - \mathbf{x}\|_2 + \|\mathbf{x}\|_2 > \|\mathbf{b}\|_2$, and therefore $\|\mathbf{b} - \mathbf{x}\|_2$ actually exceeds 1, except if $\mathbf{b} - \mathbf{x}$ is a nonnegative multiple $\alpha\mathbf{x}$. From $\mathbf{b} - \mathbf{x} = \alpha\mathbf{x}$, we have

$$1 = \|\mathbf{x}\|_2 = \left\| \frac{1}{1+\alpha}\mathbf{b} \right\|_2 = \frac{2}{1+\alpha}.$$

Hence $\alpha = 1$ and $\mathbf{x} = \frac{1}{2}\mathbf{b} = (1, 0)$. We conclude that there is a unique shortest segment: The segment from $(2, 0)$ to $(1, 0)$ is (strictly) shorter than all the others.

(b) The same (triangle) inequality applies with maxnorm:

$$\|\mathbf{b} - \mathbf{x}\|_0 \ge \|\mathbf{b}\|_0 - \|\mathbf{x}\|_0 = 2 - 1.$$

Consequently, $\mathbf{a} := (1, 0)$ gives as short a segment as possible, since $\|\mathbf{b} - \mathbf{a}\|_0 = 1$.

But there are other such segments. Simply observe that the segment to $(1, 1)$ is equally long, because

$$\|\mathbf{b} - (1, 1)\|_0 := \max\{|1|, |-1|\} = 1.$$

(See also Exercise 3.)

Exercises

1. Let $\mathbf{a} := (1, 2, 4)$, $\mathbf{b} := (-3, 4, 5)$.

   (a) Calculate $\|\mathbf{a}\|_2$, $\|\mathbf{b}\|_2$, $\|\mathbf{a} + \mathbf{b}\|_2$.

   (b) Check the triangle inequality for these vectors.

   (c) Calculate $\|3\mathbf{a}\|_2$ and $\|3\mathbf{a} + 3\mathbf{b}\|_2$.

2. Let $\mathbf{c} := (4, -2, 8)$, $\mathbf{d} := (-6, 3, -12)$.

   (a) Calculate $\|\mathbf{c}\|_2$, $\|\mathbf{d}\|_2$, $\|\mathbf{c} + \mathbf{d}\|_2$.

   (b) Check the triangle inequality for these vectors. What conclusion do you draw?

   (c) Calculate $\|30\mathbf{c} + 40\mathbf{d}\|_2$.

3. In the set $\mathbf{R}^2$, using maxnorm:

   (a) Sketch the set $\{\mathbf{x}\colon \|\mathbf{x}\|_0 = 1\}$ (what we called "the unit circle").

   (b) Describe all points on the "circle" from which the segment to $(2, 0)$ has length 1.

4. Under maxnorm in $\mathbf{R}^2$, what is the longest chord on the unit circle (the longest line segment to another point of the circle) from $(1, 0)$? Is it unique?

5. Use the properties of dot product (Theorem 1.1) to prove that:

   (a) $\|\mathbf{x}\|_2 := (\mathbf{x} \bullet \mathbf{x})^{1/2} \geq 0$, with equality iff $\mathbf{x} = \mathbf{O}$.

   (b) $\|\alpha\mathbf{x}\|_2 = |\alpha|\ \|\mathbf{x}\|_2$ for scalar α, vector $\mathbf{x}$.

6. Use the law of cosines to prove the triangle inequality (in an inner product space).

7. Show that part (a) of Theorem 1.6 is equivalent to part (b).

8. Prove that maxnorm is a norm in $\mathbf{R}^n$.

1.5 Metrics

A norm measures size. Where you can measure the size of a line segment, you can call that length the "distance" between the two ends. That is the concept we need: Once we have distance, we will define limits in terms of vanishing distance; then we do calculus.

Definition. Given a normed linear space with members $\mathbf{x}$ and $\mathbf{y}$, the real number

$$d(\mathbf{x}, \mathbf{y}) := \|\mathbf{x} - \mathbf{y}\|$$

is called the **distance between x and y**.

Theorem 1.9. *In a normed linear space, the distance function has the following properties*:

(a) Symmetry	$d(\mathbf{x}, \mathbf{y}) = d(\mathbf{y}, \mathbf{x})$ *for all* $\mathbf{x}, \mathbf{y}$.
(b) Positive Definiteness	$d(\mathbf{x}, \mathbf{y}) \geq 0$ *with equality iff* $\mathbf{x} = \mathbf{y}$.
(c) Triangle Inequality	$d(\mathbf{x}, \mathbf{z}) \leq d(\mathbf{x}, \mathbf{y}) + d(\mathbf{y}, \mathbf{z})$ *for all* $\mathbf{x}, \mathbf{y}, \mathbf{z}$.

Proof. Exercise 3.

A function associating a real number to each ordered pair $(\mathbf{x}, \mathbf{y})$ from a set—not necessarily from a vector space—and satisfying Theorem 1.9 is "distance-like"; we call such a function a **metric**.

Distance and straightness are two of the most fundamental concepts in our mental picture of the world. It is remarkable that among all the properties our idea of distance has, the three of Theorem 1.9 are sufficient to form the basis for abstraction. Statements (a) and (b) are so much a part of our notion of distance that we normally have no reason to articulate them. Statement (c) joins the two concepts, since it says that the straight path is never longer than any other.

As we did with norms, we separate our discussion into some specifics about the standard metric in $\mathbf{R}^n$ and some general results.

In $\mathbf{R}^n$, the metric defined by the standard norm coincides with what we usually call the distance formula. Thus, if $\mathbf{x} = (x_1, \ldots, x_n)$ and $\mathbf{y} = (y_1, \ldots, y_n)$, then

$$d(\mathbf{x}, \mathbf{y}) := \|\mathbf{x} - \mathbf{y}\|_2 = \left([x_1 - y_1]^2 + \cdots + [x_n - y_n]^2\right)^{1/2}.$$

The distance formula makes it natural to refer to this distance function as the "standard metric" or "Pythagorean metric."

Wherever the Pythagorean theorem applies, Euclidean geometry works. In any such world, the straight path is not just "never longer"; it is strictly shorter than the others.

Theorem 1.10. *In* $\mathbf{R}^n$, *as in any inner product space, the triangle inequality for distance is strict, that is*,

$$d(\mathbf{x}, \mathbf{z}) \leqq d(\mathbf{x}, \mathbf{y}) + d(\mathbf{y}, \mathbf{z}),$$

with equality iff $\mathbf{y}$ *is on the segment from* $\mathbf{x}$ *to* $\mathbf{z}$.

Proof. This result is based on the strict norm inequality. By Theorem 1.6(b),

$$\begin{aligned} d(\mathbf{x}, \mathbf{z}) &:= \|\mathbf{x} - \mathbf{z}\|_2 = \|\mathbf{x} - \mathbf{y} + \mathbf{y} - \mathbf{z}\|_2 \\ &\leq \|\mathbf{x} - \mathbf{y}\|_2 + \|\mathbf{y} - \mathbf{z}\|_2 = d(\mathbf{x}, \mathbf{y}) + d(\mathbf{y}, \mathbf{z}), \end{aligned}$$

with equality iff one of $\mathbf{x}-\mathbf{y}$ and $\mathbf{y}-\mathbf{z}$ is a nonnegative multiple of the other. We will show that this last clause is true iff $\mathbf{y}$ is on the segment from $\mathbf{x}$ to $\mathbf{z}$.

Suppose $\mathbf{x}-\mathbf{y}=\alpha(\mathbf{y}-\mathbf{z})$ with $\alpha \geq 0$. Solving for $\mathbf{y}$, we get

$$\mathbf{y}=\frac{1}{1+\alpha}\mathbf{x}+\frac{\alpha}{1+\alpha}\mathbf{z}.$$

In this linear combination, the coefficients are nonnegative and sum to 1. Therefore, $\mathbf{y}$ is on the segment from $\mathbf{x}$ to $\mathbf{z}$ (see the paragraph above Example 3 in Section 1.2).

The same reasoning applies if $\mathbf{y}-\mathbf{z}=\alpha(\mathbf{x}-\mathbf{y})$.

Conversely, if $\mathbf{y}$ is on the segment, then

$$\mathbf{y}=(1-\alpha)\mathbf{x}+\alpha\mathbf{z} \quad \text{for some } 0 \leq \alpha \leq 1.$$

We easily verify that $\mathbf{x}-\mathbf{y}=\alpha(\mathbf{x}-\mathbf{z})$ and $\mathbf{y}-\mathbf{z}=(1-\alpha)(\mathbf{x}-\mathbf{z})$; of these, one is $\frac{\alpha}{1-\alpha}$ (or $\frac{1-\alpha}{\alpha}$, in the event that $\alpha=1$) times the other. □

Example 1. Set $\mathbf{x}:=(1,2,3)$, $\mathbf{y}:=(-5,-4,-3)$, $\mathbf{z}:=(4,5,6)$. We calculate

$$\mathbf{y}-\mathbf{x}=(-6,-6,-6), \qquad \mathbf{z}-\mathbf{x}=(3,3,3).$$

From $\mathbf{y}-\mathbf{x}=-2(\mathbf{z}-\mathbf{x})$, we may conclude that $\mathbf{x}$, $\mathbf{y}$, and $\mathbf{z}$ lie on the same line (compare Exercise 4 in Section 1.2). Also, the negative sign in -2 tells us that $\mathbf{y}$ and $\mathbf{z}$ are in opposite directions from $\mathbf{x}$, so that $\mathbf{x}$ is between the other two.

From this discussion, we expect to find that $d(\mathbf{y},\mathbf{z})=d(\mathbf{y},\mathbf{x})+d(\mathbf{x},\mathbf{z})$. (Notice the order of appearance.) We calculate $d(\mathbf{y},\mathbf{z})=\sqrt{243}=9\sqrt{3}$, $d(\mathbf{y},\mathbf{x})=6\sqrt{3}$, $d(\mathbf{x},\mathbf{z})=3\sqrt{3}$.

All the evidence says that $\mathbf{x}$ is on the segment from $\mathbf{y}$ to $\mathbf{z}$. Solving $\mathbf{y}-\mathbf{x}=-2(\mathbf{z}-\mathbf{x})$ for $\mathbf{x}$ in terms of the other two, we conclude that $\mathbf{x}$ is two-thirds of the way from $\mathbf{y}$ to $\mathbf{z}$:

$$\mathbf{x}=\frac{1}{3}\mathbf{y}+\frac{2}{3}\mathbf{z}.$$

To discuss metrics where an inner product may be lacking, we begin with a property of all metrics that are defined by the norm in a linear space.

Theorem 1.11. *The metric* $d(\mathbf{x},\mathbf{y}):=\|\mathbf{x}-\mathbf{y}\|$ *associated with the norm in a normed vector space has the following properties*:

(a) Radial Homogeneity $\quad d(\alpha\mathbf{x},\alpha\mathbf{y})=|\alpha|\,d(\mathbf{x},\mathbf{y})$.
(b) Translation Invariance $\quad d(\mathbf{x},\mathbf{y})=d(\mathbf{x}+\mathbf{z},\mathbf{y}+\mathbf{z})$.

Proof. Exercise 4.

These algebraic relationships have simple geometric meanings. For condition (a) in the theorem, draw (Figure 1.7, left) $\mathbf{x}$ and $\mathbf{y}$, extend them to $\alpha\mathbf{x}$ and $\alpha\mathbf{y}$, and draw the segments from $\mathbf{x}$ to $\mathbf{y}$ and $\alpha\mathbf{x}$ to $\alpha\mathbf{y}$. The two triangles share an angle

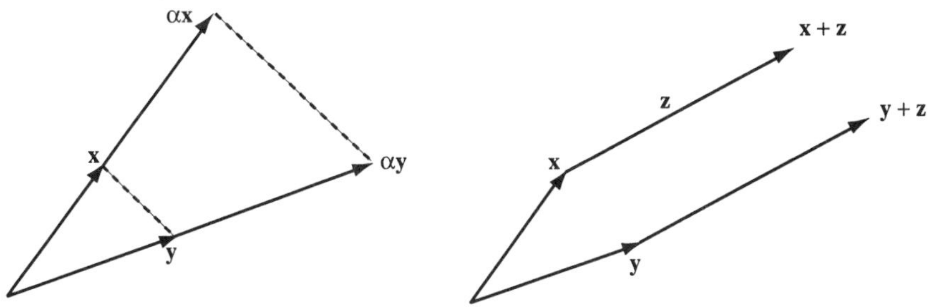

Figure 1.7.

and have two pairs of proportional sides. The condition says of the metric that in this situation, the two triangles are necessarily similar. Condition (b) says of the metric that if you move $\mathbf{x}$ and $\mathbf{y}$ the same distance in the same direction—that is, if you translate each by the same displacement $\mathbf{z}$ (Figure 1.7, right)—then the new locations are exactly as far apart as they were originally.

We next point out that a single metric is never the only possible one in a set. If you multiply a metric by a constant, such as 12 or 2.54, the resulting rescaled distance is also a metric. On the other hand, the $\mathbf{R}^n$ metric comes from a norm, and not all metrics do.

Example 2. In $\mathbf{R}^2$, define

$$D(\mathbf{x},\mathbf{y}) := \frac{d(\mathbf{x},\mathbf{y})}{1+d(\mathbf{x},\mathbf{y})},$$

where d represents the Pythagorean distance.

The function D satisfies Theorem 1.9, because:

(a) In $D(\mathbf{x},\mathbf{y})$, numerator and denominator are symmetric in $\mathbf{x}$ and $\mathbf{y}$.

(b) In $D(\mathbf{x},\mathbf{y})$, the denominator is positive, and unless $\mathbf{x} = \mathbf{y}$, so is the numerator.

(c) $f(s) := s/(1+s)$ is an increasing function of the nonnegative real variable s. Hence

$$\begin{aligned} D(\mathbf{x},\mathbf{z}) &:= \frac{d(\mathbf{x},\mathbf{z})}{1+d(\mathbf{x},\mathbf{z})} \\ &\leq \frac{d(\mathbf{x},\mathbf{y})+d(\mathbf{y},\mathbf{z})}{1+d(\mathbf{x},\mathbf{y})+d(\mathbf{y},\mathbf{z})} \\ &\qquad \text{(by triangle inequality and increasing } f\text{)} \\ &= \frac{d(\mathbf{x},\mathbf{y})}{1+d(\mathbf{x},\mathbf{y})+d(\mathbf{y},\mathbf{z})} + \frac{d(\mathbf{y},\mathbf{z})}{1+d(\mathbf{x},\mathbf{y})+d(\mathbf{y},\mathbf{z})} \\ &\leq \frac{d(\mathbf{x},\mathbf{y})}{1+d(\mathbf{x},\mathbf{y})} + \frac{d(\mathbf{y},\mathbf{z})}{1+d(\mathbf{y},\mathbf{z})} \\ &= D(\mathbf{x},\mathbf{y}) + D(\mathbf{y},\mathbf{z}). \end{aligned}$$

Thus D is a metric. But D is not related to a norm. If it were related to a norm $\|\mathbf{x}\|$, then homogeneity would give

$$D\big(\mathbf{O}, (10^6, 0)\big) = \big\|\mathbf{O} - 10^6(1, 0)\big\| = 10^6\big\|\mathbf{O} - (1, 0)\big\| = 10^6 D\big(\mathbf{O}, (1, 0)\big).$$

But $D\big(\mathbf{O}, (10^6, 0)\big) \approx .999999$ and $D(\mathbf{O}, (1, 0)) = \frac{1}{2}$.

Indeed, every distance defined by this metric is less than 1. A creature living at the origin in $\mathbf{R}^2$ and measuring this way would have the width of its entire universe as its unit of measure. Two-thirds of the width of this universe would be in what we call the unit circle. Our creature would view $\big(10^6, 0\big)$ as .999999 away, and $\big(10^{50}, 0\big)$ as just past there, at a distance of about $1 - 10^{-50}$. Also, its cousin at $\big(10^6, 0\big)$ would view both the origin and $\big(10^{50}, 0\big)$ as just under 1 unit away from *it*. Each one would consider itself to be standing at the center of the universe and its cousin to be practically at the edge.

It is clear that for the $\mathbf{R}^n$ metric $d(\mathbf{x}, \mathbf{y}) := \|\mathbf{x}-\mathbf{y}\|_2$, we have $\|\mathbf{x}\|_2 = d(\mathbf{x}, \mathbf{O})$ for every $\mathbf{x}$. That is, just as our experience dictates, the length of $\mathbf{x}$ is simply the distance between the ends of the arrow that represents it. So, given a metric D, shouldn't the same formula work? That is, it should not be automatic that $\|\mathbf{x}\| := D(\mathbf{x}, \mathbf{O})$ defines a norm?

To understand why it does not work that way, view the definition (Theorem 1.9) of a metric. Nothing in the required properties (a)–(c) refers to the vector space operations. That is the source of trouble in Example 2: The metric defined there ignores scalar multiplication. In the definition of norm (Theorem 1.5), the relation to the operations is obvious. Accordingly, to associate a metric with a norm, we must demand some kind of compatibility between the metric and the addition and scalar multiplication.

The needed connection turns out to be Theorem 1.10. We conclude this section by using Theorem 1.10 to make the following characterization of metrics that go with norms.

Theorem 1.12. *If a metric D defined on a vector space is radially homogeneous and translation invariant, then the metric is defined by a norm. Specifically, given D with properties (a) and (b) in Theorem 1.10, then the function defined by* $\|\mathbf{x}\| := D(\mathbf{x}, \mathbf{O})$:

(a) *is a norm, and*

(b) *defines D, in the sense that D is related to it by*

$$D(\mathbf{x}, \mathbf{y}) = \|\mathbf{x} - \mathbf{y}\| \text{ for every } \mathbf{x} \text{ and } \mathbf{y}.$$

Proof. Let D be a radially homogeneous, translation-invariant metric, and write

$$\|\mathbf{x}\| := D(\mathbf{x}, \mathbf{O}).$$

(a) First, $\|\mathbf{x}\|$ is positive definite: We always have $\|\mathbf{x}\| > 0$, except $\|\mathbf{O}\| = 0$. (Reason?) Second, $\|\mathbf{x}\|$ is radially homogeneous:

$$\begin{aligned}\|\alpha\mathbf{x}\| &:= D(\alpha\mathbf{x}, \mathbf{O}) \\ &= D(\alpha\mathbf{x}, \alpha\mathbf{O}) && \text{(vector algebra)} \\ &= |\alpha|\, D(\mathbf{x}, \mathbf{O}) && (D \text{ is radially homogeneous}) \\ &= |\alpha|\, \|\mathbf{x}\|.\end{aligned}$$

Third, $\|\mathbf{x}\|$ satisfies the triangle inequality:

$$\begin{aligned}\|\mathbf{x}+\mathbf{y}\| &:= D(\mathbf{x}+\mathbf{y}, \mathbf{O}) \\ &= D(\mathbf{x}+\mathbf{y}-\mathbf{y}, \mathbf{O}-\mathbf{y}) && (D \text{ is translation invariant}) \\ &= D(\mathbf{x}, -\mathbf{y}) \\ &\le D(\mathbf{x}, \mathbf{O}) + D(\mathbf{O}, -\mathbf{y}) && (D \text{ satisfies the triangle inequality}) \\ &= D(\mathbf{x}, \mathbf{O}) + D(\mathbf{O}, \mathbf{y}) && \text{(Reason?)} \\ &= D(\mathbf{x}, \mathbf{O}) + D(\mathbf{y}, \mathbf{O}) && \text{(Reason?)} \\ &= \|\mathbf{x}\| + \|\mathbf{y}\|.\end{aligned}$$

We have proved that $\|\mathbf{x}\|$ is a norm.

(b) Under the definitions we have made,

$$\|\mathbf{x}-\mathbf{y}\| := D(\mathbf{x}-\mathbf{y}, \mathbf{O}) = D(\mathbf{x}-\mathbf{y}+\mathbf{y}, \mathbf{O}+\mathbf{y}) = D(\mathbf{x}, \mathbf{y}). \qquad \square$$

Exercises

1. Set $\mathbf{x} := (1.234567, 1.234568, 1.234569)$, $\mathbf{y} := (53.234567, 53.234568, 53.234569)$, $\mathbf{z} := (17.234567, 17.234568, 17.234569)$. Which, if any, of $d(\mathbf{x}, \mathbf{y})$, $d(\mathbf{x}, \mathbf{z})$, and $d(\mathbf{y}, \mathbf{z})$ equals the sum of the other two?

2. In the set $\mathbf{R}^2$:

   (a) With the Pythagorean distance, describe the set of all points that are equidistant from the points $(1, 0)$ and $(-1, 0)$.

   (b) Answer the same question using the "maxdistance" $d_0(\mathbf{x}, \mathbf{y}) := \|\mathbf{x}-\mathbf{y}\|_0$ associated with the maxnorm.

3. Prove Theorem 1.9.

4. Prove Theorem 1.11.

1.6 Infinite-Dimensional Spaces

We have defined an inner product, norm, and metric in terms of coordinates in $\mathbf{R}^n$. This explicit use of the coordinates in an n-tuple might suggest that the finiteness of their number is essential. In this section we show that the three notions can be defined in some vector spaces of infinite dimension, where they maintain the properties we have described so far in $\mathbf{R}^n$.

In each subsequent chapter we will devote the last section to treating the chapter's material in the infinite-dimensional case. We will generalize results that carry over from finite dimensions, and give counterexamples for those that do not. This practice reflects the author's predilection for displaying ideas and results in the context of more general ones.

We are going to work with two normed spaces, meaning two vector-space-plus-norm combinations. Both use the same underlying vector space, namely, the set $C[0, 1]$ of real-valued functions that are continuous on the closed interval $[0, 1]$. We assume that the student knows that $C[0, 1]$ is a vector space under the "usual" operations defined by

$$(f + g)(x) := f(x) + g(x)$$

and

$$(\alpha f)(x) := \alpha f(x),$$

and that it is of infinite dimension.

Since we like to think geometrically, it makes sense for us to look in any space for an inner product. There is a natural extension of the Euclidean space dot product to $C[0, 1]$. The dot product in $\mathbf{R}^n$ is the sum of the products of the coordinates. With continuous functions, we would like to sum the products of the values, except of course that summing is not precisely legal.

Definition. If f and g are in $C[0, 1]$, then

$$f \bullet g := \int_0^1 f(x)g(x)\,dx$$

is the **dot product of f and g**.

Theorem 1.13. *The dot product is an inner product on* $C[0, 1]$.

Proof. Before looking at the properties demanded of an inner product (Theorem 1.1), we have to note that the expression $f \bullet g$ actually defines something. Thus, suppose f and g are in $C[0, 1]$. Then the product function fg is also continuous, and therefore integrable. Consequently, $\int_0^1 f(x)g(x)\,dx$ is a meaningful expression.

Symmetry ($f \bullet g = g \bullet f$) follows from the commutativity of multiplication on $\mathbf{R}$: $f(x)g(x) = g(x)f(x)$, so the associated integrals are the same. Distributivity and homogeneity follow from the linearity of the Riemann integral:

$$\int_0^1 [\alpha f(x) + \beta g(x)]h(x)\,dx = \alpha \int_0^1 f(x)h(x)\,dx + \beta \int_0^1 g(x)h(x)\,dx.$$

As for positive definiteness, if $f \equiv 0$, then $\int_0^1 f(x)f(x)\,dx = 0$ by a property of the integral. If instead f is not the zero function, then f^2 is a *nonnegative*, *continuous* function that is not identically zero. Such a function has, by a theorem about integrals of continuous functions [Ross, Section 34; specifically, Exercise 11 there], a positive integral. We conclude that $f \bullet f \geq 0$, with equality precisely for $f := 0$. □

Every inner product creates a norm, and we acknowledge the connection to $\mathbf{R}^n$ by employing the same notation. Thus, we write

$$\|f\|_2 := \left(\int_0^1 f(x)^2\,dx\right)^{1/2}.$$

We will refer to it as the **2-norm**, for the obvious reason. We will also call it the **mean-norm**, because it represents the root-mean-square average of f, a characteristic that adds to its significance. We will use $C_2[0,1]$ to signify the normed linear space obtained by using $\|f\|_2$ as norm in $C[0,1]$.

Example 1. The family of functions

$$\begin{aligned} s_j(x) &:= \sin 2\pi jx, \quad & j &= 1, 2, \ldots, \\ c_k(x) &:= \cos 2\pi kx, \quad & k &= 0, 1, 2, \ldots, \end{aligned}$$

is an important subset of $C_2[0,1]$. We will return to it at several stages.

(a) The functions in this family are mutually orthogonal.

If we take two different sines, we have

$$\begin{aligned} s_j \bullet s_k &:= \int_0^1 \sin 2\pi jx \sin 2\pi kx\,dx \\ &= \int_0^1 \frac{1}{2}[\cos 2\pi(j-k)x - \cos 2\pi(j+k)x]\,dx \\ &= 0 \quad \text{as long as } j \neq k. \end{aligned}$$

A similar thing happens if we take two different cosines.

If we use one sine and one cosine, then

$$\begin{aligned} s_j \bullet c_k &:= \int_0^1 \sin 2\pi jx \cos 2\pi kx\,dx \\ &= \int_0^1 \frac{1}{2}[\sin 2\pi(j+k)x + \sin 2\pi(j-k)x]\,dx \\ &= 0 \quad \text{even if } j = \pm k. \end{aligned}$$

(b) The functions are all the same size. We have

$$\begin{aligned} s_j \bullet s_j &= \int_0^1 \sin^2 2\pi jx\,dx \\ &= \int_0^1 \frac{1}{2}(1 - \cos 4\pi jx)\,dx \\ &= \frac{1}{2}; \end{aligned}$$

similarly with $c_k \bullet c_k$. Each of these functions has root-mean-square of $\sqrt{2}/2$. In our language,

$$\|s_j\|_2 = \|c_k\|_2 = \sqrt{2}/2.$$

(c) As with any orthogonal set of equally long vectors, these functions are uniformly spaced. Thus, the distance from (say) s_j to c_k is

$$\|s_j - c_k\|_2 = \left(\|s_j\|_2^2 + \|c_k\|_2^2\right)^{1/2} = 1. \qquad \text{(Reason?)}$$

We know that a finite-dimensional space does not admit an infinite family of orthogonal vectors, as in (a). (Why?) As we will see later, the same goes for (c): You cannot have an infinite family of equinormed, equispaced vectors in finite dimensions.

Is there a norm on $C[0, 1]$ that does not come from an inner product? Just as we found for the Pythagorean norm, maxnorm on $\mathbf{R}^n$ has an analogue in the space of functions. We will give this big brother the same name and symbol.

Definition. Assume that f is a continuous real-valued function on $[0, 1]$. Then the **maxnorm** of f is the real number $\|f\|_0$ defined by

$$\|f\|_0 := \sup\{|f(x)| : 0 \le x \le 1\}.$$

If f is continuous, then $|f|$ is, too. By the extreme value theorem [check Ross, Theorem 18.1], $|f|$ has a maximum value on $[0, 1]$. Hence the definition does produce a real $\|f\|_0$. Also, the "max" part of the name is justified. We next justify the "norm" part.

Theorem 1.14. *In the space* $C[0, 1]$:

(a) *Maxnorm is a norm.*

(b) *Maxnorm does not come from an inner product.*

Proof. (a) We leave the positive definiteness and radial homogeneity to Exercise 4.

There is one subtlety in the proof of the triangle inequality that we want to address. Assume that f and g are members of $C[0, 1]$. We have

$$\|f + g\|_0 := \sup\{|f(x) + g(x)| : x \in [0, 1]\}.$$

By the triangle inequality for real numbers,

$$|f(x) + g(x)| \le |f(x)| + |g(x)|$$

for each individual x. The quantity $|f(x)|$ does not exceed $\|f\|_0$, because by definition the latter is the sup of $\{|f(x)|\}$. Similarly, $|g(x)| \le \|g\|_0$ for each x. Hence for all x,

$$|f(x) + g(x)| \le \|f\|_0 + \|g\|_0.$$

Therefore, the constant $\|f\|_0 + \|g\|_0$ is an upper bound for $\{|f(x)+g(x)|\}$, forcing

$$\|f+g\|_0 := \sup\{|f(x)+g(x)|\} \leqq \|f\|_0 + \|g\|_0.$$

(b) As in Theorem 1.7(b) for the $\mathbf{R}^n$ maxnorm, we try to invoke Theorem 1.6; that is, we look for a triangle equality. Indeed, the example that proved Theorem 1.7(b) suggests what we can use here: two functions whose maxima occur at the same place. Thus, define

$$f(x) := \sin 2\pi x,$$
$$g(x) := 2 - \left(x - \frac{1}{4}\right)^2.$$

Then $\sup|f| = \sup f = f\left(\frac{1}{4}\right) = 1$, and $\sup|g| = g\left(\frac{1}{4}\right) = 2$. Therefore,

$$3 = \|f+g\|_0 = \|f\|_0 + \|g\|_0.$$

Since f and g are independent vectors (Exercise 3), it cannot be that maxnorm is associated with an inner product. □

We use $C_0[0, 1]$ to denote the normed linear space $C[0, 1]$ under maxnorm.

Example 2. Examine the functions given in Example 1 as members of $C_0[0, 1]$.

(a) They are of equal size. Each has a maximum value 1 and (except for c_0) minimum -1 on $[0, 1]$. Therefore,

$$\|s_j\|_0 = \|c_k\|_0 = 1 \quad \text{for every } j \text{ and } k.$$

(b) While not uniformly spaced (Exercise 2), they are pairwise more than 1 apart. Take s_j and c_k. Recall (Example 1(c)) that

$$\int_0^1 (s_j(x) - c_k(x))^2\, dx = \|s_j - c_k\|_2^2 = 1.$$

Since $s_j - c_k$ has root-mean-square 1 despite having zeros in $[0, 1]$, it easily follows that

$$\|s_j - c_k\|_0 := \sup\{|s_j(x) - c_k(x)|\} > 1.$$

We end the first chapter with a result that is outside our main business but that lends a kind of symmetry to our treatment. We have shown that a dot product always leads to a norm, which leads to a metric. We established that not every metric comes from a norm, nor every norm from a dot product. We went further with metrics: We characterized (Theorem 1.12) which ones do come from norms. We now add the missing piece by characterizing which norms come from dot products.

The characterization is worth pursuing because it reflects a simple geometric idea. Look at the parallelogram in Figure 1.8 determined by $\mathbf{a}$ and $\mathbf{b}$. We write $\mathbf{c} = \mathbf{a} - \mathbf{b}$ for one diagonal, $\mathbf{d} = \mathbf{a} + \mathbf{b}$ for the other, and a, b, c, d for the

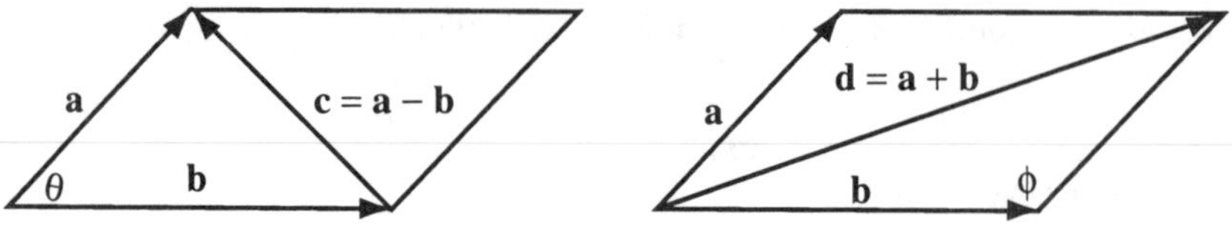

Figure 1.8.

corresponding lengths. The law of cosines (the one from geometry, not our abstract one) says of the left-hand figure that

$$c^2 = a^2 + b^2 - 2ab\cos\theta.$$

For the right-hand figure,

$$d^2 = a^2 + b^2 - 2ab\cos\phi.$$

Adding the equations and remembering that θ and ϕ, being supplementary, have opposite cosines, we get

$$c^2 + d^2 = 2a^2 + 2b^2.$$

This relationship is called the **parallelogram identity**. In the next two theorems we show that a norm comes from a dot product iff the norm leads to this equation.

Theorem 1.15. *In a space with an inner product* $\mathbf{x} \bullet \mathbf{y}$, *the associated norm* $\|\mathbf{x}\|_2 := (\mathbf{x} \bullet \mathbf{x})^{1/2}$ *satisfies*:

(a) *The parallelogram identity* $\|\mathbf{x}+\mathbf{y}\|_2^2 + \|\mathbf{x}-\mathbf{y}\|_2^2 = 2\|\mathbf{x}\|_2^2 + 2\|\mathbf{y}\|_2^2$.

(b) *The identity* $\|\mathbf{x}+\mathbf{y}\|_2^2 - \|\mathbf{x}-\mathbf{y}\|_2^2 = 4(\mathbf{x} \bullet \mathbf{y})$.

Proof. Exercise 5.

Theorem 1.16. *Suppose the norm* $\|\mathbf{x}\|$ *in a space satisfies the parallelogram identity. Then there exists a dot product that defines* $\|\mathbf{x}\|$. *Specifically, under the stated hypothesis*:

(a) *The operation* DOT *defined by*

$$\mathbf{x} \text{ DOT } \mathbf{y} := \frac{1}{4}\left(\|\mathbf{x}+\mathbf{y}\|^2 - \|\mathbf{x}-\mathbf{y}\|^2\right)$$

is an inner product.

(b) DOT *defines* $\|\mathbf{x}\|$, *in the sense that for every* $\mathbf{x}$,

$$\|\mathbf{x}\| = (\mathbf{x} \text{ DOT } \mathbf{x})^{1/2}.$$

Proof. (a) We must show that the definition given makes DOT symmetric, distributive, homogeneous, and positive definite.

(i) Symmetry is immediate, because $\|\mathbf{x}+\mathbf{y}\| = \|\mathbf{y}+\mathbf{x}\|$ and $\|\mathbf{x}-\mathbf{y}\| = \|\mathbf{y}-\mathbf{x}\|$.

(ii) Positive definiteness is also immediate, because $\mathbf{x}$ DOT $\mathbf{x} := \frac{1}{4}\|2\mathbf{x}\|^2 = \|\mathbf{x}\|^2$.

(iii) To prove distributivity we begin with

$$4\big([\mathbf{x}+\mathbf{y}] \text{ DOT } \mathbf{z}\big) := \|\mathbf{x}+\mathbf{y}+\mathbf{z}\|^2 - \|\mathbf{x}+\mathbf{y}-\mathbf{z}\|^2.$$

Applying the parallelogram identity to $\mathbf{x}+\mathbf{z}$ and $\mathbf{y}$, then to $\mathbf{x}$ and $\mathbf{z}-\mathbf{y}$, we have

$$\begin{aligned}\|\mathbf{x}+\mathbf{y}+\mathbf{z}\|^2 &= 2\|\mathbf{x}+\mathbf{z}\|^2 + 2\|\mathbf{y}\|^2 - \|\mathbf{x}+\mathbf{z}-\mathbf{y}\|^2 \\ &= 2\|\mathbf{x}+\mathbf{z}\|^2 + 2\|\mathbf{y}\|^2 - 2\|\mathbf{x}\|^2 - 2\|\mathbf{z}-\mathbf{y}\|^2 + \|\mathbf{x}-\mathbf{z}+\mathbf{y}\|^2.\end{aligned}$$

Therefore,

$$\|\mathbf{x}+\mathbf{y}+\mathbf{z}\|^2 - \|\mathbf{x}+\mathbf{y}-\mathbf{z}\|^2 = 2\|\mathbf{x}+\mathbf{z}\|^2 + 2\|\mathbf{y}\|^2 - 2\|\mathbf{x}\|^2 - 2\|\mathbf{z}-\mathbf{y}\|^2.$$

Interchanging the roles of $\mathbf{x}$ and $\mathbf{y}$, we get

$$\|\mathbf{x}+\mathbf{y}+\mathbf{z}\|^2 - \|\mathbf{x}+\mathbf{y}-\mathbf{z}\|^2 = 2\|\mathbf{y}+\mathbf{z}\|^2 + 2\|\mathbf{x}\|^2 - 2\|\mathbf{y}\|^2 - 2\|\mathbf{z}-\mathbf{x}\|^2.$$

Adding the two equations, we get

$$\begin{aligned}8([\mathbf{x}+\mathbf{y}] \text{ DOT } \mathbf{z}) &= 2\|\mathbf{x}+\mathbf{z}\|^2 - 2\|\mathbf{z}-\mathbf{y}\|^2 + 2\|\mathbf{y}+\mathbf{z}\|^2 - 2\|\mathbf{z}-\mathbf{x}\|^2 \\ &= 2\big[\|\mathbf{x}+\mathbf{z}\|^2 - \|\mathbf{x}-\mathbf{z}\|^2\big] + 2\big[\|\mathbf{y}+\mathbf{z}\|^2 - \|\mathbf{y}-\mathbf{z}\|^2\big] \\ &= 8(\mathbf{x} \text{ DOT } \mathbf{z}) + 8(\mathbf{y} \text{ DOT } \mathbf{z}).\end{aligned}$$

This proves that $[\mathbf{x}+\mathbf{y}]$ DOT $\mathbf{z} = \mathbf{x}$ DOT $\mathbf{z} + \mathbf{y}$ DOT $\mathbf{z}$. Thus DOT is distributive.

(iv) To establish homogeneity fix $\mathbf{x}$ and $\mathbf{y}$, and consider the real-valued function $f(s) := (s\mathbf{x})$ DOT $\mathbf{y}$ of the real number s. The function is additive, because

$$\begin{aligned}f(s+t) &:= ([s+t]\mathbf{x}) \text{ DOT } \mathbf{y} \\ &= (s\mathbf{x}+t\mathbf{x}) \text{ DOT } \mathbf{y} \\ &= (s\mathbf{x}) \text{ DOT } \mathbf{y} + (t\mathbf{x}) \text{ DOT } \mathbf{y} \qquad \text{(distributivity)} \\ &:= f(s) + f(t).\end{aligned}$$

The function is also continuous, because

$$\begin{aligned}f(s) - f(t) &= f(s-t) \qquad \text{(Why?)} \\ &:= (s-t)\mathbf{x} \text{ DOT } \mathbf{y} \\ &:= \frac{1}{4}\Big[\|(s-t)\mathbf{x}+\mathbf{y}\|^2 - \|(s-t)\mathbf{x}-\mathbf{y}\|^2\Big] \\ &= \frac{1}{4}\big[\|(s-t)\mathbf{x}+\mathbf{y}\| - \|(s-t)\mathbf{x}-\mathbf{y}\|\big] \\ &\quad \times \big[\|(s-t)\mathbf{x}+\mathbf{y}\| + \|(s-t)\mathbf{x}-\mathbf{y}\|\big].\end{aligned}$$

The last factor is no more than

$$2\|(s-t)\mathbf{x}\| + 2\|\mathbf{y}\| \le 2\,(|s|+|t|)\,\|\mathbf{x}\| + 2\|\mathbf{y}\|.$$

The second factor, being the difference of norms, is smaller in value (Theorem 1.8) than the norm of the difference; that is,

$$\begin{aligned} \big|\|(s-t)\mathbf{x}+\mathbf{y}\| - \|(s-t)\mathbf{x}-\mathbf{y}\|\big| &= \big|\|(s-t)\mathbf{x}+\mathbf{y}\| - \|\mathbf{y}-(s-t)\mathbf{x}\|\big| \\ &\le \|2(s-t)\mathbf{x}\| = 2\|\mathbf{x}\|\,|s-t|. \end{aligned}$$

It follows that as $s \to t$, $|f(s) - f(t)|$ tends to zero. Therefore, f is continuous.

A theorem from the single-variable case (Exercise 7) states that every additive continuous real function satisfies $f(s) = sf(1)$ for all s. Hence

$$(s\mathbf{x})\text{ DOT }\mathbf{y} = s(\mathbf{x}\text{ DOT }\mathbf{y}).$$

That is, DOT is a homogeneous operation. This completes the proof that DOT is an inner product.

(b) This was demonstrated in the third paragraph of the proof in part (a). □

Exercises

1. Let $f(x) := 1 - 2x^2$, $g(x) := 1 + x^2$. Evaluate $\|f\|_0$, $\|g\|_0$, and $\|f - g\|_0$. Do $\|f\|_2$ and $\|g\|_2$ also add up to $\|f - g\|_2$?

2. From the functions in Example 1, viewed as members of $C_0[0, 1]$, find two whose distance is 2 and two whose distance is less than 2. (Example 2(b) shows that each such distance exceeds 1. Because the range of each function is $[-1, 1]$, it is clear that no distance can exceed 2.)

3. (a) Show that the functions $F(x) := 1$, $G(x) := \sin^2 x$, and $H(x) := \cos 2x$ are linearly dependent vectors in $C[0, 1]$.

   (b) Show that $f(x) := \sin 2\pi x$ and $g(x) := 2 - \left(x - \frac{1}{4}\right)^2$ are independent.

4. Show that $\|f\|_0$ is positive definite and radially homogeneous ($\|\alpha f\|_0 = |\alpha|\,\|f\|_0$).

5. Prove Theorem 1.15.

6. Let $I[0, 1]$ represent the set of functions (defined and) *integrable* on the unit interval.

   (a) Is $f \bullet\bullet g := \int_0^1 f(x)g(x)\,dx$ defined for all f, g in I?

   (b) Is $f \bullet\bullet g$ symmetric? linear? positive definite?

7. Assume that f is an additive function of one real variable: $f(s + t) = f(s) + f(t)$ for all s, t.

(a) Show that $f(k) = k\, f(1)$ for every natural number k.

(b) Extend part (a) to $f(j) = j\, f(1)$ for every integer j.

(c) Extend (b) to $f(k/m) = (k/m)\, f(1)$ for every rational k/m.

Add the assumption that $f(x)$ is continuous at some fixed value $x = b$.

(d) Show that f is continuous at $x = 0$.

(e) Show that f is continuous everywhere.

(f) Show that $f(x) = x\, f(1)$ for every real x.

2

Sequences in Normed Spaces

Introduction

Our mission is to study continuous functions. To discuss continuity, we must introduce the concept of limit. Our tools will be sequences, whose convergence we define in terms of distance tending to zero. Since we defined (Section 1.5) distance in terms of length in $\mathbf{R}^n$—and more generally, in terms of a norm in a vector space—we will cast our material in the language of norms.

In this chapter, then, we define and prove some important results about convergence of sequences.

2.1 Neighborhoods in a Normed Space

Assume that we are in a vector space in which $\|\mathbf{x}\|$ denotes the length of $\mathbf{x}$ and $\|\mathbf{x} - \mathbf{y}\|$ the distance from $\mathbf{x}$ to $\mathbf{y}$. The sets of points surrounding a given point are used to define closeness, convergence, and related concepts.

Definition. Given a real number $r > 0$ and vector $\mathbf{x}$, we call:

(a) $B(\mathbf{x}, r) := \{\mathbf{y}\colon \|\mathbf{x} - \mathbf{y}\| \leq r\}$ the **ball of radius r centered at $\mathbf{x}$.**

(b) $S(\mathbf{x}, r) := \{\mathbf{y}\colon \|\mathbf{x} - \mathbf{y}\| = r\}$ the **sphere of radius r centered at $\mathbf{x}$.**

(c) $N(\mathbf{x}, r) := \{\mathbf{y}\colon \|\mathbf{x} - \mathbf{y}\| < r\}$ the **neighborhood of radius r centered at $\mathbf{x}$.**

A neighborhood is sometimes called an "open ball." We will save that name until we define "open."

Notice that "sphere" refers to what we think of as the *surface* of the ball. Peel off that skin and you are left with the neighborhood.

In $\mathbf{R}$, the norm is the absolute-value function. Accordingly, we write $\mathbf{x}$ and $\|\mathbf{x}\|$ as x and $\|x\|$. We have

$$\|\mathbf{x}-\mathbf{y}\| := |x-y| < r \Leftrightarrow -r < x-y < r \Leftrightarrow x-r < y < x+r.$$

Therefore, $N(x,r)$ is just the open interval $(x-r, x+r)$ centered at x. Similarly, $B(x,r)$ is the closed interval $[x-r, x+r]$, and $S(x,r)$ is the two-element set $\{x-r, x+r\}$.

In $\mathbf{R}^2$, where our figures reside, we will refer to spheres and balls by their normal two-dimensional names, "circles" and "disks."

The results in this section are useful in a number of our later proofs. What these results do is confirm that certain statements that look obvious in our figures really are justified by our definitions. Their truth will allow us to give very natural geometric arguments in situations where the analytic arguments are opaque, knowing that the pictures can be translated to valid analysis.

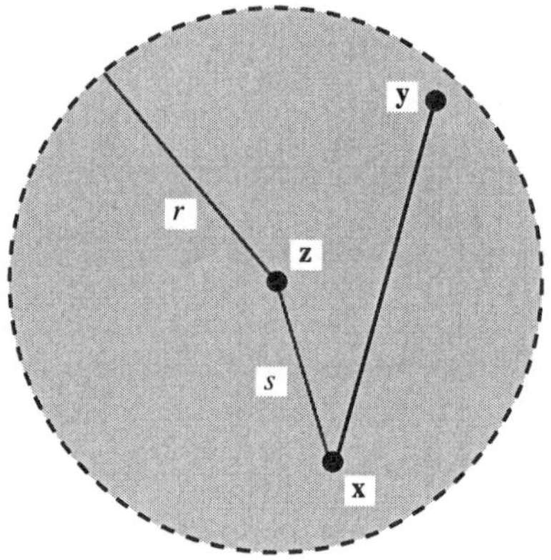

Figure 2.1.

Theorem 2.1. *Suppose* $\mathbf{x}$ *and* $\mathbf{y}$ *are in the neighborhood* $N(\mathbf{z}, r)$, *with* $\mathbf{x}$ *at distance* s *from* $\mathbf{z}$. *Then (see Figure 2.1):*

(a) $\mathbf{x}$ *and* $\mathbf{y}$ *are less than* $r+s$ *apart. In particular, the distance from* $\mathbf{x}$ *to* $\mathbf{y}$ *is less than* $2r$.

(b) *The entire line segment from* $\mathbf{x}$ *to* $\mathbf{y}$ *is contained in* $N(\mathbf{z}, r)$.

Proof. (a) By the triangle inequality, the distances satisfy

$$\|\mathbf{x}-\mathbf{y}\| \le \|\mathbf{x}-\mathbf{z}\| + \|\mathbf{z}-\mathbf{y}\|.$$

Here $\|\mathbf{x}-\mathbf{z}\| = s$, and $\|\mathbf{z}-\mathbf{y}\| < r$ because $\mathbf{y}$ is in $N(\mathbf{z}, r)$. Hence $\|\mathbf{x}-\mathbf{y}\| < s+r$.

(b) The segment from $\mathbf{x}$ to $\mathbf{y}$ consists of the vectors

$$\mathbf{u} = (1-\alpha)\mathbf{x} + \alpha\mathbf{y}, \qquad 0 \le \alpha \le 1.$$

Since $\mathbf{z} = (1-\alpha)\mathbf{z} + \alpha\mathbf{z}$, we have

$$\begin{aligned} \|\mathbf{u}-\mathbf{z}\| &= \|(1-\alpha)\mathbf{x} + \alpha\mathbf{y} - (1-\alpha)\mathbf{z} - \alpha\mathbf{z}\| \\ &\le \|(1-\alpha)\mathbf{x} - (1-\alpha)\mathbf{z}\| + \|\alpha(\mathbf{y}-\mathbf{z})\| \\ &= (1-\alpha)\|\mathbf{x}-\mathbf{z}\| + \alpha\|\mathbf{y}-\mathbf{z}\| && \text{(Why not } |1-\alpha|, |\alpha|\text{?)} \\ &< (1-\alpha)r + \alpha r = r. && \text{(Why } < \text{ instead of } \le\text{?)} \end{aligned}$$

The inequality says that $\mathbf{u}$ is in $N(\mathbf{z}, r)$. This being true for every member of the line segment, we conclude that the segment is a subset of $N(\mathbf{z}, r)$. □

Theorem 2.2. *Let $d := \|\mathbf{x}-\mathbf{y}\| > 0$. View two neighborhoods $R := N(\mathbf{x}, r)$ and $S := N(\mathbf{y}, s)$.*

(a) *If $r + s \le d$, then R and S have no common points. In particular, small enough neighborhoods of one point are disjoint from small enough neighborhoods of any other point.*

(b) *If $r + s > d$, then R and S overlap (their intersection contains some neighborhood).*

(c) *If $r \ge d + s$, then R is so big that it contains S.*

Proof. The situations (a), (b), (c) are illustrated left, middle, and right in Figure 2.2.

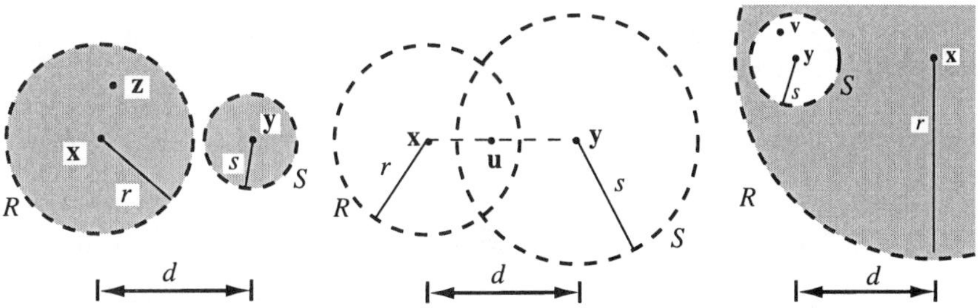

Figure 2.2.

(a) As in the figure on the left let $\mathbf{z}$ be in R; in other words, $\|\mathbf{x}-\mathbf{z}\| < r$. In the triangle $\mathbf{xyz}$,

$$\|\mathbf{x}-\mathbf{y}\| \le \|\mathbf{x}-\mathbf{z}\| + \|\mathbf{z}-\mathbf{y}\| < r + \|\mathbf{z}-\mathbf{y}\|.$$

Therefore,

$$\|\mathbf{z}-\mathbf{y}\| > d - r \ge s.$$

This relation says that $\mathbf{z}$ is not in S.

Similarly, no member of S is in R. We have proved that R and S are disjoint sets.

(b) It is a useful general principle that if α and $\beta := 1 - \alpha$ are nonnegative, then the point $\mathbf{u} := \beta\mathbf{x} + \alpha\mathbf{y}$ splits the segment from $\mathbf{x}$ to $\mathbf{y}$ into pieces whose lengths are in the ratio $\alpha : \beta$. (Exercise 1; compare Example 3 in Section 1.2.)

In the middle figure, to find a vector $\mathbf{u}$ in the intersection we need to go rightward from $\mathbf{x}$, more than the distance $d - s$ (out of the total d) and less than r. Thus, let α be any real number with $\max\{0, (d-s)/d\} < \alpha < \min\{r/d, 1\}$. This choice is possible because

$$0 < r/d;$$
$$(d - s)/d < 1 \qquad (s \text{ being positive});$$

and

$$(d - s)/d < r/d \qquad (\text{owing to } d < r + s).$$

Then

$$\mathbf{u} := (1 - \alpha)\mathbf{x} + \alpha\mathbf{y}$$

is a point in both R and S. It is in R because

$$\|\mathbf{u} - \mathbf{x}\| = \| - \alpha\mathbf{x} + \alpha\mathbf{y}\| = \alpha d < r.$$

It is in S because

$$\begin{aligned}\|\mathbf{u} - \mathbf{y}\| &= \|(1 - \alpha)\mathbf{x} + (\alpha - 1)\mathbf{y}\| = (1 - \alpha)\|\mathbf{x} - \mathbf{y}\| \\ &< (1 - [d - s]/d)\|\mathbf{x} - \mathbf{y}\| = s.\end{aligned}$$

This shows that $R \cap S$ is not empty; we leave the rest of the claim (that the intersection contains a whole neighborhood) to Exercise 4.

(c) In the right-hand figure, if $\mathbf{v} \in S$ ($\mathbf{v}$ is in the smaller circle), then

$$\|\mathbf{v} - \mathbf{x}\| \le \|\mathbf{v} - \mathbf{y}\| + \|\mathbf{y} - \mathbf{x}\| < s + d \le r.$$

Hence $\mathbf{v} \in S$ implies $\mathbf{v} \in N(\mathbf{x}, r)$; that is, $S \subseteq N(\mathbf{x}, r)$. The proof is complete.

Exercises

1. Prove that if α and β are positive and sum to 1, then the point $\beta\mathbf{x} + \alpha\mathbf{y}$ splits the segment from $\mathbf{x}$ to $\mathbf{y}$ into two lengths whose ratio is $\alpha : \beta$.

2. Suppose $\mathbf{y}$ is in a neighborhood $N := N(\mathbf{x}, r)$ of $\mathbf{x}$. Show that:

   (a) Some neighborhood of $\mathbf{y}$ is contained in N.

   (b) Any neighborhood of $\mathbf{y}$ of radius $2r$ or more contains N.

3. This exercise shows that for disjoint balls in an inner product space, the two closest points are located where the segment joining their centers crosses the balls' surfaces. Assume that $B(\mathbf{x}, r)$ and $B(\mathbf{y}, t)$ are disjoint balls in *any normed space*. Let $d := \|\mathbf{x} - \mathbf{y}\|$.

   (a) Show that $r + t < d$.

   (b) Show that if $\mathbf{z}$ is in $B(\mathbf{x}, r)$ ($\|\mathbf{x} - \mathbf{z}\| \leq r$) and $\mathbf{w}$ in $B(\mathbf{y}, t)$, then $\|\mathbf{z} - \mathbf{w}\| \geq d - r - t$.

   (c) In (b), show that the inequality is strict if either $\|\mathbf{x} - \mathbf{z}\| < r$ or $\|\mathbf{y} - \mathbf{w}\| < t$.

   What do (b) and (c) together prove?

   (d) Express in terms of $\mathbf{x}$ and $\mathbf{y}$ the point $\mathbf{u}$ where the segment from $\mathbf{x}$ to $\mathbf{y}$ meets the sphere $S(\mathbf{x}, r)$. (Hint: $\mathbf{u}$ is r/d of the way from $\mathbf{x}$ to $\mathbf{y}$. That fact will help you to express $\mathbf{u}$. You still have to show that your expression is on both the line segment and the sphere.)

   (e) Express $\mathbf{v}$, where the segment meets the sphere $S(\mathbf{y}, t)$, in terms of $\mathbf{x}$ and $\mathbf{y}$.

   (f) Show that $\|\mathbf{u} - \mathbf{v}\| = d - r - t$.

   What have (b)–(f) together established?

   (g) Now assume that we are in an inner product space, where we write $\|\mathbf{x}\|_2$ for the norm. Show that if $\mathbf{z}$ is on the sphere $S(\mathbf{x}, r)$ and $\mathbf{w}$ on $S(\mathbf{y}, t)$, then $\|\mathbf{z} - \mathbf{w}\| > d - r - t$, unless $\mathbf{z}$ and $\mathbf{w}$ are on the segment joining $\mathbf{x}$ and $\mathbf{y}$. (Hint: Theorem 1.10.)

   (h) Show that for $\mathbf{z}$ in $B(\mathbf{x}, r)$ and $\mathbf{w}$ in $B(\mathbf{y}, t)$, $\|\mathbf{z} - \mathbf{w}\| = d - r - t$ iff $\mathbf{z} = \mathbf{u}$ and $\mathbf{w} = \mathbf{v}$.

4. Suppose $N(\mathbf{x}, r)$ and $N(\mathbf{y}, t)$ are overlapping neighborhoods and $\mathbf{u}$ is in their intersection. (See the middle figure in Figure 2.2.) Show that some neighborhood of $\mathbf{u}$ is contained in the intersection.

2.2 Sequences and Convergence

In **R** we define convergence in terms of distance becoming small. We will do the same in any normed space. The definitions and results of the next two sections will be analogues of what we found in the set of real numbers.

Definition.

(a) A **sequence** in a space V is a mapping $\mathbf{f}$ of the natural numbers $\mathbf{N}$ into V. We write $\mathbf{x}_i$ in place of $\mathbf{f}(i)$ and $(\mathbf{x}_i)$ for $\mathbf{f}$. If $j : \mathbf{N} \to \mathbf{N}$ is strictly increasing, then $(\mathbf{x}_{j(i)})$ is a **subsequence** of $(\mathbf{x}_i)$.

(b) We say that $(\mathbf{x}_i)$ **converges to x** if $\|\mathbf{x}_i - \mathbf{x}\| \to 0$ as $i \to \infty$; in this case, we write $(\mathbf{x}_i) \to \mathbf{x}$ or $\mathbf{x}_i \to \mathbf{x}$, and we call $\mathbf{x}$ the **limit** of the sequence.

Notice that convergence of a sequence of vectors means convergence of a related sequence of reals: $(\mathbf{x}_i) \to \mathbf{x}$ means $(\|\mathbf{x}_i - \mathbf{x}\|) \to 0$. The latter means, by definition, that for each $\varepsilon > 0$ there is a corresponding I such that $-\varepsilon < \|\mathbf{x}_i - \mathbf{x}\| < \varepsilon$ for all terms having index $i \geq I$. We can drop the part $-\varepsilon < \|\mathbf{x}_i - \mathbf{x}\|$, since it is automatic.

Example 1. For any $\mathbf{x}$,

$$\mathbf{x}_i := \frac{\mathbf{x}}{i} \to \mathbf{O},$$

because as long as $\mathbf{x}$ is fixed,

$$\|\mathbf{x}_i - \mathbf{O}\| = \left\| \frac{1}{i}\mathbf{x} \right\| = \frac{1}{i}\|\mathbf{x}\| \to 0 \quad \text{as } i \to \infty.$$

Theorem 2.3. *Limits of sequences are*:

(a) *Unique: If* $(\mathbf{x}_i) \to \mathbf{x}$ *and* $(\mathbf{x}_i) \to \mathbf{y}$, *then* $\mathbf{x} = \mathbf{y}$.

(b) *Linear: If* $(\mathbf{x}_i) \to \mathbf{x}$ *and* $(\mathbf{y}_i) \to \mathbf{y}$, *then* $(\alpha\mathbf{x}_i + \beta\mathbf{y}_i) \to \alpha\mathbf{x} + \beta\mathbf{y}$.

Proof. (a) Assume that $(\mathbf{x}_i)$ converges to both $\mathbf{x}$ and $\mathbf{y}$. We have

$$0 \leq \|\mathbf{x} - \mathbf{y}\| \leq \|\mathbf{x} - \mathbf{x}_i\| + \|\mathbf{x}_i - \mathbf{y}\| \quad \text{for all } i.$$

Since by definition $\|\mathbf{x} - \mathbf{x}_i\|$ and $\|\mathbf{x}_i - \mathbf{y}\|$ must vanish as $i \to \infty$, we conclude that $0 = \|\mathbf{x} - \mathbf{y}\|$. Of necessity, $\mathbf{x} = \mathbf{y}$.

(b) Exercise 3a. □

In view of Theorem 2.3, we are justified in saying "the" limit of a convergent sequence. We write $\lim \mathbf{x}_i$ for such a limit.

Theorem 2.4. *In a normed space, if a sequence* $(\mathbf{x}_i)$ *converges to* $\mathbf{x}$, *then*:

(a) *All of its subsequences converge to* $\mathbf{x}$.

(b) *The limit of the norms is the norm of the limit*: $\|\mathbf{x}_i\| \to \|\mathbf{x}\|$.

Proof. (a) This is an easy proof, but it uses a general principle that is worth mentioning explicitly: If $(\mathbf{x}_{j(i)})$ is a subsequence of $(\mathbf{x}_i)$, then $j(i) \geq i$ for every i. (The statement may be proved by induction.)

With that in mind, assume that $(\mathbf{x}_i) \to \mathbf{x}$ and $(\mathbf{x}_{j(i)})$ is a subsequence. Given $\varepsilon > 0$, there exists I such that $i \geq I$ causes $\|\mathbf{x} - \mathbf{x}_i\| < \varepsilon$. By the principle cited,

$$i \geq I \Rightarrow j(i) \geq I \Rightarrow \|\mathbf{x} - \mathbf{x}_{j(i)}\| < \varepsilon.$$

We have proved that $(\mathbf{x}_{j(i)}) \to \mathbf{x}$.

(b) Exercise 4a. □

Example 2. Theorem 2.4 gives us some ammunition for denying convergence. It says that a sequence is divergent if it has either two subsequences that fail to act the same—for example, that converge to different limits—or badly behaved norms.

Define $\mathbf{x}_i := (\cos i, 0)$ in $\mathbf{R}^2$. Since $2\pi = 6.283\ldots$, each of the integers $j = 6$, $62, 628, \ldots$ is just under, but within 1 of, a multiple of 2π. Therefore,

$$\cos(-1) \leq \cos j \leq \cos 2\pi = 1 \qquad \text{for each such integer,}$$

and

$$\|\mathbf{x}_j\|_2 \geq \cos(-1) \geqq .54.$$

It immediately follows that $(\mathbf{x}_i)$ does not converge to $\mathbf{O}$. In fact, for any $\mathbf{x} = (a, b)$ with $a \leq 0$, we have

$$\|\mathbf{x} - \mathbf{x}_j\|_2 = \left([a - \cos j]^2 + b^2\right)^{1/2} \geq \cos j - a \geq .54,$$

so that the subsequence $\mathbf{x}_6, \mathbf{x}_{62}, \mathbf{x}_{628}, \ldots$ does not converge to $\mathbf{x}$. Therefore, the original sequence cannot converge to $\mathbf{x}$ (Theorem 2.4(a)).

Similarly, if $\mathbf{y} = (c, d)$ with $c \geq 0$, then $k = j - 3$ defines the sequence $k = 3, 59, 625, \ldots$, whose cosines are between $\cos(-4)$ and -1. We then have

$$\|\mathbf{y} - \mathbf{x}_k\|_2 \geqq c - \cos(-4) \geqq .65,$$

making it impossible for $(\mathbf{x}_i)$ to converge to $\mathbf{y}$.

We conclude that $(\mathbf{x}_i)$ does not have a limit.

Theorem 2.5. *The sequence $(\mathbf{x}_i)$ converges to $\mathbf{x}$ iff for every neighborhood N of $\mathbf{x}$ there is a corresponding index I with the property that $\mathbf{x}_i$ is in N for each $i \geq I$.*

Proof. Exercise 5.

There is one other kind of convergence that we will find it useful to define.

Definition. We say that $(\mathbf{x}_i)$ **converges to infinity**, and we write $(\mathbf{x}_i) \to \infty$ or $\lim \mathbf{x}_i = \infty$, if for every $M > 0$ there exists I such that $i \geq I$ implies $\|\mathbf{x}_i\| \geq M$.

Clearly, $(\mathbf{x}_i)$ converges to infinity iff the real sequence $(\|\mathbf{x}_i\|)$ converges to $+\infty$.

Example 3. (a) In $\mathbf{R}$ we call the sequence defined by $x_i := (-i)^i$ divergent because it oscillates indefinitely widely. But under our definition, $\|x_i\| = i^i \to \infty$ means that (x_i) converges to infinity in $\mathbf{R}^1$. Here is one area where we distinguish $\mathbf{R}^1$ from $\mathbf{R}$.

(b) Similarly, in $\mathbf{R}^2$, the sequence $(1, 2), (-3, 1), (-1, -4), (5, -1), \ldots$ (switch coordinates, reversing the first sign and incrementing the larger absolute value) changes direction—even quadrant—from term to term, but has

$$\|\mathbf{y}_i\|_2 = \left(1 + [i+1]^2\right)^{1/2} \to \infty.$$

(c) For any $\mathbf{x} \neq \mathbf{O}$ in an arbitrary space, $\|i\mathbf{x}\| = i\,\|\mathbf{x}\| \to \infty$, and so $(i\mathbf{x}) \to \infty$.

Theorem 2.6. *The sequence* $(\mathbf{x}_i)$ *converges to* ∞ *iff for every neighborhood* N *of any vector* $\mathbf{x}$ *there exists a corresponding* I *such that* $\mathbf{x}_i$ *is outside* N *for each* $i \geq I$.

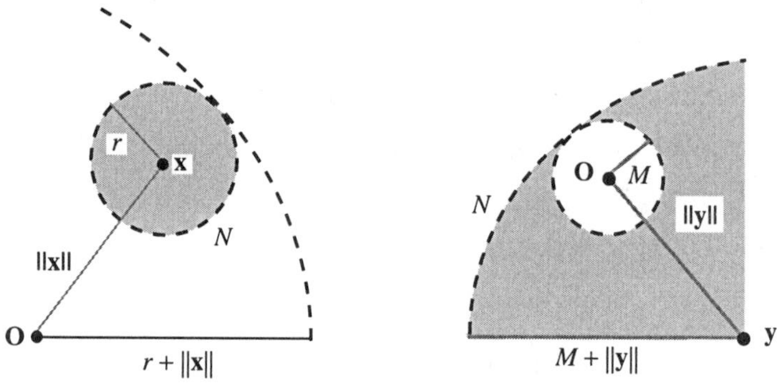

Figure 2.3.

Proof. $\Rightarrow$ Assume $\mathbf{x}_i \to \infty$. Let $N := N(\mathbf{x}, r)$ be any neighborhood. The left half of Figure 2.3 (compare Theorem 2.2(c)) shows us that $N(\mathbf{O}, r + \|\mathbf{x}\|)$ contains N. Set $M := r + \|\mathbf{x}\|$. By definition, there exists I such that

$$i \geq I \Rightarrow \|\mathbf{x}_i\| \geq M.$$

Therefore, $\mathbf{x}_i$ is not in $N(\mathbf{O}, M)$; since $N \subseteq N(\mathbf{O}, M)$, $i \geq I$ forces $\mathbf{x}_i \notin N$.

$\Leftarrow$ Assume that there is even a single $\mathbf{y}$ with the stated property: For every neighborhood $N(\mathbf{y}, s)$ there exists a corresponding I such that $\mathbf{x}_I, \mathbf{x}_{I+1}, \ldots$ are outside $N(\mathbf{y}, s)$.

Choose any positive M. We know that each neighborhood $N(\mathbf{O}, M)$ of the origin is contained in the neighborhood $N := N(\mathbf{y}, M + \|\mathbf{y}\|)$ of $\mathbf{y}$ as in the right half of Figure 2.3. By assumption, corresponding to this last neighborhood, there is I

such that $i \geq I$ implies $\mathbf{x}_i$ is outside it. Therefore, $i \geq I \Rightarrow \mathbf{x}_i$ is outside $N(\mathbf{O}, M)$, equivalent to $\|\mathbf{x}_i\| \geq M$. We have proved that $(\mathbf{x}_i) \to \infty$. □

These same arguments show that $(\mathbf{x}_i)$ approaches ∞ iff its terms go outside $B(\mathbf{x}, r)$ for every $\mathbf{x}$ and r. Hence to be near infinity is to be in the complement of $B(\mathbf{x}, r)$, that is, in $\{\mathbf{y}\colon \|\mathbf{x} - \mathbf{y}\| > r\}$. For this reason, we will refer to this last set as a **neighborhood of infinity**.

In most situations we will allow "$(\mathbf{x}_i)$ converges" to mean that $(\mathbf{x}_i)$ converges either to a vector $\mathbf{x}$ or to ∞. If we need to, we will specify "$(\mathbf{x}_i)$ converges to a vector" or "to a finite limit." Most theorems in this section stay valid, with minor modifications, if "converges to $\mathbf{x}$" is replaced by "converges to infinity" or just "converges." Thus, if the sequence $(\mathbf{x}_i)$ converges, then:

Theorem 2.3(a)*. *Its limit is unique.*

Theorem 2.4(a)*. *All its subsequences also converge, and to the same limit.*

Theorem 2.4(b)*. *The limit of the norms is the norm of the limit.*

Theorem 2.5*. *For every neighborhood N of its limit, there exists I such that $i \geq I$ implies $\mathbf{x}_i$ is in N; and conversely.*

We have one final definition.

Definition. The sequence $(\mathbf{x}_i)$ is **bounded** if $\sup\{\|\mathbf{x}_i\|\}$ is finite.

Theorem 2.7. *A sequence $(\mathbf{x}_i)$ is:*

(a) *bounded iff there exists a neighborhood N of the origin such that $\{\mathbf{x}_i\} \subseteq N$.*

(b) *bounded iff there exists some neighborhood $N(\mathbf{x}, r)$ of some point $\mathbf{x}$ such that $\{\mathbf{x}_i\} \subseteq N(\mathbf{x}, r)$.*

(c) *bounded if it is convergent to a vector, but not conversely.*

(d) *unbounded if it converges to infinity, but not conversely.*

Proof. (a) Exercise 6a.

(b) $\Rightarrow$ Assume that $(\mathbf{x}_i)$ is bounded. By (a), $\{\mathbf{x}_i\}$ is contained in a neighborhood of the origin. We simply take $\mathbf{x} = \mathbf{O}$.

$\Leftarrow$ Assume that $\{\mathbf{x}_i\}$ is contained in $N(\mathbf{x}, r)$. We know (left part of Figure 2.3) that the neighborhood $N(\mathbf{O}, r + \|\mathbf{x}\|)$ is big enough to encompass $N(\mathbf{x}, r)$. Hence $\{\mathbf{x}_i\}$ is contained in $N(\mathbf{O}, r + \|\mathbf{x}\|)$. By (a) the sequence is bounded.

(c) Assume that $\mathbf{x}_i \to \mathbf{x}$. By Theorem 2.5, if we choose a neighborhood $N(\mathbf{x}, r)$, then the terms of $(\mathbf{x}_i)$ are eventually in it. Choose $N(\mathbf{x}, 1)$. There exists I such that

$$i \geq I \Rightarrow \mathbf{x}_i \in N(\mathbf{x}, 1);$$

that is, $\mathbf{x}_I, \mathbf{x}_{I+1}, \ldots$ are in $N(\mathbf{x}, 1)$. Let

$$M := \|\mathbf{x}_1\| + \cdots + \|\mathbf{x}_{I-1}\| + \|\mathbf{x}\| + 1.$$

Then the neighborhood $N(\mathbf{O}, M)$ encloses each of $\mathbf{x}_1, \ldots, \mathbf{x}_{I-1}$, as well as $N(\mathbf{x}, 1)$. Therefore, every member of $\{\mathbf{x}_i\}$ is in $N(\mathbf{O}, M)$. By (a), the sequence is bounded.

We leave the "not conversely" in (c) to Exercise 7a, and part (d) to Exercises 6b, 7b.

Exercises

1. Which of these sequences have finite limits in $\mathbf{R}^2$?

   (a) $\mathbf{x}_i := (i \sin[1/i], [\cos i]/i)$

   (b) $\mathbf{y}_i := (\cos[i\pi/2], \sin[i\pi/2])$

   (c) $\mathbf{z}_i := (i \cos[i\pi/2], i \sin[i\pi/2])$

   (d) $\mathbf{u}_i := (2^i + [-2]^i, 2^i + [-2]^i)$.

2. Which sequences in Exercise 1 converge to infinity?

3. (a) Prove Theorem 2.3(b).

   (b) Is Theorem 2.3(b) still valid if $(\mathbf{x}_i)$ and $(\mathbf{y}_i)$ converge to infinity?

4. (a) Prove Theorem 2.4(b).

   (b) Disprove the converse, by finding an example of a sequence $(\mathbf{x}_i)$ with $\|\mathbf{x}_i\| \to \|\mathbf{x}\|$ but $(\mathbf{x}_i)$ not converging to $\mathbf{x}$.

   (c) Show that $\mathbf{x}_i \to \mathbf{O}$ iff $\|\mathbf{x}_i\| \to 0$.

5. Prove Theorem 2.5.

6. (a) Prove part (a) of Theorem 2.7.

   (b) Prove that if a sequence converges to infinity, then it is unbounded.

7. Give counterexamples in $\mathbf{R}^2$ to show that:

   (a) A bounded sequence need not be convergent.

   (b) An unbounded sequence need not converge to infinity.

2.3 Convergence in Euclidean Space

In this section we give three theorems essential to our work in Euclidean space. The second of them leads to an idea that is useful in general normed spaces.

In $\mathbf{R}^n$ it is possible to reduce the question of convergence of a sequence $(\mathbf{x}_i)$ to the question of convergence in each of its coordinates. To avoid double subscripts, we define the **coordinate projection mappings** ("projections"): π_j is defined by

$$\pi_j((x_1, x_2, \ldots, x_n)) := x_j, \qquad 1 \le j \le n.$$

Thus, $\pi_j(\mathbf{x})$ is the jth coordinate of $\mathbf{x}$.

Theorem 2.8. *In* $\mathbf{R}^n$, $(\mathbf{x}_i)$ *converges to a finite limit iff for each fixed* j, *the real sequence* $(\pi_j(\mathbf{x}_i))$ *converges to a real number; in either case,* $\pi_j(\lim_{i\to\infty}\mathbf{x}_i) = \lim_{i\to\infty}\pi_j(\mathbf{x}_i)$.

Proof. $\Rightarrow$ Assume that $(\mathbf{x}_i)$ converges to $\mathbf{y}$.

The convergence of the projected sequences comes from the fact that projections preserve or reduce distances:

$$\begin{aligned}|\pi_j(\mathbf{y}) - \pi_j(\mathbf{z})| &:= |y_j - z_j| = \left[(y_j - z_j)^2\right]^{1/2}\\ &\le \left[(y_1 - z_1)^2 + \cdots + (y_n - z_n)^2\right]^{1/2}\\ &= \|\mathbf{y} - \mathbf{z}\|_2.\end{aligned}$$

In particular,

$$|\pi_j(\mathbf{y}) - \pi_j(\mathbf{x}_i)| \le \|\mathbf{y} - \mathbf{x}_i\|_2.$$

By assumption, the right-hand side approaches zero. Hence so does the left. To say that

$$|\pi_j(\mathbf{y}) - \pi_j(\mathbf{x}_i)| \to 0$$

is to say that $(\pi_j(\mathbf{x}_i))$ converges to $\pi_j(\mathbf{y})$.

$\Leftarrow$ Assume that each $\pi_j(\mathbf{x}_i)$ converges as $i \to \infty$, and denote the limit by y_j. By definition of such convergence, for each $\varepsilon > 0$ there exist

$$I_1 \text{ such that } i \ \ge\ I_1 \text{ implies } |\pi_1(\mathbf{x}_i) - y_1| < \varepsilon/n;$$

$$\ldots$$

$$I_n \text{ such that } i \ \ge\ I_n \text{ implies } |\pi_n(\mathbf{x}_i) - y_n| < \varepsilon/n.$$

Set $I := I_1 + \cdots + I_n$. Then $i \ge I$ implies

$$\begin{aligned}\|\mathbf{x}_i - (y_1, \ldots, y_n)\|_2 &:= \left[(\pi_1(\mathbf{x}_i) - y_1)^2 + \cdots + (\pi_n(\mathbf{x}_i) - y_n)^2\right]^{1/2}\\ &< \left[n\varepsilon^2/n^2\right]^{1/2} \le \varepsilon.\end{aligned}$$

We have shown that $\mathbf{x}_i \to (y_1, \ldots, y_n)$. □

Example 1. The effect of Theorem 1 is to allow us to employ in $\mathbf{R}^n$ our knowledge of limits of sequences in $\mathbf{R}$. Thus, in $\mathbf{R}^3$,

(a)
$$\mathbf{x}_i := \left(\left[1 + \frac{1}{i}\right]^i, \left[1 + \frac{2}{i}\right]^{2i}, \left[1 + \frac{3}{i}\right]^{3i}\right)$$

converges to (e, e^4, e^9), because the three coordinate sequences separately converge to the indicated limits.

(b)
$$\mathbf{y}_i := \left(\cos \pi i, \cos[\pi i]^2, \cos[\pi i]^3\right)$$

diverges, because the first-coordinate sequence is $-1, 1, -1, 1, \ldots$, making convergence to a finite limit impossible (Theorem 2.8); and all three coordinate sequences are bounded, making convergence to infinity impossible (Reason?).

(c)
$$\mathbf{z}_i := \left(\frac{i}{\ln 2i}, \frac{i}{i+1}, \frac{i}{e^i} \right)$$

has $\pi_1(\mathbf{z}_i) \to \infty$. This alone guarantees that $(\mathbf{z}_i) \to \infty$. (Compare Exercise 6.)

In the first half of the proof of Theorem 2.8 we noted that a projection π_j satisfies

$$|\pi_j(\mathbf{y}) - \pi_j(\mathbf{z})| \le \|\mathbf{y} - \mathbf{z}\|_2.$$

In words, the distance between the real numbers $\pi_j(\mathbf{y})$ and $\pi_j(\mathbf{z})$ is no more than the distance between the vectors $\mathbf{y}$ and $\mathbf{z}$. We will slightly abuse the term **contractive** by using it to describe any function with the property that its images have the same or smaller separation than their preimages.

Definition. A sequence $(\mathbf{x}_i)$ in a normed space is **Cauchy** if

$$\lim_{i,j\to\infty} \|\mathbf{x}_i - \mathbf{x}_j\| = 0.$$

To say that the limit of $\|\mathbf{x}_i - \mathbf{x}_j\|$ is 0 is to say that whenever $\varepsilon > 0$ is specified, we can find a corresponding $I(\varepsilon)$ such that $\|\mathbf{x}_i - \mathbf{x}_j\| < \varepsilon$ for the terms with both i and $j \ge I(\varepsilon)$. For a sequence to converge to a vector, its terms must get close to that vector. For a sequence to be Cauchy, its terms must get close to each other. Our next theorem says that in $\mathbf{R}^n$, as in $\mathbf{R}$, convergence to a vector and Cauchy-ness are equivalent.

Theorem 2.9. (Cauchy's Criterion) *In* $\mathbf{R}^n$, *a sequence has a finite limit iff it is Cauchy.*

Proof. $\Rightarrow$ Exercise 7. It is worth noting that nothing about this proof need be limited to $\mathbf{R}^n$, or even to a normed space. In any set where there is a metric, every sequence convergent to a member of the set has to be Cauchy.

$\Leftarrow$ Assume that $(\mathbf{x}_i)$ is Cauchy. We will produce a limit for $(\mathbf{x}_i)$.

For each fixed $k = 1, \ldots, n$, look at the kth coordinates of the vectors $\mathbf{x}_i$. These make up the sequence $(\pi_k(\mathbf{x}_i))$, where π_k is the projection map. Because projections are contractive, we have

$$|\pi_k(\mathbf{x}_i) - \pi_k(\mathbf{x}_j)| \le \|\mathbf{x}_i - \mathbf{x}_j\|_2.$$

Since the right-hand side approaches 0 as $i, j \to \infty$, the same is true of the left. That is, $(\pi_k(\mathbf{x}_i))$ is a Cauchy sequence of real numbers. There must therefore exist a real y_k such that $\pi_k(\mathbf{x}_i) \to y_k$. By Theorem 2.8, $(\mathbf{x}_i) \to (y_1, \ldots, y_n)$. □

Theorem 2.10 (The Bolzano–Weierstrass Theorem) *In $\mathbf{R}^n$, every bounded sequence has a subsequence that converges to a finite limit.*

Proof. Assume that M is a constant such that each $\mathbf{x}_i$ satisfies $\|\mathbf{x}_i\|_2 < M$.

The first coordinates $\pi_1(\mathbf{x}_i)$ satisfy

$$|\pi_1(\mathbf{x}_i)| < M \quad \text{for all } i. \text{ (Reason?)}$$

In other words, $(\pi_1(\mathbf{x}_i))$ is a bounded sequence of real numbers. By the Bolzano–Weierstrass theorem in $\mathbf{R}$ [Ross, Section 11], there is a subsequence $(\pi_1(\mathbf{x}_{j(i)}))$ converging to a real number y_1.

Look next at the second coordinates $(\pi_2(\mathbf{x}_{j(i)}))$ *in the subsequence.* From

$$|\pi_2(\mathbf{x}_{j(i)})| \le \|\mathbf{x}_{j(i)}\|_2 < M,$$

we infer that $(\pi_2(\mathbf{x}_{j(i)}))$ is bounded. There is therefore a subsequence $(\pi_2(\mathbf{x}_{k(j(i))}))$ converging to a real y_2. Also, since $(\pi_1(\mathbf{x}_{k(j(i))}))$ is a subsequence of $(\pi_1(\mathbf{x}_{j(i)}))$, we know (Theorem 2.4(a)) that $(\pi_1(\mathbf{x}_{k(j(i))}))$ converges to y_1.

Continuing in this fashion, we arrive at a sub-sub-$\cdots$-subsequence $(\mathbf{x}_{p(\ldots(i)\ldots)})$ of the original $(\mathbf{x}_i)$ having the property that

$$\begin{aligned}
\pi_1(\mathbf{x}_{p(\ldots(i)\ldots)}) &\to y_1, \\
\pi_2(\mathbf{x}_{p(\ldots(i)\ldots)}) &\to y_2, \\
&\cdots \\
\pi_n(\mathbf{x}_{p(\ldots(i)\ldots)}) &\to y_n.
\end{aligned}$$

Again using Theorem 2.8, we conclude that $(\mathbf{x}_{p(\ldots(i)\ldots)}) \to (y_1, y_2, \ldots, y_n)$. We have found a convergent subsequence. □

Observe that in this proof of the Bolzano–Weierstrass theorem we did a kind of filtering to produce subsequences of subsequences, in a process that would have failed if the number of coordinates (= the number of projections) had been infinite. Moreover, in the proofs of both the Bolzano–Weierstrass theorem and Cauchy's criterion, we invoked Theorem 2.8. Look back at that proof. It, too, depends on exhausting the dimension. Therefore, our *arguments* fail in an infinite-dimensional space. That leaves the question of whether the theorems can be proved via some other line of reasoning. We treat that question in the next section.

Since we have mentioned convergent subsequences, we formalize the notion.

Definition. In any normed space, let $(\mathbf{x}_{j(i)})$ be a convergent subsequence of $(\mathbf{x}_i)$. Then the limit of $(\mathbf{x}_{j(i)})$ (which may be infinite) is a **sublimit** of $(\mathbf{x}_i)$.

["Sublimit" may not be a standard name, but it captures the meaning. I first heard it from a wonderfully erudite future economist named Ted Fernandez.]

Example 2. In $\mathbf{R}^2$:

(a)
$$\mathbf{x}_i := (\cos[i\pi/2], \sin[i\pi/2])$$

oscillates among $(\pm 1, 0)$ and $(0, \pm 1)$. Each of these is a sublimit. For example, $(-1, 0) = \mathbf{x}_2 = \mathbf{x}_6 = \cdots$ is the limit of the subsequence $(\mathbf{x}_{4i-2})$.

(b)
$$\mathbf{y}_i := \left(\exp([-1]^i i), \exp([-1]^i 2i)\right)$$

is always on the parabola $y = x^2$ (Verify!), with the odd terms tending toward $(0, 0)$ and the even terms tending to ∞. Hence $\mathbf{O}$ and ∞ are sublimits of $(\mathbf{y}_i)$.

Theorem 2.11. *In any normed linear space*:

(a) *A vector* $\mathbf{x}$ *(or* ∞*) is a sublimit of sequence* $(\mathbf{x}_i)$ *iff every neighborhood of* $\mathbf{x}$ *(respectively* ∞*) contains infinitely many terms of the sequence.*

(b) *Infinity is a sublimit of* $(\mathbf{x}_i)$ *iff the sequence is unbounded.*

Proof. (a) (Note that "infinitely many terms" does not mean infinitely many vectors. In Example 2(a), the single vector $(-1, 0)$ *is* infinitely many terms of the sequence.) We deal with the infinite case and leave the proof for the finite case as Exercise 9.

$\Rightarrow$ Assume that ∞ is a sublimit of $(\mathbf{x}_i)$ and that N is a neighborhood of ∞. By definition, some subsequence $(\mathbf{x}_{j(i)})$ goes to ∞. By Theorem 2.5*, there is I such that $i \geq I$ implies $\mathbf{x}_{j(i)} \in N$. Hence N has the terms $\mathbf{x}_{j(I)}, \mathbf{x}_{j(I+1)}, \ldots$.

$\Leftarrow$ Assume that each neighborhood of ∞ contains infinitely many terms of $(\mathbf{x}_i)$. First, $N_1 := \{\mathbf{x}\colon \|\mathbf{x}\| > 1\}$ is such a neighborhood. Pick some term of the sequence there, and label it $\mathbf{x}_{j(1)}$; note that $\|\mathbf{x}_{j(1)}\| > 1$. Let

$$M_1 := \|\mathbf{x}_1\| + \|\mathbf{x}_2\| + \cdots + \|\mathbf{x}_{j(1)}\| + 1.$$

Next, the neighborhood $N_2 := \{\mathbf{x}\colon \|\mathbf{x}\| > M_1\}$ contains terms from the sequence, but none of $\mathbf{x}_1, \ldots \mathbf{x}_{j(1)}$ is that far from the origin. Hence there is an $\mathbf{x}_{j(2)}$ out there that has

$$j(2) > j(1) \quad \text{and} \quad \|\mathbf{x}_{j(2)}\| > M_1 > 2.$$

Continuing the pattern, we end up with a string of vectors $\mathbf{x}_{j(i)}$ satisfying

$$j(i+1) > j(i), \quad \text{so that the string is a subsequence, and}$$
$$\|\mathbf{x}_{j(i)}\| > i, \quad \text{so that the subsequence converges to } \infty.$$

Therefore, ∞ is a sublimit of $(\mathbf{x}_i)$.

(b) If ∞ is a sublimit of $(\mathbf{x}_i)$, then by (a) each neighborhood $\{\mathbf{x}\colon \|\mathbf{x}\| > M\}$ contains terms of the sequence. That is, for each real M, there are terms $\mathbf{x}_i$ with $\|\mathbf{x}_i\| > M$. By definition, $(\mathbf{x}_i)$ is unbounded.

If ∞ is not a sublimit, then by (a) some neighborhood $\{\mathbf{x}\colon \|\mathbf{x}\| > M_0\}$ has only finitely many terms of the sequence. Call them $\mathbf{x}_{j(1)}, \ldots, \mathbf{x}_{j(k)}$. Each vector from this list satisfies

$$\|\mathbf{x}\| < T := M_0 + \|\mathbf{x}_{j(1)}\| + \cdots + \|\mathbf{x}_{j(k)}\| + 1,$$

because its norm is a term on the right; and each vector $\mathbf{x}_i$ not on the list also satisfies the inequality, because they are all in $\{\mathbf{x}\colon \|\mathbf{x}\| \le M_0\}$. Hence $\|\mathbf{x}_i\| < T$ for every i, and $(\mathbf{x}_i)$ is bounded. □

Theorem 2.11(b) says that if $(\mathbf{x}_i)$ is unbounded, then some subsequence of $(\mathbf{x}_i)$ converges to infinity. The Bolzano–Weierstrass theorem adds that if $(\mathbf{x}_i)$ in $\mathbf{R}^n$ is bounded, then some subsequence converges to a finite limit. Therefore, we may state that every sequence in $\mathbf{R}^n$ has a convergent subsequence (possibly with an infinite limit), and thereby a sublimit.

Exercises

1. Which of the following sequences converge?

   (a) $\big(i/[2i-1], \cos[1/i], \tan^{-1} i\big)$.

   (b) $\big(i \sin[1/i], i^2 - i^2 \cos[1/i], \tan[\pi(2i-1)/4]\big)$.

   (c) $(\cos i, e^i, 1/i)$.

2. (a) Find an example of a sequence that has no Cauchy subsequences.

   (b) Characterize all of the sequences in $\mathbf{R}^n$ that have the property described in (a).

3. In the two parts of Example 2, are there any sublimits other than the ones named?

4. For each sequence in Exercise 1, list all the sublimits.

5. In $\mathbf{R}^1$, does $\big([-1]^i i\big)$ have a unique sublimit?

6. Given a sequence $(\mathbf{x}_i)$ in $\mathbf{R}^n$:

   (a) Prove that if any one coordinate sequence $(\pi_j(\mathbf{x}_i))$ has infinite limit, then $(\mathbf{x}_i) \to \infty$.

   (b) Prove that the converse of (a) is false.

7. (a) Prove that a convergent sequence in a normed space is necessarily Cauchy.

   (b) Prove that a Cauchy sequence in a normed space is necessarily bounded. [This is Lemma 10.10 in Ross.]

8. (a) Prove that if a sequence converges, then it has a unique sublimit.

   (b) In $\mathbf{R}^n$, prove that if a sequence has a unique sublimit, then it converges.

9. Prove that a vector $\mathbf{x}$ is a sublimit of a sequence $(\mathbf{x}_i)$ iff every neighborhood of $\mathbf{x}$ contains infinitely many terms of the sequence.

10. Prove that a vector $\mathbf{x}$ (or ∞) is a sublimit of a sequence $(\mathbf{x}_i)$ iff for every integer I and positive ε, there is an $i > I$ such that $\|\mathbf{x}_i - \mathbf{x}\| < \varepsilon$ (respectively $\|\mathbf{x}_i\| > 1/\epsilon$). (One consequence of this is that a term of a sequence is not automatically a sublimit; the sequence must come back to it repeatedly or toward it to make it a sublimit.)

2.4 Convergence in an Infinite-Dimensional Space

We have remarked that our proofs of Cauchy's criterion and the Bolzano–Weierstrass theorem depend on the finite dimension of $\mathbf{R}^n$ over $\mathbf{R}$. In this section we show that it would not do any good to look for better arguments; in the infinite-dimensional spaces that we address, the two theorems actually fail.

For both $C_0[0, 1]$ and $C_2[0, 1]$, we consider first the Bolzano–Weierstrass theorem. This choice simplifies matters for two reasons. First, the Bolzano–Weierstrass theorem implies Cauchy's criterion. That is the import of the next result. Therefore, if you find that the Bolzano–Weierstrass theorem holds, then you know that Cauchy's criterion does as well. Second—or, the bad news is—in a space of infinite dimension, the Bolzano–Weierstrass theorem never holds. The proof of that has to wait for three chapters, but we will illustrate the statement for our two spaces.

Theorem 2.12. *In any space where the Bolzano–Weierstrass theorem holds, Cauchy's criterion does likewise. In detail: Assume that in a given normed space, every bounded sequence has a subsequence that converges (necessarily to a finite limit); then, a sequence in the space has a finite limit iff it is Cauchy.*

Proof. Exercise 5. (The exercise guides the reader through a coordinate-free proof. The coordinatelessness avoids the problem of our earlier arguments.)

We start with $C_2[0, 1]$. (Refer to Section 1.6.) With respect to the Bolzano–Weierstrass theorem, we already have enough evidence.

Theorem 2.13. *In* $C_2[0, 1]$ *there are bounded sequences with no convergent subsequences.*

Proof. In Example 1 of Section 1.6, we argued that

$$s_i(x) := \sin 2\pi i x$$

defines a sequence of functions that are equally distant from each other.

Whenever that happens, the sequence involved has no convergent subsequences. For let $(s_{k(i)})$ be a subsequence. Then it is impossible to have $\|s_{k(i)} - s_{k(j)}\|_2 \to 0$, since each such distance is 1. Therefore, $(s_{k(i)})$ is not Cauchy. As we now know (Exercise 7 in Section 2.3), it follows that this subsequence is not convergent to a vector. Since the functions are also of equal norm, $(s_{k(i)})$ cannot converge to infinity either. □

The situation in this last proof can be duplicated in any infinite-dimensional inner product space. It is a principle of linear algebra—the Gram–Schmidt process [Lay, Section 6.4]—that in a space with a dot product, it is possible to find orthonormal sets with as many vectors as the dimension allows. If the dimension is infinite, then it is possible to pick an orthonormal sequence $(\mathbf{x}_i)$. In such a sequence, the intervector distances $\|\mathbf{x}_i - \mathbf{x}_j\|_2$, $i \neq j$, are all the same (How big?), and the proof carries.

The failure of the Bolzano–Weierstrass theorem means that we have to decide Cauchy's criterion some other way. Toward settling the question, we build a Cauchy sequence that does not converge.

Example 1. For $i \geq 2$, write

$$F_i(x) := \begin{cases} 0 & \text{for} \quad x \leq \frac{1}{2} - \frac{1}{i} \\ \frac{i}{2}\left(x - \frac{1}{2} + \frac{1}{i}\right) & \frac{1}{2} - \frac{1}{i} \leq x \leq \frac{1}{2} + \frac{1}{i} \\ 1 & x \geq \frac{1}{2} + \frac{1}{i}. \end{cases}$$

In words—of which Figure 2.4 is worth two thousand, since it shows two of the graphs—each F_i has value 0 until just before $x = \frac{1}{2}$, has value 1 beginning just after $x = \frac{1}{2}$, and increases linearly from 0 to 1 in the remaining interval surrounding $x = \frac{1}{2}$.

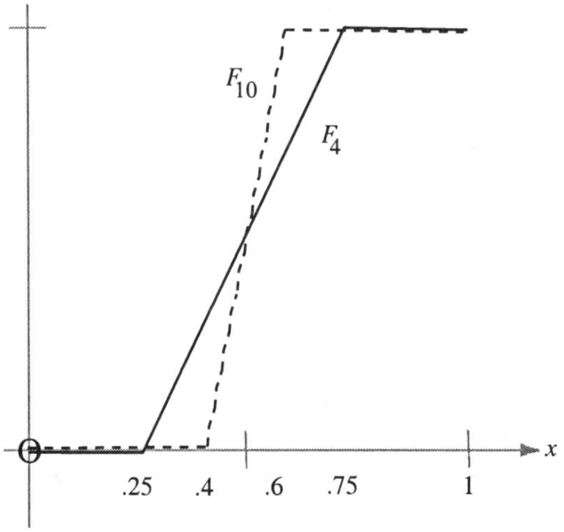

Figure 2.4.

The example illustrates both a general class of examples and a technique for their description. The general class comprises functions whose values are mostly 0 and 1. The other values of the functions are needed only because a continuous function on [0, 1] cannot have exactly two values (Exercise 4a). The technique is to use either a graph or words to specify the function. Henceforth we will skip the analytical definition (like the one that opens Example 1), and use instead graphical or verbal descriptions.

Theorem 2.14. *In* $C_2[0,1]$ *there are divergent Cauchy sequences.*

Proof. In Example 1 we see that each F_i is continuous. Further, for $i < j$, F_i and F_j are equal outside the interval $\left[\frac{1}{2}-\frac{1}{i}, \frac{1}{2}+\frac{1}{i}\right]$, and their values are between 0 and 1 inside the interval. Therefore,

$$\|F_i - F_j\|_2^2 := \int_0^1 (F_i - F_j)^2\, dx \le \int_{1/2-1/i}^{1/2+1/i} 1\, dx = \frac{2}{i},$$

so that (F_i) is Cauchy.

But (F_i) is not convergent. That is, there is no vector F in $C_2[0,1]$—no continuous function on the interval—such that $\|F - F_i\|_2 \to 0$.

To see that, assume that F is such a function. Take a fixed $y > \frac{1}{2}$. For big enough i, the function F_i is identically 1 on $[y, 1]$. Specifically, if

$$\frac{1}{2} + \frac{1}{i} < y,$$

then $F_i(x)$ is already 1 at $x = y$. Therefore,

$$\begin{aligned}\int_y^1 (1 - F(x))^2\, dx &= \int_y^1 (F_i(x) - F(x))^2\, dx \\ &\le \int_0^1 (F_i(x) - F(x))^2\, dx = \|F_i - F\|_2^2.\end{aligned}$$

Since the first quantity is fixed and nonnegative, and the last has zero limit as $i \to \infty$, we conclude that

$$\int_y^1 (1 - F(x))^2\, dx = 0.$$

So $(1-F)^2$ is a continuous, nonnegative function with zero integral on $[y, 1]$. By the integration principle cited earlier [Ross, Exercise 34.11], $1 - F \equiv 0$ on $[y, 1]$. Since this last is true for all $y > \frac{1}{2}$, we deduce that $F(y) = 1$ for all $y > \frac{1}{2}$.

Similarly, we establish that $F(z) = 0$ for all $z < \frac{1}{2}$. But this is a contradiction, because a continuous function cannot be constantly 0 on $\left[0, \frac{1}{2}\right)$ and constantly 1 on $\left(\frac{1}{2}, 1\right]$ (Exercise 4(b)). Hence no function in $C_2[0,1]$ is the limit of (F_i). □

We have found that $C_2[0,1]$, which like $\mathbf{R}^n$ has a dot product, is unlike $\mathbf{R}^n$ in that it has divergent Cauchy sequences. A normed vector space in which every Cauchy sequence is convergent is called **complete**. Completeness is a fundamental property of $\mathbf{R}$; indeed, it is possible to *define* $\mathbf{R}$ as a complete ordered field. We now know that $C_2[0, 1]$ is an incomplete infinite-dimensional space. Our next task is to show that $C_0[0, 1]$, despite its infinite dimension, is complete.

The task would be trivial, says Theorem 2.12, if bounded sequences in $C_0[0, 1]$ could be counted on to have convergent subsequences. No such luck: This Bolzano–Weierstrass theorem property just does not apply. The sequence in Example 1 is a counterexample, but we will reach again into the class of zero–one functions to give a more transparent counterexample.

Example 2. Describe G_i by intervals: $G_i(x) := 0$ on $\left[0, \frac{1}{2^i}\right]$ and $\left[\frac{1}{2^{i-1}}, 1\right]$; $G_i(c) := 1$ at the midpoint $c = 1.5/2^i$ of the interval $\left[\frac{1}{2^i}, \frac{1}{2^{i-1}}\right]$; $G_i(x)$ is linear and continuous in the remaining two pieces of $[0, 1]$.

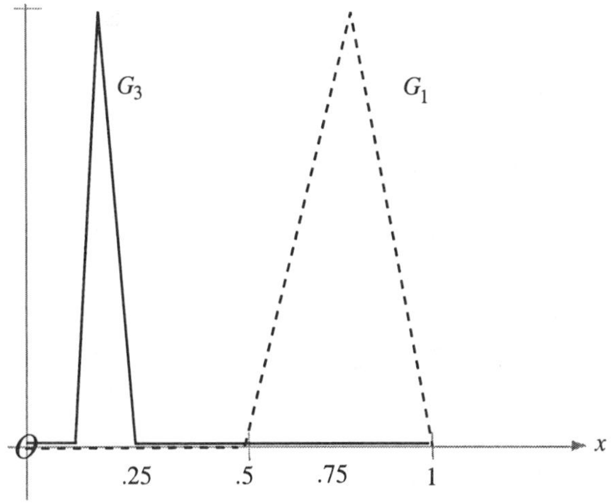

Figure 2.5.

Figure 2.5 shows the graphs of G_1 and G_3. It illustrates that each G_i is continuous and has values between 0 and 1. Hence $\|G_i\|_0 = 1$. Moreover, in the interval where G_i is positive, all the others are zero. It immediately follows that

$$i \neq j \Rightarrow \|G_i - G_j\|_0 := \sup\{|G_i(x) - G_j(x)|\} = 1.$$

Theorem 2.15. *In* $C_0[0, 1]$, *there are bounded sequences with no convergent subsequences.*

Proof. Example 2 achieves in $C_0[0, 1]$, in which we know that the norm is not induced by a dot product, what Example 1 did in $C_2[0, 1]$, namely, a sequence of equinormed, equispaced vectors. By the argument in Theorem 2.13, such a sequence is bounded but has no sublimits. □

To decide Cauchy's criterion in $C_0[0, 1]$, we first characterize convergence there.

Theorem 2.16. *In the space* $C_0[0, 1]$, *the vector sequence* (f_i) *converges to a vector* f *iff the function sequence* (f_i) *converges uniformly to the function* f *on the interval* $[0, 1]$.

Proof. $\Rightarrow$ Assume that $f_i \to f$ in the vector sense. By definition of such convergence, each $\varepsilon > 0$ has a corresponding I such that

$$i \geq I \Rightarrow \|f_i - f\|_0 < \varepsilon.$$

By definition of maxnorm, the last inequality says that

$$\max\{|f_i(x) - f(x)| : x \text{ in } [0, 1]\} < \varepsilon.$$

If the max is less than ε, then every value $|f_i(x) - f(x)|$ is less than ε. To say that for every $\varepsilon > 0$ there exists I such that

$$i \geq I \Rightarrow |f_i(x) - f(x)| < \varepsilon \text{ throughout the interval}$$

is to say that f_i converges uniformly to f.

$\Leftarrow$ Assume conversely that the functions f_i converge uniformly to f. If each f_i is in $C_0[0, 1]$, then so is f, because the uniform limit of continuous functions is continuous [Ross, Section 24]. If we choose $\varepsilon > 0$, then there is I such that

$$i \geq I \Rightarrow |f_i(x) - f(x)| < \varepsilon \text{ for every } x \text{ in the interval.}$$

In particular, $\max\{|f_i(x) - f(x)|\} < \varepsilon$. Thus, given $\varepsilon > 0$, there is I such that

$$i \geq I \Rightarrow \|f_i - f\|_0 < \varepsilon.$$

That is, $\|f_i - f\|_0 \to 0$ as $i \to \infty$, which is the definition in $C_0[0, 1]$ of $(f_i) \to f$. □

Because of Theorem 2.16, maxnorm is often called the "uniform norm".

Theorem 2.17. $C_0[0, 1]$ *is complete.*

Proof. Assume that (f_i) is a Cauchy sequence; that is, $\|f_i - f_j\|_0 \to 0$ as $i, j \to \infty$. We must prove that (f_i) is convergent; we must find f such that $\|f_i - f\|_0 \to 0$.

Let $\varepsilon > 0$ be specified. By assumption, there is $I(\varepsilon)$ such that

$$i, j \geq I(\varepsilon) \Rightarrow \|f_i - f_j\|_0 < \varepsilon \Rightarrow |f_i(x) - f_j(x)| < \varepsilon \quad \text{for all } x \in [0, 1].$$

Because $I(\varepsilon)$ depends only on ε, and not on x, the sequence (f_i) is a uniformly Cauchy function sequence.

By a theorem about convergence of functions in **R** [Ross, Theorem 25.4], there exists a function f such that $f_i \to f$ uniformly in $[0, 1]$. Because the convergence is uniform and the f_i are continuous, f is continuous also. By Theorem 2.16, $f_i \to f$ in $C_0[0, 1]$. □

Exercises

1. We will define a sequence of functions in groups ending at powers of 2—H_2, then H_3 and H_4, then H_5 to H_8, and so on—as follows. Fix an integer $k \geq 0$. Partition $[0, 1]$ into 2^k equally wide subintervals. The function H_{2^k+j}, $j = 1, \dots, 2^k$, has values

1	in subinterval number j;
increasing linearly from 0 to 1	in the subinterval number $j - 1$, if there is one;
decreasing linearly from 1 to 0	in the subinterval number $j + 1$, if any;
0	on the rest of $[0, 1]$.

   (a) Draw the graph of H_2; then H_3 and H_4; then H_5, H_6, H_7, H_8.

   (b) Show that if $2^k + 1 \leq i \leq 2^{k+1}$, then $1/2^k \leq \|H_i\|_2^2 < 1/2^{k-1}$.

   (c) Show that (H_i) has a limit in $C_2[0, 1]$.

2. Consider the functions H_i, defined in Exercise 1, as vectors in $C_0[0, 1]$.

   (a) Is the sequence (H_i) bounded in $C_0[0, 1]$?

   (b) Does (H_i) have a convergent subsequence in $C_0[0, 1]$?

3. Where does the sequence from Exercise 1 have a pointwise limit? That is, for which values of x does $\lim_{i\to\infty} H_i(x)$ exist?

4. (a) Name all the continuous functions g that have the following property: For every $x \in [0, 1]$, either $g(x) = 0$ or $g(x) = 1$.

   (b) Show that a continuous function h on $[0, 1]$ cannot satisfy both $h(x) = 0$ for all $x < \frac{1}{2}$ and $h(x) = 1$ for all $x > \frac{1}{2}$.

5. Establish the following sequence of results to prove that wherever the Bolzano–Weierstrass theorem holds, there every Cauchy sequence converges. (Recall that we have already indicated that half of Cauchy's criterion is always true: Every convergent sequence is Cauchy.) In any normed space:

   (a) Show that every Cauchy sequence ($\|\mathbf{x}_i - \mathbf{x}_j\| \to 0$) is bounded ($\|\mathbf{x}_i\|$ < constant). (This is Exercise 7b in Section 2.3.)

   (b) Show that if $(\mathbf{x}_i)$ is Cauchy and some subsequence $(\mathbf{x}_{j(i)})$ converges to a vector, then the whole sequence $(\mathbf{x}_i)$ converges to that vector.

   (c) Show that if the Bolzano–Weierstrass theorem holds, then every Cauchy sequence converges to a vector.

3
Limits and Continuity in Normed Spaces

Introduction

Having studied sequences and their convergence, we now employ those ideas to introduce continuity of functions that map vectors to vectors. We also discuss the most elementary properties of continuous functions.

The classical example of a function that assigns vectors to vectors is the "force field." A "point mass" at the origin creates a "gravitational field," which exerts a force $\mathbf{F}$ on a unit mass at any point (x, y, z) of space. The force has magnitude inversely proportional to the square of the distance from the origin, so the force vector has length $c/\left(x^2 + y^2 + z^2\right)$. It points in the direction from (x, y, z) back toward the origin, a direction specified by the unit vector $-(x, y, z)/\left(x^2 + y^2 + z^2\right)^{1/2}$. Thus,

$$\mathbf{F} = \frac{-c}{(x^2 + y^2 + z^2)^{3/2}}(x, y, z).$$

The right-hand side is a vector variable that is dependent on the independent variable $\mathbf{x} = (x, y, z)$. Here both vectors are in $\mathbf{R}^3$; we need only add to this idea the possibility that the vectors $\mathbf{F}$ and $\mathbf{x}$ might be in other spaces.

3.1 Vector-Valued Functions in Euclidean Space

Definition. Let $D \subseteq \mathbf{R}^n$ and let $\mathbf{f}$ be a function giving $\mathbf{x} \in D$ the image $\mathbf{f}(\mathbf{x}) \in \mathbf{R}^m$. We call $\mathbf{f}$ a **vector** (or **vector-valued**) **function of the vector variable** $\mathbf{x}$. If $m = 1$, then $\mathbf{f}$ is a **real** (or **real-valued, scalar, scalar-valued**) **function of the vector variable** $\mathbf{x}$.

Any vector function mapping into Euclidean space has closely related real functions. If $\mathbf{f}$ is a vector function, then for each $\mathbf{x}$, $\mathbf{f}(\mathbf{x})$ is a vector $(f_1, \ldots, f_m)$ in $\mathbf{R}^m$. Each f_j is a real number that depends on $\mathbf{x}$, so f_j is a real-valued function of $\mathbf{x}$. Using the coordinate projections as an intermediary, we have

$$f_j(\mathbf{x}) = \pi_j(\mathbf{f}(\mathbf{x})).$$

We denote the jth coordinate of $\mathbf{z}$ by $\pi_j(\mathbf{z})$, no matter whether $\mathbf{z}$ is in $\mathbf{R}^n$ or $\mathbf{R}^m$.

Example 1. We have described the force of gravity as

$$\mathbf{F}(\mathbf{x}) = \frac{-c}{\|\mathbf{x}\|_2^3}\,\mathbf{x}.$$

The natural domain (the largest set on which $\mathbf{F}$ is defined) is $D = \{\mathbf{x} \neq \mathbf{O}\}$, with $\mathbf{R}^n = \mathbf{R}^m = \mathbf{R}^3$.

The associated real functions are

$$F_1(x, y, z) := \frac{-cx}{\left(x^2 + y^2 + z^2\right)^{3/2}},$$
$$F_2(x, y, z) := \frac{-cy}{\left(x^2 + y^2 + z^2\right)^{3/2}},$$
$$F_3(x, y, z) := \frac{-cz}{\left(x^2 + y^2 + z^2\right)^{3/2}}.$$

We will adopt the name **components of f** for the f_j. The coordinates of $\mathbf{x}$ are related to $\mathbf{x}$ the same way the components of $\mathbf{f}$ are related to $\mathbf{f}$, but the separate names will serve to highlight the differences. For example, the components of $\mathbf{f}$ are functions, and those of $\mathbf{x}$ are normally not. Notice that we use $F_1(x, y, z)$ in place of the clumsier $F_1((x, y, z))$.

A vector function leads to real components, and the process can certainly be reversed. Suppose $f_1(x_1, \ldots, x_n), \ldots, f_m(x_1, \ldots, x_n)$ are functions of the variable $\mathbf{x} := (x_1, \ldots, x_n)$. If D is any set on which all the f's are defined—in other words, any subset of the intersection of their separate domains—then

$$\mathbf{f}(\mathbf{x}) := (f_1(\mathbf{x}), \ldots, f_m(\mathbf{x}))$$

defines a vector function of $\mathbf{x}$ on D.

The simplest vector function is a **constant**,

$$\mathbf{f}(\mathbf{x}) = \mathbf{b} := (b_1, \ldots, b_m).$$

The simplest nonconstant functions are first-degree functions. A **first-degree real function** has the form

$$g(\mathbf{x}) = a_1 x_1 + \cdots + a_n x_n + b.$$

If each g_j is of first degree, namely,

$$\begin{aligned} g_1(\mathbf{x}) &= a_{11}x_1 + \cdots + a_{1n}x_n + b_1, \\ &\cdots \\ g_m(\mathbf{x}) &= a_{m1}x_1 + \cdots + a_{mn}x_n + b_m, \end{aligned}$$

then

$$\mathbf{g}(\mathbf{x}) := (g_1(\mathbf{x}), \ldots, g_m(\mathbf{x}))$$

is a **first-degree vector function.**

Example 2. A first-degree vector function can be given a standard form analogous to the form for a first-degree function of one variable.

Let

$$\mathbf{G}(\mathbf{x}) := (x + 2y + 3z + 4,\ 5x + 6y + 7z + 8)$$

define an $\mathbf{R}^2$-valued function of $\mathbf{x} = (x, y, z) \in \mathbf{R}^3$. Write $\mathbf{b} = (4, 8)$. We will identify any vector with the corresponding column matrix. Thus,

$$(x, y, z) = \begin{bmatrix} x \\ y \\ z \end{bmatrix}, \qquad \mathbf{b} = \begin{bmatrix} 4 \\ 8 \end{bmatrix},$$

and

$$\mathbf{G}(\mathbf{x}) - \mathbf{b} = \begin{bmatrix} x + 2y + 3z \\ 5x + 6y + 7z \end{bmatrix} = \begin{bmatrix} 1 & 2 & 3 \\ 5 & 6 & 7 \end{bmatrix} \begin{bmatrix} x \\ y \\ z \end{bmatrix}.$$

Writing $\mathbf{A}$ for the matrix of coefficients, we have $\mathbf{G}(\mathbf{x}) - \mathbf{b} = \mathbf{A}\mathbf{x}$, allowing us to write

$$\mathbf{G}(\mathbf{x}) = \mathbf{A}\mathbf{x} + \mathbf{b}.$$

Progressing as in elementary algebra, we call the product of a real number and any number of the x_j a **term**. A sum of terms is a **polynomial**. Thus,

$$\begin{aligned} f_1(x, y, z) &:= 5, \\ f_2(x, y, z) &:= 3x + 2yz, \\ f_3(x, y, z) &:= x^2 - y^3 z \end{aligned}$$

are polynomials, and

$$\mathbf{f}(x, y, z) := \left(5, 3x + 2yz, x^2 - y^3 z\right)$$

is a **polynomial vector function** defined and valued in $\mathbf{R}^3$. A quotient of polynomials is a **rational function**, and we say that **f is rational** if each component is rational. A rational function is defined wherever all its components have nonzero denominators.

Exercises

1. Prove that $\mathbf{f}$ is of first degree iff there is a matrix $\mathbf{A}$ such that for all $\mathbf{x}$ and $\mathbf{y}$,

$$\mathbf{f}(\mathbf{x}) - \mathbf{f}(\mathbf{y}) = \mathbf{A}[\mathbf{x} - \mathbf{y}].$$

Here the right side is the matrix product of $\mathbf{A}$ and the column vector $\mathbf{x} - \mathbf{y}$.

2. Give examples of rational functions $\mathbf{g} \colon \mathbf{R}^2 \to \mathbf{R}^3$ whose domains are:

    (a) all of $\mathbf{R}^2$ except for two points.

    (b) all of $\mathbf{R}^2$ except for one line.

3. *Is there a rational function whose domain is the part of $\mathbf{R}^2$ outside the square $-1 \le x \le 1$, $-1 \le y \le 1$? [Hint: Write the denominators, which are polynomials in (x, y), as polynomials in y with coefficients that are polynomials in x; then consider roots.]

4. Suppose $\mathbf{h}(x, y) := (ax + by + c, dx + ey + f)$. Under what conditions on a, b, c, d, e, f will $\mathbf{h}$ be a one-to-one function?

5. Let $\mathbf{F}(x, y, z) = (x+2y+3z+4, 5x+6y+7z+8, 9x+10y+11z+12)$. Is (13, 14, 15) in the range of $\mathbf{F}$?

6. Let $\mathbf{i} = (j, k, \ldots, p)$ be an n-tuple of nonnegative integers. For any vector $\mathbf{x}$, define

$$\mathbf{x}^{\mathbf{i}} := x_1{}^j x_2{}^k \cdots x_n{}^p;$$

note that $\mathbf{x}^{\mathbf{i}}$ is a real number.

    (a) Show that $(\alpha\mathbf{x}^{\mathbf{i}})(\beta\mathbf{x}^{\mathbf{j}}) = (\alpha\beta)\mathbf{x}^{\mathbf{i}+\mathbf{j}}$ for real α and β.

    (b) Show that every real polynomial can be written in the form

$$f(\mathbf{x}) = \alpha\mathbf{x}^{\mathbf{i}} + \beta\mathbf{x}^{\mathbf{j}} + \cdots + \omega\mathbf{x}^{\mathbf{k}}.$$

    (c) What is the degree of $g(\mathbf{x}) := \mathbf{x}^{(2,3,4)} + 5\mathbf{x}^{(6,7,8)}$ in $\mathbf{R}^3$?

3.2 Limits of Functions in Normed Spaces

In this section, $\mathbf{f}$ represents a fixed function, with domain D contained in a normed space. The range is also a subset of some normed space, so $\mathbf{f}$ is a vector (-valued) function. However, we do not insist that the spaces involved be Euclidean.

Definition. Let $\mathbf{B}$ represent either a vector or ∞. Assume this about $\mathbf{B}$: Some sequence from D converges to $\mathbf{B}$. Next, assume this about such sequences: For every sequence $(\mathbf{x}_i)$ from D that converges to $\mathbf{B}$ and is made up of vectors that are different from $\mathbf{B}$, the corresponding sequence $(\mathbf{f}(\mathbf{x}_i))$ converges. In this situation, we say that **f has a limit at B** (or **as $\mathbf{x} \to \mathbf{B}$**).

The definition is a mouthful, so we will discuss its elements in detail, in the order of appearance there.

First, our definition allows both limits at infinity and infinite limits. As it happens, our main interest being continuous functions, the bulk of our work will deal with finite limits. Nevertheless, it is handy to have examples for the infinite case.

Example 1. (a) The gravitational force

$$\mathbf{F}(\mathbf{x}) = \frac{-c\mathbf{x}}{\|\mathbf{x}\|_2^3}$$

varies with the inverse square of distance:

$$\|\mathbf{F}(\mathbf{x})\|_2 = \frac{c}{\|\mathbf{x}\|_2^2}.$$

If $(\mathbf{x}_i)$ converges to ∞, then $\|\mathbf{x}_i\|_2 \to \infty$. Consequently, $\|\mathbf{F}(\mathbf{x}_i)\|_2 \to 0$, and so $\mathbf{F}(\mathbf{x}_i) \to \mathbf{O}$.

From $(\mathbf{x}_i) \to \infty$, we drew the conclusion that $(\mathbf{F}(\mathbf{x}_i))$ converges. Therefore, $\mathbf{F}$ has a limit at ∞.

(b) If $\mathbf{x}_i \to \mathbf{O}$, then $\|\mathbf{x}_i\|_2 \to 0$, forcing $\|\mathbf{F}(\mathbf{x}_i)\|_2 \to \infty$. That is, if $\mathbf{x}_i \to \mathbf{O}$, then $(\mathbf{F}(\mathbf{x}_i))$ converges. We see that $\mathbf{F}$, which is undefined at $\mathbf{O}$, still has a limit there.

(c) Recall that in $\mathbf{R}$, every nonconstant polynomial $p(x)$ approaches $\pm\infty$ as $\mathbf{x} \to \pm\infty$; in our current language, $p(\mathbf{x}) \to \infty$ as $\mathbf{x} \to \infty$. Our first limit surprise is that the same is not true in $\mathbf{R}^n$.

Let $p(x, y) := x^3 + y^3$. Along any vertical line $x = a$, the values $p(a, y)$ approach $\pm\infty$ as $y \to \pm\infty$. But along the line $y = -x$, $p(x, y) = p(x, -x)$ is constantly zero. If we take

$$\mathbf{x}_i := \left(i\left[-1 + (-1)^i\right], 2i\right),$$

then $(\mathbf{x}_i)$ approaches ∞, with its odd terms on $y = -x$, the even ones on $x = 0$. Hence $(p(\mathbf{x}_i))$ does not converge. We deduce that p does not have a limit at ∞.

Two things in the last argument are worth highlighting. One is a technique that we will often use to show that a function does not have a limit at some particular place: Show that there are two paths to the place along which the function values have different limits. When this happens, we do not need to recast the argument in terms of convergent sequences; we will accept that the function lacks a limit.

It is essential to note that you cannot determine that there *is* a limit by checking out different paths. We will see this in some later examples. In particular, and this is the second highlight, you cannot decide whether $\mathbf{f}(\mathbf{x})$ has a limit by finding the separate limits of $\mathbf{f}(x_1, \ldots, x_n)$ with respect to the individual x_j.

Example 2. For $\mathbf{x} = (x, y)$ in $\mathbf{R}^2$, let $g(\mathbf{x}) := xy/(x^2 + y^2)$.

For points $(x, 0)$ different from but approaching the origin along the x-axis, we have $g(x, 0) = 0$. Similarly, $g(0, y) = 0$ along the y-axis. Thus, $g(x, 0)$ has a limit as a function of x, and $g(0, y)$ has a limit as a function of y. But along $x = y$, $g(x, y) \equiv \frac{1}{2}$. Therefore, g has no limit at $(0, 0)$.

In the example, g has a limit as a function of x with y fixed, and it has a limit as a function of just y. But that information does not allow us to infer that g has a limit. We cannot, in other words, study multivariable functions one variable at a time.

Moving to the second element, suppose now that $\mathbf{B} = \mathbf{b}$ is finite. Nothing in the definition requires $\mathbf{b}$ to be in D. Indeed, in all of our examples above, the function was undefined at the place where we looked for a limit. What *is* demanded is that some sequence of vectors from D converge to $\mathbf{b}$.

Definition. Let S be a subset of a normed space. If some sequence of vectors from S converges to a point $\mathbf{c}$ (which need not be in S) then $\mathbf{c}$ is a **closure point** of S.

The name "limit point" is common for this idea. The reason for our choice of this particular term will become evident later. The reason for avoiding "limit point" is that the same name is sometimes applied to what we call "sublimit." Using one name for two things is always risky, but is particularly so in this context. There is an important difference between the related topics of sublimit of a sequence and closure point of a set: A term of a sequence is not automatically one of the sublimits (Exercise 10 in Section 2.3); a member of a set *is* automatically one of the closure points, because if $\mathbf{d}$ is a member of S, then the trivial sequence $\mathbf{d}, \mathbf{d}, \ldots$ is a sequence from S converging to $\mathbf{d}$.

Theorem 3.1. *A vector* $\mathbf{c}$ *is a closure point of a set* S *iff every neighborhood of* $\mathbf{c}$ *possesses at least one point (possibly* $\mathbf{c}$ *itself) of* S.

Proof. $\Rightarrow$ Exercise 5.

$\Leftarrow$ We will demonstrate a technique useful in many proofs of this kind.

Assume that every neighborhood of $\mathbf{c}$ has points from S. Thus, there exist $\mathbf{x}_1$ from S belonging to $N(\mathbf{c}, 1/1)$, $\mathbf{x}_2$ from S belonging to $N(\mathbf{c}, 1/2), \ldots$. We know that each $\mathbf{x}_i$ is in S. Because $\|\mathbf{c} - \mathbf{x}_i\| < 1/i$, we have $(\mathbf{x}_i) \to \mathbf{c}$. Therefore, $\mathbf{c}$ is a closure point of S. □

Example 3. For $\mathbf{x} = (x, y)$ in $\mathbf{R}^2$, let

$$h(\mathbf{x}) := \sqrt{\|\mathbf{x}\|_2 \frac{(1-x)(y-1)}{(x-1)^2 + (y-1)^2}}.$$

We see that the radicand is negative if x and y are both greater than 1 or both less than 1, except that it is zero at $\mathbf{O}$. It is undefined at $(1, 1)$. Consequently, the

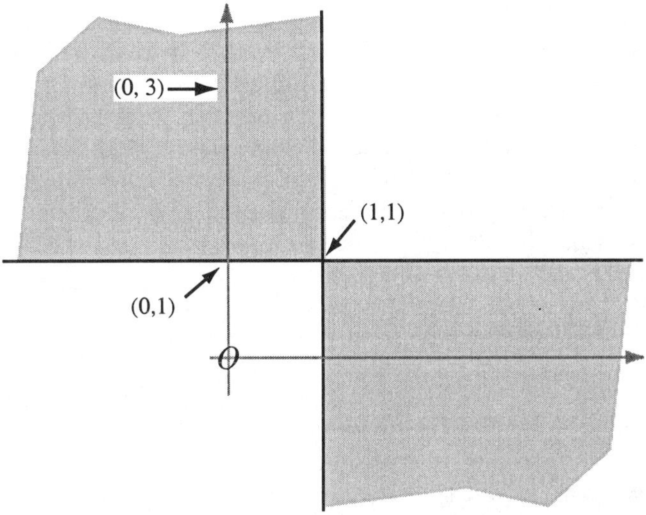

Figure 3.1.

domain T of h is the shaded region pictured in Figure 3.1 plus the origin, including the two lines $x = 1$ and $y = 1$ but not their intersection.

(a) It makes sense to discuss the limit of h at $(0, 3)$. Informally, we may say that $(0, 3)$ is surrounded by points of D, so we can talk about nearby values of h. In terms of the definition and Theorem 3.1, every neighborhood $N((0, 3), \varepsilon)$ with $\varepsilon \leq 1$ is contained in T. Therefore, $(0, 3)$ is a closure point of T, and the definition of limit applies.

(b) For $(0, 1)$, a neighborhood $N((0, 1), \varepsilon)$ is never entirely within T, but its upper half is if $\varepsilon \leq 1$. Therefore, $(0, 1)$ is a closure point of T, and the definition again applies.

A similar statement goes for $(1, 1)$, even though $(1, 1) \notin T$. A neighborhood $N((1, 1), \varepsilon)$ has points of D in its upper-left and lower-right quarters. Hence the point falls within the scope of the definition. It also makes sense to us; there are nearby points at which we may examine the values of h.

(c) The same may not be said at $\left(\frac{1}{2}, 0\right)$. In the neighborhood $N\left(\left(\frac{1}{2}, 0\right), .25\right)$, there is no point of T. Therefore, $\left(\frac{1}{2}, 0\right)$ is not a closure point of T, and we do not even talk about a limit of h there.

(d) The oddest case is $(0, 0)$. According to Theorem 3.1 and the comment that precedes it, $(0, 0)$ is a closure point of T. On the other hand, in $\mathbf{R}$ we always think of limits in terms of values of h at *other* nearby points. For h, there are no values for other points close to the origin. To resolve this case, we proceed to the next protocol.

The third element of the definition of a limit is a requirement on those sequences from D that converge to $\mathbf{b}$ without actually reaching it. At a given $\mathbf{b}$, either no such sequences exist, and the requirement is empty; or else some do, and the requirement must be met.

Suppose that in some neighborhood N of **b**, **b** is the only point of D. If $(\mathbf{x}_i)$ is a sequence from D converging to **b**, then by Theorem 2.5, the terms $\mathbf{x}_i$ are eventually (for large enough i) in N. Therefore, the terms of $(\mathbf{x}_i)$ are eventually **b** itself; it is false that the terms of $(\mathbf{x}_i)$ differ from **b**. Consequently, there are no sequences from D that converge to **b** without reaching it. Whatever the definition of limit demands is vacuously true at **b**.

Definition. If in some neighborhood of a vector **c** the only point of a set S is **c** itself, then we call **c** an **isolated point** of S.

The previous paragraph says that if **b** is an isolated point of D, then **f** trivially has a limit at **b**. This may seem like an undesirable way to look at things, but it is unavoidable from a logical standpoint. For **f** to lack a limit at **b**, there must be two approaches to **b** along which **f** produces something other than a common limit. No such approaches can be made to an isolated point of the domain.

Suppose now that in no neighborhood of **b** is **b** the only point of D. Thus, in every neighborhood of **b** there is at least one point of D different from **b**.

Definition. We call **c** an **accumulation point** of S if every neighborhood of **c** has points of S different from **c**.

Notice that by definition, an isolated point of S is necessarily a member of S. An accumulation point of S may or may not be in S. The origin of $\mathbf{R}^2$ is an accumulation point of all four quadrants, but is not a member of any.

At an accumulation point of D, we can use the argument from the proof of Theorem 3.1 to produce (Exercise 6(a)) a sequence $(\mathbf{x}_i)$ consisting of vectors from D all different from **b** and converging to **b**. In this case, the definition of limit gives **f** a test to pass. The test is this: For vector sequences that approach but do not reach **b** via D, the definition demands that the sequences of function values converge.

Example 4. Return to (Example 3)

$$h(\mathbf{x}) := \sqrt{\|\mathbf{x}\|_2 \frac{(1-x)(y-1)}{(x-1)^2+(y-1)^2}}\,.$$

(a) The origin is an isolated point of the domain T (Example 3(d)). Therefore, it is automatic that h has a limit at **O**.

(b) $(0, 3)$ is an accumulation point of T. Let us apply the test.

Assume that $(\mathbf{z}_i)$ is a sequence from T converging to $(0, 3)$. Write $\mathbf{z}_i = (x_i, y_i)$. For large enough i, $\mathbf{z}_i$ is close enough to $(0, 3)$ to make $x_i < 1$ and $y_i > 1$ (Reason?). Hence we may write

$$h(\mathbf{z}_i) = \frac{\|\mathbf{z}_i\|_2^{1/2}(1-x_i)^{1/2}(y_i-1)^{1/2}}{\|\mathbf{z}_i-(1,1)\|_2}\,.$$

Since the limit of the norms is the norm of the limit (Theorem 2.4(b)), we have

$$\|\mathbf{z}_i\|_2^{1/2} \to \sqrt{3} \quad \text{and} \quad \|\mathbf{z}_i - (1, 1)\|_2 \to \sqrt{5}.$$

Since the limit of the projection x_i is the projection $\pi_1((0, 3))$ of the limit (Theorem 2.8), we have $x_i \to 0$; similarly, $y_i \to 3$. By properties of sequences of real numbers, we conclude that

$$h(\mathbf{z}_i) \to \sqrt{3}(\sqrt{1})(\sqrt{2})/\sqrt{5}.$$

We have shown that $(\mathbf{z}_i) \to (0, 3)$ implies that $(h(\mathbf{z}_i))$ converges. This proves that h has a limit at $(0, 3)$.

(c) $(1, 1)$ is also an accumulation point of T. But near there h behaves something like the function of Example 2.

Along the line $y = 1$, the values $h(x, 1)$ for $x \neq 1$ are identically 0. Along the line

$$y - 1 = -(x - 1),$$

which crosses $(1, 1)$ going down from left to right, and is within the domain except for $(1, 1)$ itself, we have

$$h(\mathbf{x}) = \frac{\|\mathbf{x}\|_2^{1/2}|y - 1|}{[2(y - 1)^2]^{1/2}} = \frac{\|\mathbf{x}\|_2^{1/2}}{\sqrt{2}},$$

whose limit is $2^{-1/4}$. Different paths yield different limits of the function values. We infer that h does not have a limit at $(1, 1)$.

Notice within Example 4 that in part (c), finding two appropriate paths of approach sufficed to prove that the function does not possess a limit. By contrast, in part (b) it was necessary to consider a sequence approaching the point in question, making no assumption about the sequence except that it converges to the point without reaching it. As we noted previously, it is always that way: Path counterexamples can show that there is no limit, but not establish that there is one.

Within the context of the third element, it is important to note the provisions "sequence from D" and "different from $\mathbf{b}$." The latter requirement can be softened to "eventually different from $\mathbf{b}$." In other words, we can test with sequences $(\mathbf{x}_i)$ for which some term $\mathbf{x}_I$ and all its followers $\mathbf{x}_{I+1}, \mathbf{x}_{I+2}, \ldots$ are different from $\mathbf{b}$. But some requirement like that is essential. If $\mathbf{f}$ is defined near $\mathbf{b}$, then we want the limit of $\mathbf{f}$ to be determined by its behavior near $\mathbf{b}$, unaffected by the existence or value of $\mathbf{f}(\mathbf{b})$. For instance, our notion of behavior "nearby" tells us that

$$K(x, y) := 0 \quad \text{for } (x, y) \neq \mathbf{O}, \; := 1 \text{ at } \mathbf{O},$$

has a limit as $(x, y) \to \mathbf{O}$. The proposal fits the limit definition, because $K(\mathbf{x}_i)$ is identically zero for any sequence $(\mathbf{x}_i)$ of nonzero vectors. The same would not be true if the definition allowed a sequence like $(1, 0)$, $(0, 0)$, $(\frac{1}{2}, 0)$, $(0, 0)$, $(\frac{1}{3}, 0)$, $(0, 0), \ldots$ that converges to $\mathbf{O}$ but includes infinitely many $\mathbf{O}$'s.

The provision "sequence from D" makes D a very important part of any question regarding the limit. In talking about functions, we are not always careful to specify a domain. Typically, the domain is somehow understood, possibly as the largest set in which some formula is defined. It is essential to remember that a description (such as a formula) is just part of a function. The same formula applied within two different domains necessarily defines two different functions, which may then answer differently as to whether a limit exists. (Compare Exercise 8.)

The final element we study is actually missing from the definition of limit. At a point $\mathbf{b}$ that can be approached by vectors $\mathbf{x}_i \neq \mathbf{b}$ from D, the definition demands that the sequences of values $\mathbf{f}(\mathbf{x}_i)$ have limits. It does not insist that these sequences have the same limit. We now show that a common limit is guaranteed.

Theorem 3.2. *Suppose* $\mathbf{b}$ *is an accumulation point of* D *and* $\mathbf{f}$ *has a limit at* $\mathbf{b}$. *For any two sequences of vectors from* D *approaching* $\mathbf{b}$ *without reaching it, the two sequences of function values have the same limit.*

Proof. We are assuming that $\mathbf{f}$ has a limit at an accumulation point $\mathbf{b}$ of the domain. That is, whenever a vector sequence $(\mathbf{x}_i)$ avoids but approaches $\mathbf{b}$—and there are such sequences—the value sequence $(\mathbf{f}(\mathbf{x}_i))$ converges. We must show that if $(\mathbf{y}_i)$ is another such vector sequence, then $(\mathbf{f}(\mathbf{y}_i))$ has the same limit as $(\mathbf{f}(\mathbf{x}_i))$.

Suppose $(\mathbf{x}_i)$ and $(\mathbf{y}_i)$ are as described. Consider the sequence $\mathbf{x}_1, \mathbf{y}_1, \mathbf{x}_2, \mathbf{y}_2, \ldots$. This sequence converges to $\mathbf{b}$ and avoids it. By hypothesis, the value sequence $\mathbf{f}(\mathbf{x}_1), \mathbf{f}(\mathbf{y}_1), \mathbf{f}(\mathbf{x}_2), \mathbf{f}(\mathbf{y}_2), \ldots$ must converge. Since subsequences of a convergent sequence must converge to the parent sequence's limit (Theorem 2.4(a)), the subsequences $(\mathbf{f}(\mathbf{x}_i))$ and $(\mathbf{f}(\mathbf{y}_i))$ must have the same limit. □

This result tells us that if $\mathbf{b}$ is an accumulation point of D and $\mathbf{f}$ has a limit there, then there is one and only one limit of function values $\mathbf{f}(\mathbf{x}_i)$ for $\mathbf{b} \neq \mathbf{x}_i \to \mathbf{b}$. We call this limiting value, which could be infinite, the **limit of f at b** and denote it by $\lim_{\mathbf{x}\to\mathbf{b}} \mathbf{f}(\mathbf{x})$.

If $\mathbf{b}$ is in D but not an accumulation point, then $\mathbf{b}$ is an isolated point of D. In this case, we are compelled to admit that $\mathbf{f}$ has a limit, and we need a value for it. We therefore agree to this convention: At an isolated point $\mathbf{b}$ of D, set $\lim_{\mathbf{x}\to\mathbf{b}} \mathbf{f}(\mathbf{x}) := \mathbf{f}(\mathbf{b})$.

Exercises

1. Let $f(x, y) := x^2 + y^2$.
   (a) Does f have a limit at ∞?
   (b) Does f have a limit at $\mathbf{O}$?
2. Let $g(x, y) := x^4 + y^4$.
   (a) Does g have a limit at ∞?
   (b) Does g have a limit at $\mathbf{O}$?

(c) Does $1/g$ (the reciprocal) have a limit at $\mathbf{O}$?

3. Let $G(x, y) := xy$.

   (a) Does G have a limit at ∞?

   (b) Does G have a limit at $\mathbf{O}$?

   (c) Does $1/G$ (the reciprocal) have a limit at $\mathbf{O}$?

   (d) Does $\sqrt{G}$ have a limit at $\mathbf{O}$?

4. Find the limits, if they exist, at $\mathbf{O}$ and at ∞:

   (a) $f_1(x, y) := 5/(xy)^{1/2}$

   (b) $f_2(x, y) := (\sin xy)/(xy)$

   (c) $\mathbf{f}_3(x, y) := (x^2 - y^2, 2xy)$

   (d) $\mathbf{f}_4(x, y) := \left([x^4 - y^4]/[x^2 + y^2], 2xy/[x^2 + y^2]\right)$

5. Prove that if $\mathbf{c}$ is a closure point of S, then every neighborhood of $\mathbf{c}$ contains at least one point (possibly $\mathbf{c}$ itself) of S.

6. (a) Prove that $\mathbf{d}$ is an accumulation point of S iff there is a sequence $(\mathbf{x}_i)$ from S converging to $\mathbf{d}$ with every $\mathbf{x}_i$ different from $\mathbf{d}$.

   (b) Prove that $\mathbf{d}$ is an accumulation point of S iff every neighborhood of $\mathbf{d}$ has infinitely many points of S.

7. Let $H(x, y) := y^2/x$.

   (a) Show that along any nonvertical line through the origin we have $H(x, y) \to 0$ as $(x, y) \to (0, 0)$. (Therefore, H has a limit, indeed the same limit, for approach to the origin along any line in the domain.)

   (b) Show that H does not have a limit at $(0, 0)$.

8. (a) Define $K(x) := x/|x|$ on the domain $\mathbf{R}^\# := \{x \in \mathbf{R} : x \neq 0\}$. Does K have a limit at 0?

   (b) Let K^+ denote the restriction of K to the set $\mathbf{R}^+$ of positive real numbers. Does K^+ have a limit at $x = 0$?

3.3 Finite Limits

Keeping an eye on our future preoccupation with multivariable calculus, we now focus on finite limits at finite places. We are still in the general context of functions mapping one normed vector space to another.

We first establish that some familiar algebraic combinations of functions have limits under our definition.

Theorem 3.3. *At a normal place, if two functions have finite limits, then the limit of a linear combination of the two exists, and is the combination of the limits; and the limit of the product exists, and is the product of limits. In detail: Assume* $\mathbf{f}$ *and* $\mathbf{g}$ *both have finite limits at* $\mathbf{b}$. *Assume further that* $\mathbf{b}$ *is an accumulation point of the set in which* $\mathbf{f}$ *and* $\mathbf{g}$ *are both defined. Then:*

(a) *For any real constants* α *and* β, *the linear combination* $\alpha\mathbf{f} + \beta\mathbf{g}$ *has a limit at* $\mathbf{b}$, *and*

$$\lim_{\mathbf{x}\to\mathbf{b}} [\alpha\mathbf{f}(\mathbf{x}) + \beta\mathbf{g}(\mathbf{x})] = \alpha \lim_{\mathbf{x}\to\mathbf{b}} \mathbf{f}(\mathbf{x}) + \beta \lim_{\mathbf{x}\to\mathbf{b}} \mathbf{g}(\mathbf{x}).$$

(b) *If* $\mathbf{f} = f$ *is real, then the product* $f\mathbf{g}$ *has a limit at* $\mathbf{b}$, *and*

$$\lim_{\mathbf{x}\to\mathbf{b}} [f(\mathbf{x})\mathbf{g}(\mathbf{x})] = \left[\lim_{\mathbf{x}\to\mathbf{b}} f(\mathbf{x})\right]\left[\lim_{\mathbf{x}\to\mathbf{b}} \mathbf{g}(\mathbf{x})\right].$$

Proof. (a) Exercise 6.

(b) Write $L := \lim_{\mathbf{x}\to\mathbf{b}} f(\mathbf{x})$ and $\mathbf{y} := \lim_{\mathbf{x}\to\mathbf{b}} \mathbf{g}(\mathbf{x})$.

By hypothesis, there exist sequences $(\mathbf{x}_i)$ converging to $\mathbf{b}$, not reaching $\mathbf{b}$, and consisting of points at which both f and $\mathbf{g}$ are defined. For any such sequence,

$$\begin{aligned} \|f(\mathbf{x}_i)\mathbf{g}(\mathbf{x}_i) - L\mathbf{y}\| &\le \|f(\mathbf{x}_i)\mathbf{g}(\mathbf{x}_i) - L\mathbf{g}(\mathbf{x}_i)\| + \|L\mathbf{g}(\mathbf{x}_i) - L\mathbf{y}\| \\ &= |f(\mathbf{x}_i) - L|\, \|\mathbf{g}(\mathbf{x}_i)\| + |L|\, \|\mathbf{g}(\mathbf{x}_i) - \mathbf{y}\|. \end{aligned}$$

By the definition of the limit of a function, $f(\mathbf{x}_i) \to L$ and $\mathbf{g}(\mathbf{x}_i) \to \mathbf{y}$. Therefore, the last sum approaches $0\|\mathbf{y}\| + |L|0$. We conclude that

$$\|f(\mathbf{x}_i)\mathbf{g}(\mathbf{x}_i) - L\mathbf{y}\| \to 0.$$

We have proved simultaneously that $(f(\mathbf{x}_i)\mathbf{g}(\mathbf{x}_i))$ converges, which proves that $f\mathbf{g}$ has a limit, and that the limit is $L\mathbf{y}$. □

Example 1. The hypothesis in Theorem 3.3 that $\mathbf{b}$ is an accumulation point of the common domain is essential. A variety of things can go wrong without it.

(a) It is insufficient to assume that $\mathbf{b}$ is a closure point of the common domain. In $\mathbf{R}^n$, let

$$f(\mathbf{x}) := 1 + (\|\mathbf{x}\|_2[\|\mathbf{x}\|_2 - 1])^{1/2},$$

$$g(\mathbf{x}) := \begin{cases} 0 & \text{if } \mathbf{x} \ne \mathbf{O}, \\ 1 & \text{at } \mathbf{O}. \end{cases}$$

The origin is in the domain of f, but it is an isolated point; by contrast, g is defined everywhere. We may check that

$$\lim_{\mathbf{x}\to\mathbf{O}} f(\mathbf{x}) = f(\mathbf{O}) = 1,$$

$$\lim_{\mathbf{x}\to\mathbf{O}} g(\mathbf{x}) = 0.$$

But $f+g$ and fg are defined in $N(\mathbf{O}, 1)$ only at $\mathbf{O}$, so

$$\lim_{\mathbf{x}\to\mathbf{O}} [f+g] = [f+g](\mathbf{O}) = 2$$

and

$$\lim_{\mathbf{x}\to\mathbf{O}} [fg] = 1.$$

The limit of the sum is not the sum of the limits, and similarly with the product.

(b) It is likewise insufficient to assume that $\mathbf{b}$ is an accumulation point of the two separate domains. In $\mathbf{R}^2$,

$$f(x, y) := xy \ln(xy)$$

is defined in quadrants I and III. Therefore, $\mathbf{O}$ is an accumulation point of its domain. We find that $\lim_{\mathbf{x}\to\mathbf{O}} f(\mathbf{x}) = 0$ (Exercise 1). Similarly,

$$g(x, y) := xy \ln(-xy)$$

is defined in quadrants II and IV, and $\lim_{\mathbf{x}\to\mathbf{O}} g(\mathbf{x}) = 0$. But $f+g$ has no limit at $\mathbf{O}$, because its domain is empty.

Next we want to show that quotients have limits at the expected places. To do that, we must first establish some results that turn out to have their own significance. The first is the "ε–δ" definition of function limit. It says that $\mathbf{y}$ is the limit of $\mathbf{f}$ at $\mathbf{b}$ exactly if you can keep the values $\mathbf{f}(\mathbf{x})$ close to $\mathbf{y}$ by confining $\mathbf{x}$ to vectors close to $\mathbf{b}$. The second takes advantage of what the first says, that the values stay close to the limit.

Theorem 3.4. *Suppose* $\mathbf{b}$ *is an accumulation point of the domain* D *of* $\mathbf{f}$. *Then* $\mathbf{f}$ *has a finite limit at* $\mathbf{b}$ *iff there is* $\mathbf{y}$ *(in the range space) with this property*:

> *For every* $\varepsilon > 0$, *there is a corresponding* $\delta > 0$ *such that* $\|\mathbf{y}-\mathbf{f}(\mathbf{x})\| < \varepsilon$ *for every* $\mathbf{x}$ *in* D *with* $0 < \|\mathbf{x}-\mathbf{b}\| < \delta$.

In this situation, $\mathbf{y} = \lim_{\mathbf{x}\to\mathbf{b}} \mathbf{f}(\mathbf{x})$.

Proof. $\Rightarrow$ Suppose $\mathbf{f}$ has a finite limit at the accumulation point $\mathbf{b}$. By definition of accumulation point, there is some sequence $(\mathbf{u}_i)$ with $\mathbf{b} \neq \mathbf{u}_i \to \mathbf{b}$. By definition of the limit of a function, the values $(f(\mathbf{u}_i))$ have a limit. Let $\mathbf{y}$ be this limit.

Choose a positive real ε. Look at the places $\mathbf{x} \neq \mathbf{b}$ in D where $\|\mathbf{y}-\mathbf{f}(\mathbf{x})\| \geq \varepsilon$; that is, let

$$S = S(\varepsilon) := \{\mathbf{x} \in D \colon \mathbf{x} \neq \mathbf{b} \text{ and } \|\mathbf{y}-\mathbf{f}(\mathbf{x})\| \geq \varepsilon\}.$$

If there were a sequence $(\mathbf{x}_i)$ from S converging to $\mathbf{b}$, then by Theorem 3.2, the function values $\mathbf{f}(\mathbf{x}_i)$ would converge to the same limit as $f(\mathbf{u}_i)$. That is, we would have $\|\mathbf{y}-\mathbf{f}(\mathbf{x}_i)\| \to 0$. But every $\|\mathbf{y}-\mathbf{f}(\mathbf{x}_i)\|$ is ε or more. Hence there cannot be $(\mathbf{x}_i)$ from S approaching $\mathbf{b}$.

Thus, $\mathbf{b}$ is not a closure point of S. By Theorem 3.1, there is a neighborhood $N(\mathbf{b}, \delta)$ (depending on S, which depends on ε) that has no points of S.

Suppose now that $\mathbf{x}$ is in D with $0 < \|\mathbf{x} - \mathbf{b}\| < \delta$. First, $\mathbf{x}$ is in $N(\mathbf{b}, \delta)$, so $\mathbf{x}$ is not in S. Since $\mathbf{x} \in D$ and $\mathbf{x} \neq \mathbf{b}$, it must be that $\|\mathbf{y} - \mathbf{f}(\mathbf{x})\| \geq \varepsilon$ is false. Consequently, we have

$$\|\mathbf{y} - \mathbf{f}(\mathbf{x})\| < \varepsilon,$$

as required.

$\Leftarrow$ Suppose $\mathbf{y}$ has the stated property.

Let $(\mathbf{x}_i)$ be any sequence from D that does not reach but does approach $\mathbf{b}$. Let ε be any fixed positive real number. By hypothesis, there exists $\delta = \delta(\varepsilon)$ such that

$$0 < \|\mathbf{x} - \mathbf{b}\| < \delta \text{ implies } \|\mathbf{y} - \mathbf{f}(\mathbf{x})\| < \varepsilon.$$

From $\mathbf{x}_i \to \mathbf{b}$, we conclude that there exists $I = I(\delta(\varepsilon))$ such that

$$i \geq I \text{ implies } \|\mathbf{x}_i - \mathbf{b}\| < \delta.$$

Since also $\mathbf{x}_i \neq \mathbf{b}$, we have

$$i \geq I \Rightarrow 0 < \|\mathbf{x}_i - \mathbf{b}\| < \delta \Rightarrow \|\mathbf{y} - \mathbf{f}(\mathbf{x}_i)\| < \varepsilon.$$

This proves that $\mathbf{f}(\mathbf{x}_i) \to \mathbf{y}$.

We have established both that $\mathbf{f}$ has a limit at $\mathbf{b}$, and that the limit is $\mathbf{y}$. □

Theorem 3.5. *Suppose* $\mathbf{f}$ *has a finite limit at* $\mathbf{b}$. *Then*:

(a) $\mathbf{f}$ *is bounded near* $\mathbf{b}$. *In symbols: There exist a neighborhood N of* $\mathbf{b}$ *and a real bound M such that*

$$\|\mathbf{f}(\mathbf{x})\| < M \qquad \textit{for each } \mathbf{x} \neq \mathbf{b} \textit{ from } D \cap N.$$

(b) *If the limit is nonzero, then* $\mathbf{f}$ *is bounded away from* $\mathbf{O}$ *near* $\mathbf{b}$. *In symbols: If* $\lim_{\mathbf{x}\neq\mathbf{b}} \mathbf{f}(\mathbf{x}) \neq \mathbf{O}$, *then there exist a neighborhood N of* $\mathbf{b}$ *and an* $\varepsilon > 0$ *such that*

$$\|\mathbf{f}(\mathbf{x})\| > \varepsilon \qquad \textit{for each } \mathbf{x} \neq \mathbf{b} \textit{ from } D \cap N.$$

(c) *If* $\mathbf{f} = f$ *is real and* $\lim_{\mathbf{x}\to\mathbf{b}} f(\mathbf{x}) > 0$ *(or* < 0*), then there exist* $N(\mathbf{b}, \delta)$ *and* $\varepsilon > 0$ *such that* $f(\mathbf{x}) > \varepsilon$ *(respectively* $f(\mathbf{x}) < -\varepsilon$*) for each* $\mathbf{x} \neq \mathbf{b}$ *in* $D \cap N(\mathbf{b}, \delta)$. *In particular*:

(d) *If f is real and the limit is nonzero, then f has "persistence of sign." In symbols: If* $\lim_{\mathbf{x}\to\mathbf{b}} f(\mathbf{x}) \neq 0$, *then there is a neighborhood of* $\mathbf{b}$ *in which all the values* $f(\mathbf{x})$, *except maybe* $f(\mathbf{b})$, *have the same sign.*

Proof. All of these are trivial at an isolated point. Hence we assume that $\mathbf{b}$ is an accumulation point of D. We denote $\lim_{\mathbf{x}\to\mathbf{b}} \mathbf{f}(\mathbf{x})$ by $\mathbf{y}$ (or y if $\mathbf{f}$ is real).

(a) By Theorem 3.4, corresponding to $\varepsilon = 1$ there is $\delta > 0$ such that

$$\|\mathbf{y} - \mathbf{f}(\mathbf{x})\| < 1 \qquad \text{for every } \mathbf{x} \in D \text{ with } 0 < \|\mathbf{x} - \mathbf{b}\| < \delta.$$

We simply take $N := N(\mathbf{b}, \delta)$ and $M := 1 + \|\mathbf{y}\|$. Then for any $\mathbf{x} \neq \mathbf{b}$ from $D \cap N$,

$$\|\mathbf{f}(\mathbf{x})\| \leq \|\mathbf{f}(\mathbf{x}) - \mathbf{y}\| + \|\mathbf{y}\| < 1 + \|\mathbf{y}\| = M.$$

(b) With $\mathbf{y} \neq \mathbf{O}$, we simply take $\varepsilon := \|\mathbf{y}\|/2$. Corresponding to this ε there is a neighborhood $N := N(\mathbf{b}, \delta)$ in which $\|\mathbf{y} - \mathbf{f}(\mathbf{x})\| < \varepsilon$ except maybe at $\mathbf{b}$. Then

$$\|\mathbf{f}(\mathbf{x})\| \geq \|\mathbf{y}\| - \|\mathbf{y} - \mathbf{f}(\mathbf{x})\| > 2\varepsilon - \varepsilon \quad \text{for } \mathbf{x} \neq \mathbf{b} \text{ in } D \cap N.$$

(c) Let us handle the case $\lim_{\mathbf{x}\to\mathbf{b}} f(\mathbf{x}) = y > 0$. Set $\varepsilon = y/2$. By Theorem 3.4, there is a neighborhood $N(\mathbf{b}, \delta)$ in which $\mathbf{b} \neq \mathbf{x} \in D$ implies $|f(\mathbf{x}) - y| < \varepsilon$. For such $\mathbf{x}$,

$$y - \varepsilon < f(\mathbf{x}) < y + \varepsilon.$$

The relation $y - \varepsilon < f(\mathbf{x})$ says what we need. (See also Exercise 3.)

(d) is immediate from (c).

Now we can prove that quotients have limits.

Theorem 3.6. *The limit of a quotient is the quotient of the limits, provided that the denominator limit is nonzero. In symbols: Assume that f is real, that $L := \lim_{\mathbf{x}\to\mathbf{b}} f(\mathbf{x}) \neq 0$, and that $\lim_{\mathbf{x}\to\mathbf{b}} \mathbf{g}(\mathbf{x}) = \mathbf{y}$, all at a point $\mathbf{b}$ that is an accumulation point of their joint domain; then $\lim_{\mathbf{x}\to\mathbf{b}}[\mathbf{g}(\mathbf{x})/f(\mathbf{x})]$ exists and equals $\mathbf{y}/L$.*

Proof. The proof is easy once we establish that it is legal to talk about the limit of $\mathbf{g}/f$.

By hypothesis, there exists a sequence $(\mathbf{u}_i) \to \mathbf{b}$ of vectors not equal to $\mathbf{b}$ at each of which both f and $\mathbf{g}$ are defined. Because $L \neq 0$, there exist (Theorem 3.5(b)) a positive ε and a neighborhood N of $\mathbf{b}$ in which $|f(\mathbf{x})| > \varepsilon$ for $\mathbf{x} \neq \mathbf{b}$ in the domain of f. Since $\mathbf{u}_i \to \mathbf{b}$, there exists (Theorem 2.5) I such that $i \geq I$ implies $\mathbf{u}_i \in N$. Hence

$$i \geq I \Rightarrow |f(\mathbf{u}_i)| > \varepsilon.$$

Thus, $\mathbf{u}_I, \mathbf{u}_{I+1}, \ldots$ is a sequence converging to but not reaching $\mathbf{b}$ at whose points $\mathbf{g}/f$ is defined. This shows that $\mathbf{b}$ is an accumulation point of the domain of $\mathbf{g}/f$.

With the legality settled, let $(\mathbf{x}_i)$ be a sequence from the domain of $\mathbf{g}/f$ converging to but not reaching $\mathbf{b}$. We have

$$\begin{aligned}\|\mathbf{g}(\mathbf{x}_i)/f(\mathbf{x}_i) - \mathbf{y}/L\| &= \left\| \frac{L[\mathbf{g}(\mathbf{x}_i) - \mathbf{y}] + [L - f(\mathbf{x}_i)]\mathbf{y}}{Lf(\mathbf{x}_i)} \right\| \\ &\leq \|\mathbf{g}(\mathbf{x}_i) - \mathbf{y}\|/|f(\mathbf{x}_i)| + |L - f(\mathbf{x}_i)|\,\|\mathbf{y}\|/|Lf(\mathbf{x}_i)|.\end{aligned}$$

For the real sequences on the right, the numerators approach zero but the denominators have strictly positive limits. Therefore,

$$\|\mathbf{g}(\mathbf{x}_i)/f(\mathbf{x}_i) - \mathbf{y}/L\| \to 0.$$

This establishes that $\mathbf{g}(\mathbf{x})/f(\mathbf{x})$ has limit $\mathbf{y}/L$. □

To take some of the mystery from limits, we return to Euclidean space. We will prove a result that can be used to simplify limit questions, then use it to give us a large class of known limits.

We have indicated that you cannot study limits of a vector function $\mathbf{f} = (f_1, \ldots, f_m)$ of a vector variable $\mathbf{x} = (x_1, \ldots, x_n)$ just by looking at $\mathbf{f}$ as a function of each individual coordinate x_i. One simplification we can make, however, is to study the individual components $f_j(\mathbf{x})$.

Theorem 3.7. *The coordinates of the limit are the limits of the components: Thus, the vector function* $\mathbf{f} = (f_1, \ldots, f_m)$ *has a finite limit at* $\mathbf{b}$ *iff each real function* f_j *has a finite limit at* $\mathbf{b}$, *and* $\lim_{\mathbf{x}\to\mathbf{b}} \mathbf{f}(\mathbf{x}) = (\lim_{\mathbf{x}\to\mathbf{b}} f_1(\mathbf{x}), \ldots, \lim_{\mathbf{x}\to\mathbf{b}} f_m(\mathbf{x}))$.

Proof. Bear in mind the implicit assumption that we are viewing $\mathbf{f}$ and all the f_j on a common domain D. We need to assume also that $\mathbf{b}$ is a closure point of D.

If $\mathbf{b}$ is isolated, then the result is trivial; we therefore assume that $\mathbf{b}$ is an accumulation point of D.

⇒ Suppose

$$\mathbf{y} := (y_1, \ldots, y_m) = \lim_{\mathbf{x}\to\mathbf{b}} \mathbf{f}(\mathbf{x}).$$

For any sequence $(\mathbf{x}_i)$ from D with $\mathbf{b} \neq \mathbf{x}_i \to \mathbf{b}$, we know that $(\mathbf{f}(\mathbf{x}_i))$ converges to $\mathbf{y}$. By Theorem 2.8, the projection-value sequence $(\pi_1(\mathbf{f}(\mathbf{x}_i)))$ converges to $\pi_1(\mathbf{y})$. That is, the sequence of values $f_1(\mathbf{x}_i)$ converges to y_1. This being true for every test sequence $(\mathbf{x}_i)$, we conclude that $\lim_{\mathbf{x}\to\mathbf{b}} f_1(\mathbf{x})$ exists and is y_1, and similarly for f_2 through f_m. We have shown that each f_j has a limit, and that

$$\mathbf{y} = \left(\lim_{\mathbf{x}\to\mathbf{b}} f_1(\mathbf{x}), \ldots, \lim_{\mathbf{x}\to\mathbf{b}} f_m(\mathbf{x})\right).$$

⇐ Suppose conversely that each $f_j(\mathbf{x})$ has a finite limit y_j at $\mathbf{b}$. For each $(\mathbf{x}_i)$ from D with $\mathbf{b} \neq \mathbf{x}_i \to \mathbf{b}$, we have $f_j(\mathbf{x}_i) \to y_j$ as $i \to \infty$. Therefore (same theorem)

$$\mathbf{f}(\mathbf{x}_i) = (f_1(\mathbf{x}_i), \ldots, f_m(\mathbf{x}_i)) \to (y_1, \ldots, y_m).$$

It follows that $\mathbf{f}$ has the limit $(y_1, \ldots, y_m)$. □

Example 2. We have looked at arithmetic combinations, but not at composites. The reason is that in the context of limits, composites allow too many odd possibilities. We would want something like this: If the limit at $\mathbf{b}$ of $\mathbf{f}$ is $\mathbf{c}$, and the limit at $\mathbf{c}$ of $\mathbf{g}$ is $\mathbf{d}$, then the limit at $\mathbf{b}$ of $\mathbf{g}(\mathbf{f})$ ought to be $\mathbf{d}$. This can fail in numerous ways. (Justifications are left to Exercise 4.)

(a) Let

$$\mathbf{f}(x, y, z) := \left(4 - [x-1]^2, 5 - [y-2]^2 - [z-3]^2\right)$$

and

$$g(u, v) := 6 + [u-4]\ln[u-4] + [v-5]\ln[v-5].$$

Then at $(1, 2, 3)$, $\mathbf{f}$ has limit $(4, 5)$; at $(4, 5)$, g has limit 6; but $g(\mathbf{f})$ is never defined.

(b) Extend g by setting

$$g^+(x, y) := \begin{cases} 7 & \text{if } (x, y) = (4, 5), \\ g(x, y) & \text{otherwise.} \end{cases}$$

Then g^+ has the same limit 6 as g at $(4, 5)$, and $g^+(\mathbf{f})$ has a limit there, but that limit is not 6.

(c) Suppose

$$\mathbf{F}(x, y, z) := (4 + [x-1][y-2][z-3], 5 + [x-1][y-2][z-3])$$

and

$$G(u, v) := \begin{cases} 8 & \text{at } (4, 5), \\ 6 & \text{elsewhere.} \end{cases}$$

Then at $(1, 2, 3)$, $\mathbf{F}$ has limit $(4, 5)$; at $(4, 5)$, G has limit 6; and $G(\mathbf{F})$ is always defined. But $G(\mathbf{F})$ does not have a limit at $(1, 2, 3)$.

(d) Finally, reduce $\mathbf{F}$ to the constant $\mathbf{F}^\# \equiv (4, 5)$. Then at $(1, 2, 3)$, $\mathbf{F}^\#$ has limit $(4, 5)$; at $(4, 5)$, G has limit 6; $G(\mathbf{F})$ is always defined and has a limit at $(1, 2, 3)$. But the limit is again different from the limit of G.

Theorem 3.8. *For a rational function, the limit is the value. In symbols: If* $\mathbf{f} = (f_1, \ldots, f_m)$ *is a rational function of* $\mathbf{x} = (x_1, \ldots, x_n)$, *then* $\lim_{\mathbf{x}\to\mathbf{b}} \mathbf{f}(\mathbf{x}) = \mathbf{f}(\mathbf{b})$ *at any point* $\mathbf{b}$ *in the domain of* $\mathbf{f}$.

Proof. It is easy to construct a proof from the ground up. This choice is also instructive, because it reminds us that certain basic functions have the stated property.

The limit of a projection is the projection of the limit, because (Theorem 2.8) if $\mathbf{x}_i \to \mathbf{b}$, then $\pi_j(\mathbf{x}_i) \to \pi_j(\mathbf{b})$. The same holds for a term, by Theorem 3.3, because a term is a multiple of the product of projections. The same theorem extends the result to a polynomial, which is a linear combination of terms.

To say that $\mathbf{f}$ is rational is to say that each component is rational. Thus, each $f_k(\mathbf{x})$ is a quotient $p_k(\mathbf{x})/q_k(\mathbf{x})$ of polynomials. By hypothesis, $q_k(\mathbf{b})$ is nonzero. We argued in the previous paragraph that

$$\lim_{\mathbf{x}\to\mathbf{b}} q_k(\mathbf{x}) = q_k(\mathbf{b}).$$

Hence Theorem 3.6 applies:

$$\lim_{\mathbf{x}\to\mathbf{b}} f_k(\mathbf{x}) = \lim_{\mathbf{x}\to\mathbf{b}} p_k(\mathbf{x}) / \lim_{\mathbf{x}\to\mathbf{b}} q_k(\mathbf{x}),$$

which by the last paragraph is $p_k(\mathbf{b})/q_k(\mathbf{b})$. By Theorem 3.7,

$$\lim_{\mathbf{x}\to\mathbf{b}} \mathbf{f}(\mathbf{x}) = \left(\lim_{\mathbf{x}\to\mathbf{b}} f_1(\mathbf{x}), \ldots, \lim_{\mathbf{x}\to\mathbf{b}} f_m(\mathbf{x})\right)$$
$$= (p_1(\mathbf{b})/q_1(\mathbf{b}), \ldots, p_m(\mathbf{b})/q_m(\mathbf{b})) = \mathbf{f}(\mathbf{b}). \qquad \square$$

Example 3. Rational functions can also have limits at points where they are not defined. Thus

$$g(x, y) := \frac{x^4 - y^4}{x^2 - y^2}$$

has limit 2 at (1, 1). At such a point, Theorem 3.8 does not deny that there is a limit; it simply gives no information.

Exercises

1. Show that $\lim_{(x,y)\to(0,0)}[xy\ln(xy)] = 0$.

2. Let $\mathbf{f}(x, y) := \left(\sin x/[x - \pi], [\cos xy - 1]/x^2y^2\right)$. Does $\mathbf{f}(x, y)$ have a limit at $(\pi, 0)$?

3. Prove that if f is a real function with $\lim_{\mathbf{x}\to\mathbf{b}} f(\mathbf{x}) < 0$, then there is a positive ε and a neighborhood of $\mathbf{b}$ in which all the values $f(\mathbf{x})$, except maybe $f(\mathbf{b})$, are less than $-\varepsilon$.

4. In Example 2, justify the corresponding statements:

   (a) $\lim_{\mathbf{x}\to(1,2,3)} \mathbf{f}(\mathbf{x}) = (4, 5)$; $\lim_{\mathbf{y}\to(4,5)} g(\mathbf{y}) = 6$; and $g(\mathbf{f})$ has empty domain.

   (b) $\lim_{\mathbf{y}\to(4,5)} g^+(\mathbf{y}) = 6$; and $\lim_{\mathbf{x}\to(1,2,3)} g^+(\mathbf{f}(\mathbf{x}))$ exists but equals 7.

   (c) $\lim_{\mathbf{x}\to(1,2,3)} \mathbf{F}(\mathbf{x}) = (4, 5)$; $\lim_{\mathbf{y}\to(4,5)} G(\mathbf{y}) = 6$; $G(\mathbf{F}(\mathbf{x}))$ is always defined; but $\lim_{\mathbf{x}\to(1,2,3)} G(\mathbf{F}(\mathbf{x}))$ does not exist.

   (d) $\lim_{\mathbf{x}\to(1,2,3)} \mathbf{F}^\#(\mathbf{x}) = (4, 5)$; $G(\mathbf{F}^\#(\mathbf{x}))$ is always defined; and $\lim_{\mathbf{x}\to(1,2,3)} G(\mathbf{F}^\#(\mathbf{x}))$ exists but equals 8.

5. Prove "Cauchy's criterion" for limits of functions: Assume that $\mathbf{f}$ maps $D \subseteq \mathbf{R}^n$ to $\mathbf{R}^m$, and that $\mathbf{b}$ is a closure point of D; then $\mathbf{f}$ has a finite limit at $\mathbf{b}$ iff for every $\varepsilon > 0$, there is a corresponding δ such that

$$\|\mathbf{f}(\mathbf{x}) - \mathbf{f}(\mathbf{y})\|_2 < \varepsilon \quad \text{for } \mathbf{x}, \mathbf{y} \neq \mathbf{b} \text{ in } D$$
$$\text{with } 0 < \|\mathbf{x} - \mathbf{b}\|_2 < \delta, 0 < \|\mathbf{y} - \mathbf{b}\|_2 < \delta.$$

6. Prove part (a) of Theorem 3.3.

7. Give an example of a function defined for all $\mathbf{x}$ (in, say, $\mathbf{R}^2$) that never has a limit.

3.4 Continuity

In this section we define continuous functions and give the properties that follow immediately from the definition. For these results, we have already done most of the work in the sections on limits.

Except as stated near the end, in this section $\mathbf{f}$ represents a function with domain D contained in a normed space V and vector values belonging to a normed space W.

Definition. Suppose $\mathbf{b}$ is a point of D. We say **f is continuous at b** to mean that for every sequence $(\mathbf{x}_i)$ of points from D converging to $\mathbf{b}$, $(\mathbf{f}(\mathbf{x}_i))$ converges to $\mathbf{f}(\mathbf{b})$. We say that **f is continuous on** D if $\mathbf{f}$ is continuous at each individual point of D.

It is essential to remember the role of the domain in this definition.

Example 1. Write

$$f(\mathbf{x}) := \begin{cases} 1 & \text{if } \|\mathbf{x}\| \le 1, \\ 0 & \text{if } \|\mathbf{x}\| > 1. \end{cases}$$

Then f is continuous at each point of the set $D^\infty = \{\mathbf{x}\colon \|\mathbf{x}\| > 1\}$, but discontinuous at some points of the unit ball $D^0 = \{\mathbf{x}\colon \|\mathbf{x}\| \le 1\}$ (Exercise 1).

(a) The restriction f^∞ of f to D^∞ is continuous "on" its domain. For suppose that $(\mathbf{x}_i)$ is a sequence from D^∞ converging to $\mathbf{x} \in D^\infty$. Then

$$f^\infty(\mathbf{x}_i) = 0 \to 0 = f^\infty(\mathbf{x}).$$

This illustrates a general principle: The restriction of f to a subset in which f is continuous is itself continuous.

(b) By a similar calculation, the restriction f^0 of f to D^0 is also continuous on its domain, even though f is discontinuous at some places in D^0. Clearly, restricting the domain eliminates some test sequences, so that the restriction might be continuous where f is not.

To avoid ambiguity, if S is a subset of D, then we will reserve the phrase "f is continuous on S" to mean that the restriction of f to the domain S is continuous.

We write two general properties of continuous functions.

Theorem 3.9. *Assume that* $\mathbf{b}$ *is in* D. *Then the following are equivalent*:

(a) $\mathbf{f}$ *is continuous at* $\mathbf{b}$.

(b) $\lim_{\mathbf{x}\to\mathbf{b}} \mathbf{f}(\mathbf{x})$ *exists and equals* $\mathbf{f}(\mathbf{b})$.

(c) *To every* $\varepsilon > 0$ *there corresponds* $\delta > 0$ *such that* $\|\mathbf{f}(\mathbf{x}) - \mathbf{f}(\mathbf{b})\| < \varepsilon$ *for every* $\mathbf{x}$ *in* D *for which* $\|\mathbf{x} - \mathbf{b}\| < \delta$.

(d) *To every neighborhood* $N(\mathbf{f}(\mathbf{b}), \varepsilon)$ *in* W *there corresponds a neighborhood* $N(\mathbf{b}, \delta)$ *in* V *such that* $\mathbf{f}(\mathbf{x})$ *is in* $N(\mathbf{f}(\mathbf{b}), \varepsilon)$ *for every* $\mathbf{x}$ *in* $N(\mathbf{b}, \delta) \cap D$.

Proof. Exercise 5.

As was the case for functions in $\mathbf{R}$, our first (sequence-related) definition gives us convenient ways to establish the properties of continuous functions; that definition will be our normal one. Nevertheless, parts (b), (c), and (d) of Theorem 3.9 give us alternative characterizations of continuity that will help our work.

Example 2. In $\mathbf{R}^n$ we will apply the name **Dirichlet's function** to

$$F(\mathbf{x}) := \begin{cases} 1 & \text{if every coordinate of } \mathbf{x} \text{ is rational,} \\ 0 & \text{if some coordinate of } \mathbf{x} \text{ is irrational.} \end{cases}$$

From experience in $\mathbf{R}$, we should expect this function to be discontinuous everywhere. Such is actually the case.

Let $\mathbf{x} = (x_1, \dots, x_n)$, and choose a neighborhood $N(\mathbf{x}, \varepsilon)$. The rational numbers are dense in $\mathbf{R}$, so there exist rationals q_1 within ε/n of x_1, q_2 within ε/n of $x_2, \dots, q_n$ within ε/n of x_n. Setting $\mathbf{q} := (q_1, \dots, q_n)$, we easily see that $\mathbf{q} \in N(\mathbf{x}, \varepsilon)$:

$$\|\mathbf{x} - \mathbf{q}\|_2 = \left((x_1 - q_1)^2 + \cdots + (x_n - q_n)^2\right)^{1/2} < \left(n\,\varepsilon^2/n^2\right)^{1/2} \leq \varepsilon.$$

In words, every neighborhood of $\mathbf{x}$ includes members of $\mathbf{Q}^n$. Consequently, $\mathbf{x}$ is a closure point of $\mathbf{Q}^n$, and there exists a sequence $(\mathbf{q}_i)$ from $\mathbf{Q}^n$ converging to $\mathbf{x}$.

By similar reasoning, we can find a sequence of vectors $\mathbf{p}_i$ with (some) irrational coordinates and $\mathbf{p}_i \to \mathbf{x}$.

Since all the terms of $(F(\mathbf{q}_i))$ are 1 and all the terms of $(F(\mathbf{p}_i))$ are 0, it is not true that all sequences approaching $\mathbf{x}$ yield the same function limit. Therefore, F does not have a limit anywhere (answering Exercise 7 in Section 3.3). By Theorem 3.9(b), F is discontinuous everywhere in $\mathbf{R}^n$.

Example 3. Modify Dirichlet's function as follows. If $\mathbf{q} = (q_1, \dots, q_n)$ is in $\mathbf{Q}^n$, write

$$q_1 = i_1/j_1, \dots, q_n = i_n/j_n,$$

each of these fractions being in lowest terms with positive denominator. At such a point, set

$$G(\mathbf{q}) := 1/[j_1 j_2 \cdots j_n].$$

If $\mathbf{x}$ is in $\mathbf{R}^n$ but not in $\mathbf{Q}^n$, then set $G(\mathbf{x}) := 0$. This function has the interesting property of being discontinuous at the points with (all) rational coordinates and continuous everywhere else.

To give some evidence, consider first $\mathbf{b} = \mathbf{O}$. For the sequence

$$\mathbf{x}_i := (1/i, \dots, 1/i),$$

which converges to $\mathbf{O}$, it is clear that the reduced denominators are all i, so that

$$G(\mathbf{x}_i) = i^{-n}.$$

Therefore, $G(\mathbf{x}_i) \to 0$, whereas $G(\mathbf{O}) = 1$. Hence G is discontinuous at $\mathbf{O}$. A similar construction works at any $\mathbf{b} \in \mathbf{Q}^n$.

Next take

$$\mathbf{c} := \left(\sqrt{2}, 0, \ldots, 0\right),$$

at which $G(\mathbf{c}) = 0$. Let J be any natural number. There is a positive real number δ such that every fraction with reduced denominator J or less is at least δ away from $\sqrt{2}$ (Exercise 7c). In other words, for fractions q with $\left|q - \sqrt{2}\right| < \delta$, even the reduced denominator of q is more than J. Let $\mathbf{x} = (x_1, \ldots, x_n)$ be any vector in $N(\mathbf{c}, \delta)$. If any coordinate of $\mathbf{x}$ is irrational, then $G(\mathbf{x}) = 0$. If instead every coordinate of $\mathbf{x}$ is rational, say

$$\mathbf{x} = (i_1/j_1, \ldots, i_n/j_n),$$

then

$$\left|i_1/j_1 - \sqrt{2}\right| \le \|\mathbf{x} - \mathbf{c}\|_2 < \delta.$$

That is, i_1/j_1 is a fraction within δ of $\sqrt{2}$, forcing $j_1 > J$. Therefore,

$$G(\mathbf{x}) = 1/(j_1 \cdots j_n) < 1/J.$$

We have shown that for every integer J there exists δ such that $\|\mathbf{x} - \mathbf{c}\|_2 < \delta$ implies

$$|G(\mathbf{x}) - G(\mathbf{c})| = |G(\mathbf{x})| < 1/J.$$

It follows (Theorem 3.9(c)) that G is continuous at $\mathbf{c}$.

The next result has to do with "persistence of value." It says that if $\mathbf{f}$ is continuous at $\mathbf{b}$, then near $\mathbf{b}$, $\mathbf{f}$ stays away from any value other than $\mathbf{f}(\mathbf{b})$, including ∞. Staying away from infinity means staying bounded. Staying away from finite values $\mathbf{y}$ means $\mathbf{f}(\mathbf{x}) - \mathbf{y}$ stays away from $\mathbf{O}$, and we know numerous reasons why zero is important to avoid.

Theorem 3.10. *Assume that* $\mathbf{f}$ *is continuous at* $\mathbf{b}$. *Then*:

(a) $\mathbf{f}$ *is bounded near* $\mathbf{b}$. *That is, there exist a bound* M *and a neighborhood* $N(\mathbf{b}, \delta)$ *in which* $\|\mathbf{f}(\mathbf{x})\| < M$ *for every* $\mathbf{x}$ *in* D.

(b) *If* $\mathbf{y} \neq \mathbf{f}(\mathbf{b})$, *then near* $\mathbf{b}$, $\mathbf{f}$ *is bounded away from* $\mathbf{y}$. *That is, there are a positive* ε *and a neighborhood* $N(\mathbf{b}, \delta)$ *in* V *such that* $\|\mathbf{f}(\mathbf{x}) - \mathbf{y}\| > \varepsilon$ *for any* $\mathbf{x}$ *from* $D \cap N(\mathbf{b}, \delta)$. *In particular*:

(c) (Persistence of Sign) *If* $\mathbf{f} = f$ *is real and* $f(\mathbf{b}) > 0$ *(alternatively,* $f(\mathbf{b}) < 0$*), then in some neighborhood of* $\mathbf{b}$, f *stays not only positive (respectively, negative), but stays greater than* $f(\mathbf{b})/2$ *(respectively, less than* $f(\mathbf{b})/2$*).*

Proof. Exercise 6.

In the next two results we take care of the arithmetic combinations of functions and of the most important algebraic combination.

Theorem 3.11. *Assume that* $\mathbf{f}$ *and* $\mathbf{g}$ *are continuous at* $\mathbf{b}$. *Then*:

(a) *Every linear combination* $\alpha\mathbf{f} + \beta\mathbf{g}$ *is continuous at* $\mathbf{b}$.

(b) *If* f *is real, then* $f\mathbf{g}$ *is continuous at* $\mathbf{b}$.

(c) *If* f *is real and* $f(\mathbf{b}) \neq 0$, *then* $\mathbf{g}/f$ *is continuous at* $\mathbf{b}$.

Proof. Notice first that the hypothesis guarantees that $\mathbf{b}$ is in the domain of every function listed, so that it makes sense to talk about limits and values.

That said, we observe that all the parts follow easily from Theorems 3.3 and 3.6 and the characterization of continuity in Theorem 3.9(b).

Theorem 3.12. *If* $\mathbf{f}$ *is continuous at* $\mathbf{b}$, *and* $\mathbf{g}$ *(which maps* W *to a third normed space) is continuous at* $\mathbf{f}(\mathbf{b})$, *then* $\mathbf{g}(\mathbf{f})$ *is continuous at* $\mathbf{b}$.

Proof. Observe that the hypothesis guarantees that $\mathbf{g}(\mathbf{f}(\mathbf{b}))$ is defined.

Suppose $(\mathbf{x}_i)$ is a sequence of vectors *in the domain of* $\mathbf{g}(\mathbf{f})$ with $(\mathbf{x}_i) \to \mathbf{b}$. Then $(\mathbf{x}_i)$ comes from the domain of $\mathbf{f}$, and so $\mathbf{f}(\mathbf{x}_i) \to \mathbf{f}(\mathbf{b})$ because $\mathbf{f}$ is continuous at $\mathbf{b}$. Moreover, each $\mathbf{f}(\mathbf{x}_i)$ is in the domain of $\mathbf{g}$. By the continuity of $\mathbf{g}$ at $\mathbf{f}(\mathbf{b})$, the convergence $\mathbf{f}(\mathbf{x}_i) \to \mathbf{f}(\mathbf{b})$ implies that $\mathbf{g}(\mathbf{f}(\mathbf{x}_i)) \to \mathbf{g}(\mathbf{f}(\mathbf{b}))$. The conclusion follows.

We end the section with results specific to Euclidean space.

Theorem 3.13. *Suppose* $\mathbf{f} = (f_1, \ldots, f_m)$ *maps* D *to* $\mathbf{R}^m$ *and* $\mathbf{b} \in D$.

(a) $\mathbf{f}$ *is continuous at* $\mathbf{b}$ *iff each component* f_j *is continuous at* $\mathbf{b}$.

(b) *If* $D \subseteq \mathbf{R}^n$ *and* $\mathbf{f}$ *is a rational vector function, then* $\mathbf{f}$ *is continuous at* $\mathbf{b}$.

Proof. (a) Let $(\mathbf{x}_i)$ be a sequence from D converging to $\mathbf{b}$. Then $\mathbf{f}(\mathbf{x}_i) \to \mathbf{f}(\mathbf{b}) \Leftrightarrow$ (Theorem 2.8) each $\pi_j(\mathbf{f}(\mathbf{x}_i)) \to \pi_j(\mathbf{f}(\mathbf{b})) \Leftrightarrow$ each $f_j(\mathbf{x}_i) \to f_j(\mathbf{b})$. If $\mathbf{f}(\mathbf{x}_i) \to \mathbf{f}(\mathbf{b})$ for every $(\mathbf{x}_i)$ converging to $\mathbf{b}$, then the same is true with f_j in place of $\mathbf{f}$, and conversely. Hence $\mathbf{f}$ is continuous $\Rightarrow$ each f_j is, and conversely.

(b) Theorem 3.8 tells us that a rational function defined at $\mathbf{b}$ has limit equal to the value of the function. By Theorem 3.9(b), such a function is continuous at $\mathbf{b}$.

Example 4. Theorems 3.11–3.13 give us a large store of vector functions whose continuity we can verify, based on our knowledge of functions in $\mathbf{R}$.

Let $\mathbf{H}(x, y) := \left(e^{-x}, \cos[x + y]/xy^2\right)$. The first component is a composite of $x = \pi_1(x, y)$, which is a continuous function of the vector (x, y), and the continuous function e^{-s} of the real variable s. The second is the quotient of a continuous composite and a polynomial. Each component is continuous for $xy \neq 0$, and therefore so is $\mathbf{H}$.

Exercises

1. Prove that in a normed space,

$$f(\mathbf{x}) := \begin{cases} 1 & \text{if } \|\mathbf{x}\| \le 1, \\ 0 & \text{if } \|\mathbf{x}\| > 1, \end{cases}$$

is discontinuous at the unit sphere and continuous everywhere else.

2. Let $\mathbf{b}$ be a fixed vector. Show that $\|\mathbf{x} - \mathbf{b}\|$ is a continuous function of $\mathbf{x}$; in particular, $\|\mathbf{x}\|$ is continuous.

3. Prove that $f(\mathbf{x}) := (\|\mathbf{x}\|[\|\mathbf{x}\| - 1])^{1/2}$ is continuous on its domain. Can you make a general statement about any function $g(\mathbf{x})$ that depends only on $\|\mathbf{x}\|$?

4. (a) Let $\mathbf{b}$ be a fixed vector in $\mathbf{R}^n$. Show that $\mathbf{x} \bullet \mathbf{b}$ is a continuous function of $\mathbf{x}$.

   (b) Is the same true in any inner product space?

5. Prove Theorem 3.9 by establishing the following results. In every part, $\mathbf{b}$ and all $\mathbf{x}$-vectors are assumed to be in D. Show that:

   (a) Statement (a) implies statement (b). (Is $\mathbf{b}$ a closure point of D?)

   (b) At an isolated point $\mathbf{b}$ of D, statements (b) and (c) are trivial; at an accumulation point, statement (b) implies statement (c). (Hint: Theorem 3.4.)

   (c) Statement (c) implies statement (d).

   (d) Statement (d) implies statement (a). (Hint: Theorem 2.5.)

6. Prove Theorem 3.10. (Hint: Theorem 3.9 and Theorem 3.5 or its methods.)

7. Show that if s is an irrational real number:

   (a) Among the fractions i/j whose denominator is 10 or less, there is one i_1/j_1 that is closest to s, and its distance $|i_1/j_1 - s|$ from s is a positive number δ_1. Conclude that $|i/j - s| < \delta_1$ implies $j > 10$.

   (b) Similarly for the fractions with denominator 100 or less, 1000 or less, ...

   (c) Given a positive integer J, it is always possible to find a positive δ such that for every fraction i/j with $j \le J$, we have $|i/j - s| \ge \delta$.

8. Let S be a fixed subset and $\mathbf{b}$ a fixed vector in a normed space. By the **distance from b to** S, symbolized by $d(\mathbf{b}, S)$, we mean $\inf\{\|\mathbf{b} - \mathbf{x}\| : \mathbf{x} \text{ is in } S\}$.

   (a) In $\mathbf{R}^2$, find the distance from $(3, 0)$ to the unit circle $\{\mathbf{x}\colon \|\mathbf{x}\|_2 = 1\}$.

(b) In $\mathbf{R}^2$, find the distance from $(3, 0)$ to the neighborhood $N(\mathbf{O}, 1)$.

(c) If $d(\mathbf{b}, S) = 0$, does it follow that $\mathbf{b}$ is in S? What about the converse?

(d) Prove that if S is finite, then $d(\mathbf{b}, S) = 0$ iff $\mathbf{b}$ is in S.

(e) Use part (d) to give another proof (compare Exercise 7) that there exists a positive δ such that if a fraction has reduced denominator less than 10^{100}, then that fraction is more than δ away from $\sqrt{2}$.

3.5 Continuity in Infinite-Dimensional Spaces

We have examined the infinite-dimensional situation for the material in each of the first two chapters. Here we do something similar. Specifically, we look at continuity without making a separate study of limits.

One reason we do not want to get in too deeply is that the vector variables in the infinite-dimensional spaces we have considered are themselves functions. Recall that our two examples are $C_0[0, 1]$, meaning the set of continuous real functions on $[0, 1]$, with the norm

$$\|f\|_0 := \sup |f|;$$

and $C_2[0, 1]$, which uses the same set of functions, but with the norm

$$\|f\|_2 := \left(\int_0^1 f(x)^2 \, dx \right)^{1/2}.$$

These are normed vector spaces, so we can talk about convergence in them (Section 2.4), and therefore we can talk about continuity of functions of these vectors. Our discussion will revolve around three specific examples of such functions of functions.

Example 1. Pick a fixed element of $[0, 1]$, like $\frac{1}{2}$. Write

$$F(f) := f\left(\frac{1}{2}\right).$$

This formula associates the real number $f\left(\frac{1}{2}\right)$ to each f defined in $[0, 1]$. Therefore, F is a real-valued function of the vector variable f.

Example 2. Pick three fixed continuous functions, say $\cos 2\pi x$, $\cos 4\pi x$, and $\cos 6\pi x$. Write

$$\mathbf{G}(f) := \left(\int_0^1 f(x) \cos 2\pi x \, dx, \int_0^1 f(x) \cos 4\pi x \, dx, \int_0^1 f(x) \cos 6\pi x \, dx \right).$$

Under this formula, a continuous f is matched with an ordered triple $\mathbf{G}(f)$ of real numbers. Therefore, $\mathbf{G}$ is an $\mathbf{R}^3$-valued function of f.

Example 3. Write

$$\mathbf{H}(f) := f^2.$$

In more detail, if f is a function on $[0, 1]$, then $\mathbf{H}(f)$ is the function of x defined by

$$[\mathbf{H}(f)](x) := [f(x)]^2.$$

This $\mathbf{H}$ is a function-valued function of f.

We initially consider each of these examples in $C_0[0, 1]$.

First, F is, like the projections in $\mathbf{R}^n$, contractive:

$$|F(f) - F(g)| := \left| f\left(\frac{1}{2}\right) - g\left(\frac{1}{2}\right) \right| \leq \max |f - g| = \|f - g\|_0.$$

Every contractive function is continuous on its whole domain. With F, for example, if $f_i \to f$, then

$$|F(f_i) - F(f)| \leq \|f_i - f\|_0 \to 0.$$

That is, $f_i \to f$ implies $F(f_i) \to F(f)$. We conclude that F is continuous.

Look next at $\mathbf{G}$. Each component of $\mathbf{G}$ is contractive:

$$\begin{aligned} |G_k(f) - G_k(g)| &:= \left| \int_0^1 f(x) \cos 2k\pi x \, dx - \int_0^1 g(x) \cos 2k\pi x \, dx \right| \\ &\leq \sup |[f(x) - g(x)] \cos 2k\pi x| \leq \|f - g\|_0. \end{aligned}$$

Therefore, each component is continuous. This fact is actually enough to let us conclude that $\mathbf{G}$ is continuous (Exercise 1). However, it is worth pursuing the question a little further. Observe that

$$|G_k(f)| = |G_k(f) - G_k(\mathbf{O})| \leq \|f\|_0,$$

so that

$$\begin{aligned} \|\mathbf{G}(f)\|_2 &:= \left[G_1(f)^2 + G_2(f)^2 + G_3(f)^2 \right]^{1/2} \\ &\leq \left(\|f\|_0^2 + \|f\|_0^2 + \|f\|_0^2 \right)^{1/2} = \sqrt{3}\|f\|_0. \end{aligned}$$

If $f_i \to f$, then

$$\begin{aligned} \|\mathbf{G}(f_i) - \mathbf{G}(f)\|_2 &= \|\mathbf{G}(f_i - f)\|_2 \qquad \text{(Why?)} \\ &\leq \sqrt{3}\,\|f_i - f\|_0 \to 0. \end{aligned}$$

This proves that $\mathbf{G}$ is continuous.

What were the key steps in the preceding argument? First, we said that $\mathbf{G}$ satisfies $\|\mathbf{G}(f)\|_2 \leq \sqrt{3}\|f\|_0$. We can describe this property by saying that $\mathbf{G}$ magnifies f by no more than a factor of $\sqrt{3}$. Obviously, $\sqrt{3}$ is not special here; any fixed positive number would have sufficed. If $\mathbf{G}$ maps all of one normed space to another and the norm of any image is no more than a fixed multiple of the norm of the preimage (in symbols: if $\mathbf{G}$ maps V to W and there exists a constant M such that

$$\|\mathbf{G}(\mathbf{x})\|_W \leq M\|\mathbf{x}\|_V \qquad \text{for every } \mathbf{x} \text{ in } V)$$

then we will call $\mathbf{G}$ a map of **bounded magnification**.

The second key was that $\mathbf{G}$ is additive:

$$\mathbf{G}(f+g) = \mathbf{G}(f) + \mathbf{G}(g).$$

Any additive map is also "subtractive," because from

$$\mathbf{G}(f-g) + \mathbf{G}(g) = \mathbf{G}(f),$$

we conclude that

$$\mathbf{G}(f-g) = \mathbf{G}(f) - \mathbf{G}(g);$$

and conversely.

With the information that the foregoing are the crucial data, we leave the following result as Exercise 2:

Theorem 3.14. *Suppose Φ maps (all of) one linear space V to another. If Φ is additive and has bounded magnification, then Φ is continuous everywhere in V.*

Turn now to the function $\mathbf{H}$. Squaring is almost never additive; in this case, the vectors $f(x) \equiv 1$ and $g(x) \equiv 2$ give

$$9 = \mathbf{H}(f+g) \neq \mathbf{H}(f) + \mathbf{H}(g) = 5.$$

Squaring is also generally not of bounded magnification; we have

$$\|\mathbf{H}(2g)\|_0 = 4\|2g\|_0,$$
$$\|\mathbf{H}(3g)\|_0 = 6\|3g\|_0,$$

... (Verify!), so $\mathbf{H}$ has unbounded magnification. Accordingly, $\mathbf{H}$ is not covered by Theorem 3.14.

Is $\mathbf{H}$ continuous anyway (as a function into $C_0[0,1]$)? To answer this question, we need to examine $\|\mathbf{H}(f_i) - \mathbf{H}(f)\|_0$ for a sequence (f_i) converging to f. We have

$$\begin{aligned}\|\mathbf{H}(f_i) - \mathbf{H}(f)\|_0 &:= \|f_i^2 - f^2\|_0 := \sup\{|f_i - f|\,|f_i + f|\}\\ &\leq \sup\{|f_i - f|[|f_i| + |f|]\}\\ &\leq \sup\{|f_i - f|[\|f_i\|_0 + \|f\|_0]\}. \quad \text{(Justify all!)}\end{aligned}$$

Assume $(f_i) \to f$. Then (Theorem 2.7(c)) the sequence is bounded. That is, there exists M such that every $\|f_i\|_0$ is less than M. Because the norm of the limit is the limit of the norms, we have $\|f\|_0 \leq M$ also. Therefore,

$$\|\mathbf{H}(f_i) - \mathbf{H}(f)\|_0 \leq \sup\{|f_i - f|[2M]\} = 2M\|f_i - f\|_0 \to 0.$$

From $f_i \to f$ we have reached $\mathbf{H}(f_i) \to \mathbf{H}(f)$. Hence $\mathbf{H}$ is continuous.

(Since the squaring function is continuous, we should anticipate that all polynomials are continuous. This idea is illustrated in Exercise 3.)

That finishes our discussion of these functions as members of $C_0[0, 1]$. Now, view Examples 1–3 as functions with domains in $C_2[0, 1]$.

We can still call F a projection, but one thing changes: F is not contractive. In fact, F is not even of bounded magnification. For evidence, consider the function g_k whose graph is shown in Figure 3.2. Clearly,

$$\|g_k\|_2 := \left(\int_0^1 g_k(x)^2\,dx\right)^{1/2} \leq \left(\frac{2^k\,3}{2^k}\right)^{1/2} = \sqrt{3},$$

whereas

$$F(g_k) := g_k\left(\frac{1}{2}\right) = 2^{k/2}.$$

In words: g_k has large values but not a large integral. The function is based on Exercise 1 in Section 2.4. Using the functions H_i from that exercise, the g_k shown in Figure 3.2 is

$$g_k := 2^{k/2} H_{2^k+2^{k-1}}, \qquad k \geq 2.$$

Hence

$$|F(g_k)| \geq 2^{k/2} 3^{-1/2} \|g_k\|_2.$$

This being true for arbitrary k, F has unbounded magnification.

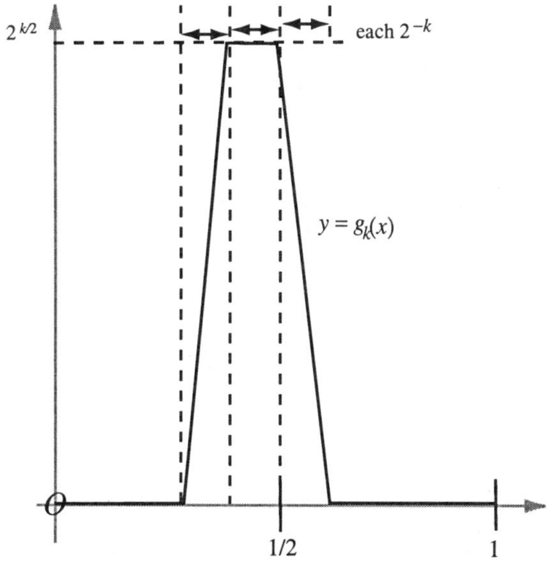

Figure 3.2.

We can now show that F is discontinuous, at least at **O**. The sequence defined by

$$h_k := 2^{-k/2}\, g_k$$

has

$$\|h_k\|_2 = 2^{-k/2}\|g_k\|_2 \leq \sqrt{3}/2^{k/2} \to 0.$$

So, it converges to $\mathbf{O}$. (Compare Exercise 1b and c in Section 2.4.) But

$$F(h_k) = 2^{-k/2} g_k\left(\frac{1}{2}\right) = 1,$$

whence $(F(h_k))$ does not converge to $F(\mathbf{O}) = 0$. Therefore, F is discontinuous at $\mathbf{O}$.

Again we ask; what were the keys in the preceding argument? First, we needed the fact that F has unbounded magnification. Second, it was essential to be able to factor scalar multipliers from F. Thus, we needed to know that F is homogeneous:

$$F(\alpha g) := (\alpha g)\left(\frac{1}{2}\right) := \alpha g\left(\frac{1}{2}\right) = \alpha F(g).$$

These properties allowed us to turn

$$\|g_k\|_2 \approx 1 \quad \text{with } F(g_k) = 2^{k/2}$$

into

$$\|2^{-k/2} g_k\|_2 \to 0 \quad \text{with } F(2^{-k/2} g_k) = 1 \text{ not approaching } 0.$$

We have again formulated the basis for a theorem–exercise (Exercise 4):

Theorem 3.15. *If* $\mathbf{U}$, *which maps (all of) one normed space to another, is homogeneous and has unbounded magnification, then* $\mathbf{U}$ *is discontinuous at* $\mathbf{O}$.

Notice that Theorem 3.14 deals with additive functions, Theorem 3.15 with homogeneous ones. A vector function $\mathbf{L}$ that is both additive and homogeneous satisfies

$$\begin{aligned} \mathbf{L}(\mathbf{x}+\mathbf{y}) &= \mathbf{L}(\mathbf{x}) + \mathbf{L}(\mathbf{y}), \\ \mathbf{L}(\alpha\mathbf{x}) &= \alpha\mathbf{L}(\mathbf{x}). \end{aligned}$$

Linear algebra calls such a function a **linear transformation** (or **linear mapping**). We can now prove the following theorem:

Theorem 3.16. *Let* $\mathbf{L}$ *be a linear map from one normed space to another. Then* $\mathbf{L}$ *is continuous everywhere iff* $\mathbf{L}$ *is of bounded magnification.*

Proof. $\Rightarrow$ Assume that $\mathbf{L}$ is continuous everywhere. We know that $\mathbf{L}$ is homogeneous. If $\mathbf{L}$ had unbounded magnification, then by Theorem 3.15 it would be discontinuous at the origin. Hence $\mathbf{L}$ must have bounded magnification.

$\Leftarrow$ Assume conversely that $\mathbf{L}$ is of bounded magnification. Since $\mathbf{L}$ is additive, Theorem 3.14 applies, and so $\mathbf{L}$ is continuous everywhere. □

It is standard to call a linear map **bounded** if it has the property we called "bounded magnification." We adopted the longer name to preserve "bounded map" and "bounded function" for a different use.

Go back for a minute to finite dimensions. Any linear map from $\mathbf{R}^n$ to $\mathbf{R}^m$ is a first-degree polynomial function (but not vice versa; see Exercise 5), so every such map is continuous. We now know two new things: Every linear map from $\mathbf{R}^n$ to $\mathbf{R}^m$ has bounded magnification, and (as we showed for F on $C_2[0, 1]$) the same *cannot* be said in infinite dimensions.

Next we face

$$\mathbf{G}(f) := \left(\int_0^1 f(x)\cos 2\pi x\,dx,\ \int_0^1 f(x)\cos 4\pi x\,dx,\ \int_0^1 f(x)\cos 6\pi x\,dx\right).$$

Recall that the integral yields an inner product in $C_2[0, 1]$. Therefore, Cauchy's inequality holds:

$$\left|\int_0^1 f(x)\cos 2k\pi x\,dx\right| \le \|f\|_2\|\cos 2k\pi x\|_2 = 2^{-1/2}\|f\|_2 \quad \text{for each } k.$$

Consequently, $\mathbf{G}$ is of bounded magnification:

$$\|\mathbf{G}(f)\|_2 := \left(G_1(f)^2 + G_2(f)^2 + G_3(f)^2\right)^{1/2} \le \sqrt{1.5}\|f\|_2.$$

Since $\mathbf{G}$ is a linear map, Theorem 3.16 tells us that $\mathbf{G}$ is continuous throughout $C_2[0, 1]$.

Finally, for the squaring operator $\mathbf{H}(f) := f^2$, we can recycle the sequence (g_k) defined in Figure 3.2. Let

$$f_k := 2^{-k/4} g_k.$$

From

$$\|g_k\|_2 \le \sqrt{3},$$

we know that $f_k \to \mathbf{O}$. But

$$\|\mathbf{H}(f_k)\|_2 = \|(2^{-k/4} g_k)^2\|_2 = \left(\int_0^1 2^{-k} g_k(x)^4\,dx\right)^{1/2} \ge \left(\frac{2^{-k}2^{2k}}{2^k}\right)^{1/2} = 1.$$

$(\mathbf{H}(f_k))$ does not approach $\mathbf{O}$, and $\mathbf{H}$ is discontinuous at the origin.

Exercises

1. Prove that if $\mathbf{G} = (G_1, \ldots, G_m)$ maps any normed vector space into $\mathbf{R}^m$, then $\mathbf{G}$ is continuous iff every component of $\mathbf{G}$ is continuous.

2. Prove Theorem 3.14.

3. Show that the "cubic polynomial" $\mathbf{K}(f) := f^3 + 5f^2 - 3f + 1$ is a continuous function from $C_0[0, 1]$ to $C_0[0, 1]$.

4. Prove Theorem 3.15. (Hint: Decide what the value of $\mathbf{U}$ at the origin is.)

5. How can a first-degree function fail to be a linear mapping? (The answer shows why we have not used "linear," whose meaning we take from linear algebra, as a synonym for "first-degree.")

6. Assume that **L** is a linear map between normed spaces. Theorem 3.16 says that "**L** is continuous everywhere" is equivalent to "**L** has bounded magnification." Prove that the following are also equivalent:

   (a) **L** is continuous everywhere.

   (b) **L** is continuous somewhere.

   (c) **L** is continuous at the origin.

7. For any continuous function f on $[0, 1]$, define $T(f) := \max f(x)$. Note that this maximum does exist, and is not the same as $\|f\|_0 := \max |f(x)|$.

   (a) Is T continuous (as a real-valued function) on $C_0[0, 1]$?

   (b) Is T continuous on $C_2[0, 1]$?

8. Pick a fixed function g_0 that is continuous on $[0, 1]$. Define $\mathbf{U}(f) := fg_0$ (the product). Is **U** continuous as:

   (a) a $C_0[0, 1]$-valued function on $C_0[0, 1]$?

   (b) a $C_2[0, 1]$-valued function on $C_2[0, 1]$?

   (c) *a $C_0[0, 1]$-valued function on $C_2[0, 1]$?

   (d) a $C_2[0, 1]$-valued function on $C_0[0, 1]$?

4

Characteristics of Continuous Functions

Introduction

In this chapter we develop fundamental results about continuous functions in normed spaces, with special attention to the specific setting of Euclidean space.

These results are generalizations of theorems about continuous functions of one variable. On appropriate subsets of **R**, a continuous function necessarily:

1. is bounded;
2. attains maximum and minimum values;
3. is uniformly continuous;
4. achieves intermediate values.

We will prove that the same statements carry over to more general spaces.

As usual, our methods suggest other ideas, in this case ideas involving neighborhoods and related sets. We will also begin to study properties of the sets themselves.

4.1 Continuous Functions on Boxes in Euclidean Space

We begin by generalizing a class of sets from **R**.

Definition. Suppose $\mathbf{a} := (a_1, \ldots, a_n)$ and $\mathbf{b} := (b_1, \ldots, b_n)$ are in $\mathbf{R}^n$.

(a) We say that **a is to the left of b**, and write $\mathbf{a} \leq \mathbf{b}$, if $a_1 \leq b_1$, $a_2 \leq b_2, \ldots, a_n \leq b_n$.

(b) If $a_1 < b_1, \ldots, a_n < b_n$, then we say that **a is strictly to the left of b**, and write $\mathbf{a} < \mathbf{b}$.

(c) If $\mathbf{a} \leq \mathbf{b}$, then we refer to $\{\mathbf{x} \colon \mathbf{a} \leq \mathbf{x} \leq \mathbf{b}\}$ as a **box**, and denote it by **[a, b]**.

It is important to see that $\leq$ is a partial order (Exercise 5). However, it is not a total order. That is, given **a** and **b**, it may be that neither $\mathbf{a} \leq \mathbf{b}$ nor $\mathbf{b} \leq \mathbf{a}$. For example, in $\mathbf{R}^2$, neither of $(1, 3)$ and $(2, 2)$ is to the left of the other. Accordingly, although it is tempting to read $\leq$ as "is smaller than or equal to," the reading suggests something that may not hold. Thus $(1, 3) \leq (2, 3)$, but neither $(1, 3) < (2, 3)$ nor $(1, 3) = (2, 3)$ is true.

If $\mathbf{a} \leq \mathbf{b}$ is false, then $\{\mathbf{x} \colon \mathbf{a} \leq \mathbf{x} \leq \mathbf{b}\}$ is empty (Exercise 6). It is for this reason that our definition of **[a, b]** assumes that $\mathbf{a} \leq \mathbf{b}$. If $\mathbf{a} \leq \mathbf{b}$ is true but it is false that $\mathbf{a} < \mathbf{b}$, then the box is squashed in at least one direction. In that case, we describe it as "flat" or "degenerate," but we still call it a box. We will continue to use the familiar names for a box in **R** (a **closed interval**) and in $\mathbf{R}^2$ (a **rectangle**).

Theorem 4.1. *Given a box* **[a, b]**:

(a) *Every vector* **x** *in* **[a, b]** *satisfies* $\|\mathbf{x}\|_2^2 \leq \|\mathbf{a}\|_2^2 + \|\mathbf{b}\|_2^2$.

(b) *If a sequence from* **[a, b]** *is convergent, then its limit is also in the box.*

Proof. (a) By definition, $\mathbf{x} \in [\mathbf{a}, \mathbf{b}]$ means

$$a_k \leq x_k \leq b_k \qquad \text{for each } k.$$

For a given k, if x_k is nonnegative, then

$$0 \leq x_k^2 \leq b_k^2 \leq a_k^2 + b_k^2.$$

If x_k is negative, then

$$0 \leq x_k^2 \leq a_k^2 \leq a_k^2 + b_k^2.$$

Either way,

$$x_k^2 \leq a_k^2 + b_k^2 \qquad \text{for every } k,$$

so that

$$\|\mathbf{x}\|_2^2 := x_1^2 + \cdots + x_n^2 \leq (a_1^2 + b_1^2) + \cdots + (a_n^2 + b_n^2) = \|\mathbf{a}\|_2^2 + \|\mathbf{b}\|_2^2.$$

(b) Assume that $(\mathbf{x}_i)$ is a convergent sequence of vectors from **[a, b]**. By (a), $(\mathbf{x}_i)$ is bounded, so the limit is finite (Theorem 2.7(d)). Call the limit **x**. For each k (Theorem 2.8),

$$\pi_k(\mathbf{x}) = \lim_{i \to \infty} \pi_k(\mathbf{x}_i).$$

Since $\mathbf{x}_i \in [\mathbf{a}, \mathbf{b}]$,

$$a_k \leq \pi_k(\mathbf{x}_i) \leq b_k \quad \text{for all } i.$$

Passing to the limit as $i \to \infty$, we conclude that

$$a_k \leq \pi_k(\mathbf{x}) \leq b_k;$$

that is, $\mathbf{x}$ is in the box. □

Definition. A function $\mathbf{f}$ (from any normed space to another) is **bounded on the set** S if there is a real M with $\|\mathbf{f}(\mathbf{x})\| < M$ for every $\mathbf{x}$ in S.

In the results that follow, we deal with a fixed function $\mathbf{f}$ whose domain is a box $[\mathbf{a}, \mathbf{b}] \subseteq \mathbf{R}^n$ and whose range is a subset of a normed linear space. For many theorems we exhibit proofs that are identical in language to the (sequence-based) proofs in $\mathbf{R}$.

Theorem 4.2. *A function continuous on a box is bounded.*

Proof. Assume that $\mathbf{f}$ is unbounded. Then no integer i can satisfy $\|\mathbf{f}(\mathbf{x})\| < i$ for all $\mathbf{x}$. That is, for every i we can find a corresponding $\mathbf{x}_i$ in the box such that $\|\mathbf{f}(\mathbf{x}_i)\| \geq i$. Thus, $(\mathbf{x}_i)$ is a sequence with $(\mathbf{f}(\mathbf{x}_i)) \to \infty$.

Because $(\mathbf{x}_i)$ comes from the box, it is a bounded sequence (Theorem 4.1(a)). By the Bolzano–Weierstrass theorem, there is a subsequence $(\mathbf{x}_{j(i)})$ converging to a finite $\mathbf{y}$. For the function-value subsequence, we have $\mathbf{f}(\mathbf{x}_{j(i)}) \to \infty$, because this subsequence of $(\mathbf{f}(\mathbf{x}_i))$ must have the parent's limit. Since $\mathbf{x}_{j(i)} \to \mathbf{y}$, $\mathbf{f}$ cannot have a finite limit at $\mathbf{y}$. By Theorem 3.9(b), $\mathbf{f}$ is discontinuous at $\mathbf{y}$. But $\mathbf{y}$ is in the box (Theorem 4.1(b)). Hence there is a place in the box at which $\mathbf{f}$ is discontinuous.

By contraposition, if $\mathbf{f}$ is continuous throughout the box, then $\mathbf{f}$ is bounded. □

Theorem 4.3 (The Extreme Value Theorem). *A continuous real-valued function on a box has a maximum value and a minimum value in the box.*

Proof. Theorem 4.2 provides us with some K such that $\|f\| < K$ throughout the box. Since f is real, the inequality becomes

$$-K < f(\mathbf{x}) < K.$$

Thus, $\{f(\mathbf{x}) : \mathbf{a} \leq \mathbf{x} \leq \mathbf{b}\}$ (the range of f) is bounded above and below. Therefore, it possesses a finite supremum and finite infimum.

Write $M := \sup\{f(\mathbf{x})\}$. By a property of suprema, corresponding to each integer i there is $\mathbf{x}_i$ in the box with

$$M - \frac{1}{i} < f(\mathbf{x}_i) \leq M.$$

Passing to the limit as $i \to \infty$, we find $f(\mathbf{x}_i) \to M$.

As in Theorem 4.2, $(\mathbf{x}_i)$ has a subsequence $(\mathbf{x}_{j(i)})$ converging to a limit $\mathbf{y}$, which by Theorem 4.1(b) is also in the box. By the continuity of f, we conclude that

$$f(\mathbf{y}) = \lim_{i \to \infty} f(\mathbf{x}_{j(i)}) = \lim_{i \to \infty} f(\mathbf{x}_i) = M.$$

The supremum M of the function values is reached at $\mathbf{y}$, so necessarily $f(\mathbf{y})$ is the maximum value of f.

An analogous argument takes care of $m := \inf\{f(\mathbf{x})\}$. □

Definition. A function $\mathbf{f}$ between normed spaces is **uniformly continuous** on a set S if to any specified ε there corresponds δ such that $\|\mathbf{f}(\mathbf{x}) - \mathbf{f}(\mathbf{y})\| < \varepsilon$ for every $\mathbf{x}$ and $\mathbf{y}$ in S having $\|\mathbf{x} - \mathbf{y}\| < \delta$.

Theorem 4.4. *A continuous function on a box is uniformly continuous.*

Proof. We must prove that every positive real ε leads to a δ such that

$$\|\mathbf{f}(\mathbf{x}) - \mathbf{f}(\mathbf{y})\| < \varepsilon \qquad \text{whenever } \|\mathbf{x} - \mathbf{y}\|_2 < \delta.$$

Let us assume the contrary: Assume that ε is a number for which no δ works. For example, $\delta = 1$ does not work. Thus, there exist $\mathbf{x}_1$ and $\mathbf{y}_1$ for which

$$\|\mathbf{x}_1 - \mathbf{y}_1\|_2 < 1 \qquad \text{but } \|\mathbf{f}(\mathbf{x}_1) - \mathbf{f}(\mathbf{y}_1)\| \geq \varepsilon.$$

Also, $\delta = \frac{1}{2}$ does not work, so there exist $\mathbf{x}_2$ and $\mathbf{y}_2$ for which

$$\|\mathbf{x}_2 - \mathbf{y}_2\|_2 < \frac{1}{2} \qquad \text{but } \|\mathbf{f}(\mathbf{x}_2) - \mathbf{f}(\mathbf{y}_2)\| \geq \varepsilon.$$

Also, $\delta = \frac{1}{3}$ does not work, and so on. That is, we have sequences $(\mathbf{x}_i)$ and $(\mathbf{y}_i)$ with

$$\|\mathbf{x}_i - \mathbf{y}_i\|_2 < 1/i \qquad \text{but } \|\mathbf{f}(\mathbf{x}_i) - \mathbf{f}(\mathbf{y}_i)\| \geq \varepsilon.$$

Since $(\mathbf{x}_i)$ comes from the box, there is a subsequence $(\mathbf{x}_{j(i)})$ converging to a finite limit $\mathbf{c}$ in the box. The sequence $(\mathbf{y}_{j(i)})$ also converges to $\mathbf{c}$, because

$$\|\mathbf{c} - \mathbf{y}_{j(i)}\|_2 \leq \|\mathbf{c} - \mathbf{x}_{j(i)}\|_2 + \|\mathbf{x}_{j(i)} - \mathbf{y}_{j(i)}\|_2,$$

and both terms on the right approach zero.

Look now at function values. If $\mathbf{f}(\mathbf{x}_{j(i)})$ and $\mathbf{f}(\mathbf{y}_{j(i)})$ had the same finite limit, then the distance $\|\mathbf{f}(\mathbf{x}_{j(i)}) - \mathbf{f}(\mathbf{y}_{j(i)})\|$ would approach 0. But we chose the original $(\mathbf{x}_i)$ and $(\mathbf{y}_i)$ to keep all these difference norms greater than or equal to ε. Therefore, the value sequences $(\mathbf{f}(\mathbf{x}_{j(i)}))$ and $(\mathbf{f}(\mathbf{y}_{j(i)}))$ do not have the same finite limit. Since the sequences $(\mathbf{x}_{j(i)})$ and $(\mathbf{y}_{j(i)})$ both converge to $\mathbf{c}$, we conclude that $\mathbf{f}$ does not have a finite limit at $\mathbf{c}$.

We have proved that if $\mathbf{f}$ is not uniformly continuous, then $\mathbf{f}$ is discontinuous someplace in the box. □

Theorem 4.5 (Intermediate Value Theorem). *A continuous real function on a box attains intermediate values. In symbols: Assume that f is continuous on* $[\mathbf{a}, \mathbf{b}]$ *and $f(\mathbf{c}) < f(\mathbf{d})$ are two values of f; if y is between $f(\mathbf{c})$ and $f(\mathbf{d})$, then there is some* $\mathbf{x} \in [\mathbf{a}, \mathbf{b}]$ **at which** $f(\mathbf{x}) = y$.

Proof. Consider the function

$$F(t) := f([1-t]\mathbf{c} + t\mathbf{d}), \qquad 0 \le t \le 1.$$

F is the composite of

$$\mathbf{G}(t) := [1-t]\mathbf{c} + t\mathbf{d}$$

and f. First, $\mathbf{G}$ is a first-degree function, so it is continuous. Second, for each t, $f(\mathbf{G}(t))$ is defined. The reason is that $\mathbf{G}(t)$ is a point on the segment joining $\mathbf{c}$ and $\mathbf{d}$, and the segment joining two points in a box lies entirely within the box: If

$$\mathbf{a} \le \mathbf{c} \le \mathbf{b}, \quad \mathbf{a} \le \mathbf{d} \le \mathbf{b}, \quad \text{and } 0 \le t \le 1,$$

then the kth coordinates satisfy

$$a_k = (1-t)a_k + ta_k \le (1-t)c_k + td_k \le (1-t)b_k + tb_k = b_k,$$

which says that $[1-t]\mathbf{c} + t\mathbf{d}$ is in the box. Third, since f is continuous in the box, $F = f(\mathbf{G})$ is continuous (Theorem 3.12).

Thus, $F(t)$ is a continuous function on $[0, 1]$. Because $F(0) = f(\mathbf{c})$, $F(1) = f(\mathbf{d})$, and y is between them, the intermediate value theorem in $\mathbf{R}$ [Ross, Theorem 18.2] tells us that there is a $u \in [0, 1]$ where $F(u) = y$. Then $\mathbf{x} := (1-u)\mathbf{c} + u\mathbf{d}$ is a point in the box (Why?) with

$$f(\mathbf{x}) = f(\mathbf{G}(u)) = F(u) = y. \qquad \square$$

Exercises

1. Which of these functions has a maximum value on the part of its domain contained in the box $[(-1, -1), (1, 1)]$?

   (a) $f(x, y) := xye^{xy}$

   (b) $g(x, y) := \left(x^2 + y^2\right)e^{1/(xy)}$

   (c) $h(x, y) := xy \sin\left(\frac{1}{xy}\right)$

2. (a) What point of the box $[(-3, 3), (1, 4)]$ is farthest from the origin?

   (b) Show that for every box in $\mathbf{R}^n$, the point that is farthest from the origin is one of the corners.

3. In Theorems 4.2, 4.3, and 4.4, it is essential that the box have finite extent. Thus, define the "infinite box" $[\mathbf{a}, \infty) := \{\mathbf{x} \colon \mathbf{a} \le \mathbf{x}\}$. Give examples to show that a continuous real function on $[\mathbf{a}, \infty)$ in $\mathbf{R}^3$ can:

   (a) be unbounded;

   (b) be bounded, but fail to reach a maximum or minimum;

   (c) fail to be uniformly continuous.

4. In Theorems 4.1–4.4, it is essential that the faces of the box be part of the domain. Write $(\mathbf{a}, \mathbf{b}) := \{\mathbf{x} \colon \mathbf{a} < \mathbf{x} < \mathbf{b}\}$. Give examples to illustrate that in such a nonempty "open box" $(\mathbf{a}, \mathbf{b})$ in $\mathbf{R}^3$, it is possible to find:

   (a) a sequence from the open box converging to a limit that is not in the open box;

   (b) an unbounded continuous function;

   (c) a bounded continuous function that does not reach a maximum or minimum;

   (d) a function that is continuous, but not uniformly.

5. Prove that $\leq$ has the three properties of a **partial order**:

   (a) reflexivity: $\mathbf{a} \leq \mathbf{a}$
   (b) antisymmetry: if both $\mathbf{a} \leq \mathbf{b}$ and $\mathbf{b} \leq \mathbf{a}$, then $\mathbf{a} = \mathbf{b}$
   (c) transitivity: if both $\mathbf{a} \leq \mathbf{b}$ and $\mathbf{b} \leq \mathbf{c}$, then $\mathbf{a} \leq \mathbf{c}$.

6. Prove that if $\mathbf{a}$ is not to the left of $\mathbf{b}$, then there are no $\mathbf{x}$ such that $\mathbf{a} \leq \mathbf{x} \leq \mathbf{b}$.

4.2 Continuous Functions on Bounded Closed Subsets of Euclidean Space

In Section 4.1 we saw that functions continuous in a box share three attributes: boundedness, attainment of extreme values, uniform continuity. For proofs, we depended on a single property of the box: Every sequence of vectors from the box has a subsequence converging to a vector in the box. In this section we will see that sets with this property belong to each of two classes that are of interest on their own, and we shall talk about characteristics of continuous functions on such sets.

That every sequence from a box has a sublimit within the box depends in turn on two features of the box. They are consequences of Theorem 4.1. The first is that every sequence from a box is a bounded sequence. This feature allows us to invoke the Bolzano–Weierstrass theorem and infer that the sequence has a sublimit. We begin by examining this attribute.

Definition. A set S of vectors in a normed linear space is **bounded** if there is a real M such that $\|\mathbf{x}\| < M$ for every $\mathbf{x}$ in S.

Example 1. (a) Every ball is bounded. If

$$\mathbf{x} \in B(\mathbf{b}, \delta) := \{\mathbf{x} \colon \|\mathbf{x} - \mathbf{b}\| \leq \delta\},$$

then

$$\|\mathbf{x}\| \leq \|\mathbf{x} - \mathbf{b}\| + \|\mathbf{b}\| \leq \delta + \|\mathbf{b}\|.$$

(b) Any subset of a bounded set is bounded (Justify!). Consequently, neighborhoods and spheres, being subsets of balls, are bounded sets.

(c) Our definition of "bounded function" ($\|\mathbf{f}(\mathbf{x})\| < M$) requires precisely that the range $\{\mathbf{f}(\mathbf{x})\}$ be a bounded set. The same holds for "bounded sequence," since a sequence is just a special kind of function.

(d) In $\mathbf{R}^n$, every box is bounded (Theorem 4.1(a)).

Theorem 4.6. *Let S be a subset of a normed space. Then the following are equivalent:*

(a) *S is bounded.*

(b) *Every sequence of vectors from S is a bounded sequence.*

(c) *S is contained in some neighborhood of the origin.*

(d) *S is contained in some neighborhood.*

If S is a subset of $\mathbf{R}^n$, then (a)–(d) are also equivalent to:

(e) *S is contained in some box.*

(f) *Each of the projections $\pi_k(S) := \{\pi_k(\mathbf{x}) : \mathbf{x} \in S\}$ is a bounded set of real numbers.*

We leave the proof as a long exercise (Exercise 7). Some of the proofs are implicitly or expressly done in Example 1.

The second feature of the box is that a sublimit of a sequence from the box is itself in the box. We turn now to that property.

Definition. A set S in a normed space is **closed** if every sequence $(\mathbf{x}_i)$ of vectors from S that converges to a finite limit has its limit in S.

It is important to understand that the definition makes no demand that sequences converge. Thus in $\mathbf{R}^n$, $\mathbf{x}_i := (i, \ldots, i)$ defines a sequence that does not converge to a limit within $\mathbf{R}^n$. Such a sequence does not pose a test for closedness. What is demanded is that *if* $(\mathbf{x}_i)$ converges to a vector, then the vector must belong to the set being tested.

Example 2. (a) In a normed space V, V itself is automatically a closed subset of V, because every finite limit is in V. The empty set is vacuously a closed subset, because we cannot find sequences to test it.

(b) No neighborhood is closed.

Consider a neighborhood $N(\mathbf{b}, \delta)$. Let

$$\mathbf{x}_i := \mathbf{b} + \delta\left(1 - \frac{1}{i}\right)\frac{\mathbf{b}}{\|\mathbf{b}\|} \qquad \text{if } \mathbf{b} \neq \mathbf{O};$$

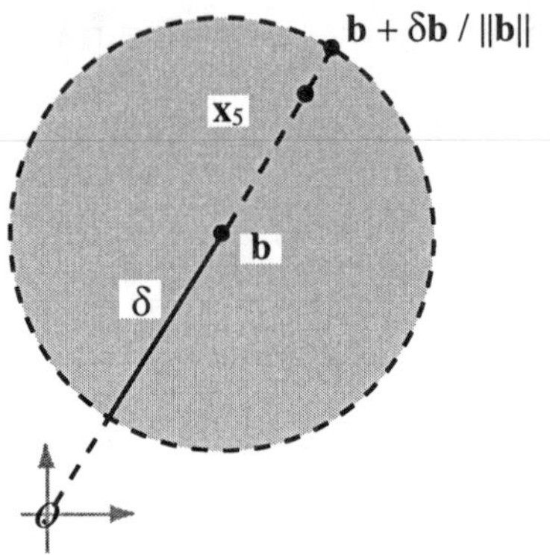

Figure 4.1.

if $\mathbf{b}$ is zero, then any unit vector will do in place of $\mathbf{b}/\|\mathbf{b}\|$. Figure 4.1 shows $\mathbf{x}_5 = \mathbf{b} + .8\delta\mathbf{b}/\|\mathbf{b}\|$, on the line from the origin through $\mathbf{b}$, $\frac{4}{5}$ of the way to the edge of the disk. Then

$$\|\mathbf{x}_i - \mathbf{b}\| = \delta(1 - 1/i) < \delta,$$

so $(\mathbf{x}_i)$ comes from the neighborhood. But

$$\mathbf{x}_i \to \mathbf{b} + \frac{\delta\mathbf{b}}{\|\mathbf{b}\|}.$$

This last vector is at the sphere; its distance to $\mathbf{b}$ is δ. That is, $(\mathbf{x}_i)$ converges to a point that is outside $N(\mathbf{b}, \delta)$. We conclude that $N(\mathbf{b}, \delta)$ is not closed.

(c) Every box is closed. Theorem 4.1(b) says just that.

(d) Every ball is closed, as is every sphere (Exercise 4).

Theorem 4.7. *For a subset S of a normed space, the following are equivalent:*

(a) *S is closed.*

(b) *Every accumulation point of S belongs to S.*

(c) *Every closure point of S belongs to S.*

(d) *The set of closure points of S equals S.*

Proof. (a) $\Rightarrow$ (b) Assume that S is closed.

Let $\mathbf{x}$ be an accumulation point of S. Then there exists a sequence $(\mathbf{x}_i)$ from S with $\mathbf{x} \neq \mathbf{x}_i \to \mathbf{x}$. By definition of closedness, the limit of $(\mathbf{x}_i)$ is in S; that is, $\mathbf{x} \in S$.

(b) $\Rightarrow$ (c) Assume that every accumulation point of S actually belongs to S. We must prove that every closure point of S is a member of S.

Let $\mathbf{x}$ be a closure point of S. By definition, some sequence $(\mathbf{x}_i)$ from S converges to $\mathbf{x}$. If one of the $\mathbf{x}_i$ equals $\mathbf{x}$, then $\mathbf{x} \in S$. If none of the $\mathbf{x}_i$ is $\mathbf{x}$, then $\mathbf{x}$ is an accumulation point of S, and by assumption $\mathbf{x} \in S$. Either way, we have the desired conclusion.

(c) $\Rightarrow$ (d) Write $\text{cl}(S)$ for the set of closure points of S. It is automatic that $S \subset \text{cl}(S)$; we stated earlier (Section 3.2) that a member of a set is trivially a closure point of it.

If we assume (c), then we are assuming that $\text{cl}(S) \subseteq S$. That forces $\text{cl}(S) = S$.

(d) $\Rightarrow$ (a) Assume $\text{cl}(S) = S$.

Let $\mathbf{x}$ be the limit of a sequence from S. By definition of closure point, $\mathbf{x}$ is in $\text{cl}(S)$. By the assumption, $\mathbf{x}$ is in S. Hence every limit of a sequence from S is in S. □

The main result of this section is that in Euclidean space, the properties of a function continuous in a box hold in any closed, bounded set.

Theorem 4.8. *Assume that* $\mathbf{f}$ *is continuous on a closed, bounded set* $S \in \mathbf{R}^n$. *Then on* S:

(a) $\mathbf{f}$ *is bounded.*

(b) *If* $\mathbf{f}$ *is real, then it reaches a maximum and a minimum value.*

(c) $\mathbf{f}$ *is uniformly continuous.*

Proof. The proofs are modifications of the arguments in Theorems 4.2, 4.3, and 4.4. We exhibit (a), then leave (b) and (c) to Exercise 8.

Assume that $\mathbf{f}$ is unbounded on S. Then to each i, there corresponds $\mathbf{x}_i$ at which $\|\mathbf{f}(\mathbf{x}_i)\| > i$. The sequence $(\mathbf{x}_i)$ is bounded, being picked from the bounded set S (Theorem 4.6(b)). By the Bolzano–Weierstrass theorem, there is a subsequence $(\mathbf{x}_{j(i)})$ converging to a finite $\mathbf{x}$. Because $(\mathbf{x}_{j(i)})$ is drawn from the closed set S, we infer that $\mathbf{x}$ is in S. Since $\mathbf{x}_{j(i)} \to \mathbf{x}$ and $\mathbf{f}(\mathbf{x}_{j(i)}) \to \infty$, it is impossible for $\mathbf{f}$ to be continuous at $\mathbf{x}$.

Of necessity, if $\mathbf{f}$ is continuous everywhere in S, then $\mathbf{f}$ is bounded. □

(There is interesting history [Kline, Chapter 40, Sections 1 and especially 2] connecting a number of things we have written. Bolzano, Cauchy, and Weierstrass helped develop the theory of the real number system, leading to the definition of a continuous function of a real variable. Bolzano showed the existence of suprema, a defining characteristic of the real numbers. We said in Section 2.4 that Cauchy's criterion—Theorem 2.9 and Exercise 5 in Section 3.3—can similarly characterize $\mathbf{R}$. The "criterion" was actually conceived by Bolzano. His work was unknown. By contrast, Cauchy was a celebrity, with contributions in almost all areas of mathematics; his investigations into limits and continuity were immediately widely known. Decades later Weierstrass, himself working in obscurity for many years, refined Bolzano's and Cauchy's descriptions of continuity to give the precise "ϵ–δ" definition, as in Theorem 3.9(c). He used Bolzano's supremum notion to prove that a bounded, infinite set of real numbers must have an accumulation point—compare Exercise 9—then used that result to prove that a continuous function on a closed interval achieves extreme values.)

Exercises

1. Which of the following sets in $\mathbf{R}^2$ is closed? Which bounded?

    (a) $\mathbf{Z}^2 := \{(x, y): x \text{ and } y \text{ are both integers}\}$.

    (b) $1/\mathbf{Z}^2 := \{(1/x, 1/y): x \text{ and } y \text{ both nonzero integers}\}$.

    (c) the x-axis.

    (d) the graph of $y = e^x$.

    (e) the graph of $y = x/|x|$.

2. Suppose T is a set of real numbers.

    (a) Under what circumstances will the set $\{x + y: x \text{ and } y \text{ both in } T\}$ be bounded?

    (b) through (d) Same questions for $\{x - y\}$, $\{xy\}$, and $\{x/y: y \neq 0\}$.

3. Prove that in $\mathbf{R}^n$, the "infinite boxes" $(-\infty, \mathbf{b}]$ and $[\mathbf{a}, \infty)$ are closed.

4. In a normed space, prove that:

    (a) Every finite set is closed. Is the converse true?

    (b) Every ball is closed.

    (c) Every sphere is closed.

    (d) The union and the intersection of two closed sets is closed.

5. Suppose $\mathbf{b}$ is a vector and S a nonempty subset of $\mathbf{R}^n$.

    (a) Assume that S is closed and bounded. Show that there must exist $\mathbf{c} \in S$ such that the distance from $\mathbf{b}$ to S (Exercise 8 in Section 3.4) is the distance from $\mathbf{b}$ to $\mathbf{c}$. (In words: Among the points of S, there must be one—or more—closest to $\mathbf{b}$.)

    (b) If S is closed and bounded, must it also have a point farthest from $\mathbf{b}$?

    (c) In $\mathbf{R}^2$, what point of the unit circle is closest to $(3, 4)$?

    (d) In $\mathbf{R}^n$, what point of the ball $B(\mathbf{a}, \delta)$ is closest to $\mathbf{b}$? (Hint: Draw the possible cases. Remember that a ball is not just a sphere.)

    (e) *Show that if S is closed, then among the points of S there is one (or more) that is closest to $\mathbf{b}$. Need there also be a farthest point?

6. Theorem 4.7 says that a closed set equals its set of closure points. Show that a closed set may not equal its set of accumulation points.

7. For the parts of Theorem 4.6, prove that:

    (a) In a normed space, (a) $\Rightarrow$ (b) $\Rightarrow$ (c) $\Rightarrow$ (d) $\Rightarrow$ (a).

    (b) In $\mathbf{R}^n$, (a) $\Rightarrow$ (e) $\Rightarrow$ (f) $\Rightarrow$ (a).

8. Prove (b) and (c) of Theorem 4.8.

9. Prove that these statements are equivalent in a normed linear space:

   (a) Every bounded infinite set has an accumulation point (possibly outside the set).

   (b) Every bounded sequence has a finite sublimit.

   (Call a set S **infinite** if there exists a one-to-one mapping of the natural numbers into S. In our sequence-oriented approach, S is infinite iff it contains a sequence whose vectors are all different. In that approach, (b) is our "Bolzano–Weierstrass theorem." But version (a) is how old Weierstrass put it. Be wary: (a) $\Rightarrow$ (b) is subtle.)

4.3 Extreme Values and Sequentially Compact Sets

The boundedness and extreme-value properties have now appeared twice, and with like justifications. The similarity leads us to turn them into a single property.

Theorem 4.9. *Let S be a subset of a normed linear space. The following are equivalent*:

(a) *Every continuous vector function on S is bounded.*

(b) *Every continuous real function on S reaches a maximum value.*

(c) *Every continuous real function on S reaches a minimum value.*

Proof. (a) $\Rightarrow$ (b) We will prove that if some continuous real function on S does not hit a maximum value, then some continuous function is unbounded.

Let f be a continuous real function on S with no maximum. If $\sup\{f(\mathbf{x}) \colon \mathbf{x} \in S\}$ were infinite, then f would be unbounded (Why?), and we would already have found an unbounded continuous function. Hence we may assume that $M := \sup\{f(\mathbf{x})\}$ is finite.

If M were one of the values of f, that is, if there were $\mathbf{y} \in S$ with $f(\mathbf{y}) = M$, then

$$f(\mathbf{x}) \le M = f(\mathbf{y})$$

would tell us that $f(\mathbf{y})$ is the maximum of f. We assumed that there is no such maximum, so it must be that $f(\mathbf{x}) < M$ for all $\mathbf{x} \in S$. Let now

$$g(\mathbf{x}) := \frac{1}{[M - f(\mathbf{x})]}.$$

The function $g_1(\mathbf{x}) := M - f(\mathbf{x})$ is continuous and positive on S, while $g_2(t) := 1/t$ is a continuous function of the real variable t as long as t is nonzero. By our theorem about continuous composites (Theorem 3.12), $g = g_2(g_1)$ is continuous on S.

By the nature of suprema, for each i there is $\mathbf{x}_i$ with

$$f(\mathbf{x}_i) > M - 1/i.$$

Hence $g(\mathbf{x}_i) > i$, which tells us that g is an unbounded continuous function on S.

(b) $\Rightarrow$ (c) Assume that every continuous real function has a maximum. Then h is continuous $\Rightarrow -h$ has a maximum $(-h)(\mathbf{y}) \Rightarrow h$ has a minimum $h(\mathbf{y})$.

(c) $\Rightarrow$ (a) Assume that $\mathbf{F}$ is a continuous vector function. Then

$$G(\mathbf{x}) := -\|\mathbf{F}(\mathbf{x})\|$$

is a continuous real function. (Reason?) By assumption, $G(\mathbf{x})$ has a minimum $-M$. Therefore,

$$-\|\mathbf{F}(\mathbf{x})\| \geq -M \quad \text{for every } \mathbf{x} \in S,$$

and it follows that $\mathbf{F}$ is bounded. □

Definition. A subset of a normed space **has the extreme value property (EVP)** if it satisfies (a)–(c) of Theorem 4.9.

Example 1. The EVP is a property of sets, not of individual functions. Clearly, any real function that has a maximum and a minimum is necessarily bounded. But a given function may have bounds without extremes, or a bound or extreme on one side without a bound on the other. In $\mathbf{R}^2$ (Exercise 1):

(a) $\tan^{-1} xy$ and $|\tan^{-1} xy|$ are both bounded, but the first reaches neither a maximum nor a minimum, and the second achieves a minimum but not a maximum.

(b) e^{xy} is bounded below, and $e^{|xy|}$ even has a minimum, but neither is bounded above.

Theorem 4.10. *If a subset of a normed space has the EVP, then it is closed and bounded.*

Proof. We have previously noted that $F(\mathbf{x}) := \|\mathbf{x}\|$ is continuous on any set. If every continuous real function on S is bounded, then F has to be. Therefore, there exists M such that

$$\|\mathbf{x}\| = F(\mathbf{x}) < M \quad \text{for every } \mathbf{x} \text{ in } S.$$

Hence S is bounded.

Take next a fixed $\mathbf{b}$ outside S, and consider

$$G(\mathbf{x}) := \frac{1}{\|\mathbf{x} - \mathbf{b}\|}.$$

The denominator $\|\mathbf{x} - \mathbf{b}\|$ is a continuous function of $\mathbf{x}$ everywhere, and is nonzero on S because $\mathbf{b}$ is not one of the vectors of S. Hence G is a continuous composite on S. By assumption, G is bounded: There is M with $G < M$ throughout S. In other words,

$$\|\mathbf{x} - \mathbf{b}\| = \frac{1}{G(\mathbf{x})} > \frac{1}{M} \quad \text{for all } \mathbf{x} \text{ in } S.$$

This says that the neighborhood $N(\mathbf{b}, 1/M)$ has no members of S. We conclude that $\mathbf{b}$ is not a closure point of S.

So, if $\mathbf{b}$ is outside S, then $\mathbf{b}$ is not a closure point of S. Equivalently, if $\mathbf{b}$ is a closure point of S, then $\mathbf{b}$ is in S. By Theorem 4.7(c), S is closed. □

Theorems 4.8 and 4.10 together tell us that in $\mathbf{R}^n$, the EVP belongs precisely to closed bounded sets. We indicated the crucial factor at the opening of Section 4.2: In a closed, bounded subset of $\mathbf{R}^n$, every sequence has a subsequence converging to a limit in the set. We will show that this last property characterizes the EVP in every normed space.

Definition. A subset of a normed space is **sequentially compact** if every sequence from the set possesses a sublimit in the set, that is, if every sequence of vectors that belong to the set has a subsequence that converges to a member of the set.

Theorem 4.11. *A subset of a normed space is sequentially compact iff it has the EVP.*

Proof. ⇒ We omit the familiar argument; compare Theorem 4.8(a).

⇐ Assume that S has the EVP and $(\mathbf{u}_i)$ is a sequence from S. We will use a function to point us to a sublimit for the sequence.

For $\mathbf{x} \in S$, write

$$f(\mathbf{x}) := \inf\{1/j + \|\mathbf{x} - \mathbf{u}_j\| : j \geq 1\}.$$

If $\mathbf{x}$ and $\mathbf{y}$ are in S and $j \geq 1$, then the triangle inequality gives

$$\frac{1}{j} + \|\mathbf{x} - \mathbf{u}_j\| \leq \frac{1}{j} + \|\mathbf{y} - \mathbf{u}_j\| + \|\mathbf{x} - \mathbf{y}\|.$$

Hence

$$f(\mathbf{x}) \leq f(\mathbf{y}) + \|\mathbf{x} - \mathbf{y}\|.$$

We can interchange $\mathbf{x}$ and $\mathbf{y}$, so we conclude that

$$|f(\mathbf{x}) - f(\mathbf{y})| \leq \|\mathbf{x} - \mathbf{y}\|;$$

the function is contractive, and therefore continuous.

Observe that for a fixed i,

$$f(\mathbf{u}_i) \leq \frac{1}{i} + \|\mathbf{u}_i - \mathbf{u}_i\| = \frac{1}{i}.$$

Therefore, f has arbitrarily small values. Since f has to be nonnegative, we conclude that

$$\inf\{f(\mathbf{x}) : \mathbf{x} \in S\} = 0.$$

By EVP, 0 has to be a value of f, say

$$0 = f(\mathbf{c}) := \inf\{1/j + \|\mathbf{c} - \mathbf{u}_j\| : j \geq 1\}.$$

Because the numbers $1/j + \|\mathbf{c} - \mathbf{u}_j\|$ constitute a sequence of positive terms, there must exist a subsequence $(1/j(k) + \|\mathbf{c} - \mathbf{u}_{j(k)}\|)$ converging to 0. That convergence forces

$$\|\mathbf{c} - \mathbf{u}_{j(k)}\| \to 0,$$

and $\mathbf{c}$ is a sublimit of $(\mathbf{u}_i)$. □

(The concise yet elementary proof in Theorem 4.11 is due to a brilliant City College student, Jan Siwanowicz.)

Notice how we have shown that in $\mathbf{R}^n$, closed-boundedness, EVP, and sequential compactness coincide, and that the latter two always coincide. Closed-boundedness does not match the others, but proof has to wait.

A less simple situation obtains with uniform continuity. Sequential compactness (and therefore EVP) guarantees that every continuous function is uniformly continuous (Exercise 2). In the opposite direction, the possible deduction is meager: We show below that a set where every continuous function is uniformly continuous has to be closed, but need not be bounded.

Theorem 4.12. *Let S be a subset of a normed space. If every continuous real function on S is uniformly continuous, then S is closed.*

Proof. Assume that S is not closed. Then there exists a sequence $(\mathbf{x}_i)$ from S converging to $\mathbf{b} \notin S$. Since $\mathbf{b}$ is outside S, $G(\mathbf{x}) := 1/\|\mathbf{x} - \mathbf{b}\|$ is defined and continuous on S. G is also unbounded, because $\|\mathbf{x}_i - \mathbf{b}\| \to 0$, implying that $G(\mathbf{x}_i) \to \infty$.

A general principle: If $(\mathbf{u}_i)$ is a Cauchy sequence and $\mathbf{f}$ a uniformly continuous function, then $(\mathbf{f}(\mathbf{u}_i))$ is Cauchy. To prove this, choose $\varepsilon > 0$. By the uniform continuity, there exists $\delta = \delta(\varepsilon)$ such that

$$\|\mathbf{f}(\mathbf{u}) - \mathbf{f}(\mathbf{v})\| < \varepsilon \qquad \text{for } \|\mathbf{u} - \mathbf{v}\| < \delta.$$

By the Cauchyness of $(\mathbf{u}_i)$, there is $I = I(\delta(\varepsilon))$ such that $i, j \geq I$ forces $\|\mathbf{u}_i - \mathbf{u}_j\| < \delta$. Hence

$$i, j \geq I \text{ implies } \|\mathbf{f}(\mathbf{u}_i) - \mathbf{f}(\mathbf{u}_j)\| < \varepsilon;$$

we conclude that $(\mathbf{f}(\mathbf{u}_i))$ is Cauchy.

Now G cannot be uniformly continuous. If it were, then the (convergent and therefore) Cauchy sequence $(\mathbf{x}_i)$ would lead to a Cauchy $(G(\mathbf{x}_i))$. But we specifically picked $(G(\mathbf{x}_i))$ to be unbounded, so (compare Exercise 7b in Section 2.3) it cannot be Cauchy.

We have shown that if S is not closed, then some continuous real function is not uniformly continuous on S. □

Example 2. Consider

$$S := \{(1, 0, \dots, 0),\ (2, 0, \dots, 0), \dots\}.$$

S is an unbounded set in $\mathbf{R}^n$, but every function on S is uniformly continuous. The reason is that if $\mathbf{f}$ is defined on S and $\varepsilon > 0$, then $\delta = \frac{1}{2}$ works: For any $\mathbf{x}$ and $\mathbf{y}$ in S with

$$\|\mathbf{x} - \mathbf{y}\|_2 < \delta,$$

necessarily

$$\|\mathbf{f}(\mathbf{x}) - \mathbf{f}(\mathbf{y})\| < \varepsilon.$$

(Why? Compare Exercise 3.)

Definition. For any set S and function $\mathbf{f}$ defined on S, the **image of S under $\mathbf{f}$** is the set $\mathbf{f}(S) := \{\mathbf{f}(\mathbf{x})\colon \mathbf{x} \in S\}$. In words, the image of a set is the set of images.

Theorem 4.13. *The continuous image of a sequentially compact set is sequentially compact; that is, if S is sequentially compact and $\mathbf{f}$ is continuous on S, then $\mathbf{f}(S)$ is a sequentially compact set.*

Proof. Assume that $\mathbf{f}$ is a continuous vector function on S and S is sequentially compact.

Let $(\mathbf{y}_i) = (\mathbf{f}(\mathbf{x}_i))$ be a sequence in $\mathbf{f}(S)$. Because $(\mathbf{x}_i)$ is a sequence from S, there is a subsequence $(\mathbf{x}_{j(i)})$ converging to a limit $\mathbf{x} \in S$. Since $\mathbf{f}$ is continuous on S, $\mathbf{f}(\mathbf{x}_{j(i)})$ converges to $\mathbf{f}(\mathbf{x}) \in \mathbf{f}(S)$. We have found in $\mathbf{f}(S)$ a sublimit of $(\mathbf{y}_i)$. This proves that $\mathbf{f}(S)$ is sequentially compact.

Exercises

1. In Example 1, why are the functions in (a) and (b) continuous? Why are statements (a) and (b) true?

2. Prove that in a sequentially compact set, every continuous function is uniformly continuous.

3. (a) Show that if all the points of S are isolated, then every function defined on S is continuous on S.

   (b) Find an example of a set T whose points are all isolated, but on which some functions are *not uniformly* continuous.

4. Let S be a nonempty sequentially compact set and $\mathbf{b}$ any vector. Prove that among the points of S, there is a closest one to $\mathbf{b}$ (one with minimal distance) and a furthest one from $\mathbf{b}$. (Compare Exercise 5a–b in Section 4.2.)

5. Show that for a nonempty S, the set-distance function $d(\mathbf{x}, S) := \inf\{\|\mathbf{x}-\mathbf{z}\|: \mathbf{z} \in S\}$ is contractive: $|d(\mathbf{x}, S) - d(\mathbf{y}, S)| \leq \|\mathbf{x} - \mathbf{y}\|$.

6. For any two sets S and T, call $d(S, T) := \inf\{\|\mathbf{x} - \mathbf{y}\|: \mathbf{x} \in S, \mathbf{y} \in T\}$ the **distance from** S **to** T (or **between** S **and** T). Show that if S and T are nonempty and sequentially compact, then there are $\mathbf{a}$ in S and $\mathbf{b}$ in T whose distance is the distance from S to T.

7. Given a nonempty set S, the **diameter** of S is the quantity $\sup\{\|\mathbf{x} - \mathbf{y}\|: \mathbf{x}, \mathbf{y} \in S\}$.

   (a) Prove that the diameter of a ball is twice its radius.

   (b) Prove that if S is sequentially compact, then there are $\mathbf{a}$ and $\mathbf{b}$ in S with $\|\mathbf{a} - \mathbf{b}\|$ equal to the diameter of S (two vectors that are as far apart as two in S can get).

8. Give examples to show that the conclusions in Exercises 4, 6, and 7(b) may fail if any of the sets involved is not sequentially compact.

9. Let S and T be nonempty subsets of a normed linear space.

   (a) Suppose S is sequentially compact. Show that there is a point $\mathbf{a}$ of S closest to T, and that $d(S, T) = d(\mathbf{a}, T)$. (Refer to Exercises 5 and 6; also, compare Exercise 4.)

   (b) Suppose S is sequentially compact, T is closed, and S and T are disjoint. Show that $d(S, T) > 0$.

   (c) Show by example(s) that (a) and (b) may fail if S is, like T, merely closed.

   (d) Show that (b) may fail if T is not closed.

   (e) Assume that S is sequentially compact, T is closed, and they are both subsets of $\mathbf{R}^n$. Show that there is a point of T closest to S.

4.4 Continuous Functions and Open Sets

Suppose S is a closed set. If $\mathbf{b}$ is not in S, then (Theorem 4.7(c)) $\mathbf{b}$ is not a closure point of S. Consequently (Theorem 3.1) there is a neighborhood $N(\mathbf{b}, \delta)$ that has no points of S. Viewing the situation from the outside in, we can say that the complement S^* of S has the following property: If $\mathbf{b}$ is in S^*, then there is a neighborhood $N(\mathbf{b}, \delta)$ having only points of S^*, so that $N(\mathbf{b}, \delta) \subseteq S^*$.

Definition. A subset T of a normed space is called **open** if for each $\mathbf{x} \in T$ there exists a neighborhood of $\mathbf{x}$ that is contained in T.

Example 1. (a) Every neighborhood is open.

Let $N := N(\mathbf{x}, r)$, and suppose $\mathbf{y} \in N$. By Theorem 2.2(c), if s satisfies

$$r \geq \|\mathbf{y} - \mathbf{x}\| + s,$$

then N encompasses the neighborhood $N(\mathbf{y}, s)$. Hence $N(\mathbf{y}, r - \|\mathbf{y} - \mathbf{x}\|)$ is a neighborhood of $\mathbf{y}$ that is contained in N. We conclude that N is open.

(b) In $\mathbf{R}^n$, if $\mathbf{a} < \mathbf{b}$, then $(\mathbf{a}, \mathbf{b}) := \{\mathbf{x} \colon \mathbf{a} < \mathbf{x} < \mathbf{b}\}$ is an open set. (If $\mathbf{a}$ is not strictly to the left of $\mathbf{b}$, then $(\mathbf{a}, \mathbf{b})$ is empty, but the statement that $(\mathbf{a}, \mathbf{b})$ is open still holds; see Exercises 3 and 4.)

Assume $\mathbf{a} < \mathbf{b}$ and $\mathbf{x} \in (\mathbf{a}, \mathbf{b})$. From $\mathbf{a} < \mathbf{x} < \mathbf{b}$,

$$a_1 < x_1 < b_1, \ldots, a_n < x_n < b_n.$$

Let r be the smallest of the numbers $x_1 - a_1, \ldots, x_n - a_n$, $b_1 - x_1, \ldots, b_n - x_n$. In Figure 4.2 it is clear that r is the distance from $\mathbf{x}$ to the nearest wall of the box. If $\mathbf{y} \in N(\mathbf{x}, r)$, then for each k,

$$|y_k - x_k| \leq \|\mathbf{y} - \mathbf{x}\|_2 < r,$$

so

$$a_k \leq x_k - r < y_k < x_k + r \leq b_k.$$

Therefore, $\mathbf{y} \in (\mathbf{a}, \mathbf{b})$. We conclude that $N(\mathbf{x}, r) \subseteq (\mathbf{a}, \mathbf{b})$, and $(\mathbf{a}, \mathbf{b})$ is open.

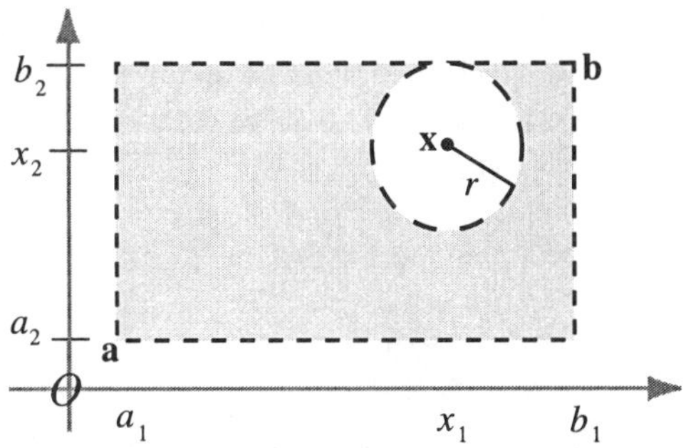

Figure 4.2.

Given what we know about $(\mathbf{a}, \mathbf{b})$ and $[\mathbf{a}, \mathbf{b}]$, it is natural to refer to $(\mathbf{a}, \mathbf{b})$ as an **open interval** and $[\mathbf{a}, \mathbf{b}]$ as a **closed interval**. These are standard names, but we will continue to call $[\mathbf{a}, \mathbf{b}]$ a "box." We can call $(\mathbf{a}, \mathbf{b})$ an "open box," but it is rare to need the name.

Along the same lines, it is common to call $N(\mathbf{x}, \delta)$ an **open ball** and $B(\mathbf{x}, \delta)$ the corresponding **closed ball**. We will continue to use "ball" as before, but will sometimes adopt "open ball" as a more descriptive term for $N(\mathbf{x}, \delta)$.

This last usage frees up "neighborhood." It is common to use "neighborhood of **x**" synonymously with "open set containing **x**." The reason is that one is typically considering whether something is true "near **x**." For instance, we next talk about a function continuous near $\mathbf{x}$. If $\mathbf{f}$ is continuous in an open set to which $\mathbf{x}$ belongs, then that set contains a neighborhood of $\mathbf{x}$, in which $\mathbf{f}$ is continuous. Conversely, if $\mathbf{f}$ is continuous in a neighborhood N of $\mathbf{x}$, then N is an open set including $\mathbf{x}$ in which $\mathbf{f}$ is continuous. Thus, continuity in an open set that has $\mathbf{x}$ as a member is equivalent to continuity in a neighborhood of $\mathbf{x}$. In any case, we will not identify "neighborhood" with "open set"; any mention of neighborhood refers, as before, to an open ball.

The introductory paragraph of this section leads to this characterization.

Theorem 4.14. *A set T in a normed space is open iff its complement T^* is closed.*

Proof. Exercise 7.

It will frequently occur that we describe an open set in terms of the open balls that make it up.

Theorem 4.15. *A set is open iff it is the union of neighborhoods.*

Proof. $\Rightarrow$ Assume that S is open. For each $\mathbf{x} \in S$ there is $N(\mathbf{x}) := N(\mathbf{x}, \delta(\mathbf{x}))$ contained in S. Let Σ be the class $\{N(\mathbf{x})\colon \mathbf{x} \in S\}$. We will show that the union U of the sets in Σ is precisely S.

Suppose $\mathbf{y} \in S$. Then $N(\mathbf{y})$ is one of the neighborhoods in the class Σ, so $\mathbf{y} \in N(\mathbf{y}) \subseteq U$. Thus, $\mathbf{y} \in S$ implies $y \in U$. That says that $S \subseteq U$.

Suppose instead that $\mathbf{y} \in U$. By definition of union, there is some $N(\mathbf{x})$ that $\mathbf{y}$ belongs to. Since each $N(\mathbf{x})$ is contained in S, we get $\mathbf{y} \in N(\mathbf{x}) \subseteq S$. From $\mathbf{y} \in U$, we derived $\mathbf{y} \in S$; it follows that $U \subseteq S$.

From the two set inclusions, we infer $S = U$.

$\Leftarrow$ Suppose Σ is a class of neighborhoods, and T is their union.

Take any $\mathbf{x} \in T$. By definition, $\mathbf{x}$ is in one of the neighborhoods; call it $N(\mathbf{b}, \varepsilon)$. Because $N(\mathbf{b}, \varepsilon)$ is an open set (Example 1), there is a neighborhood $N(\mathbf{x}, \delta)$ contained in $N(\mathbf{b}, \varepsilon)$. This $N(\mathbf{x}, \delta)$, being contained in one of the sets from Σ, is contained in the union of the sets from Σ. That is, $N(\mathbf{x}, \delta) \subseteq T$.

For an $\mathbf{x} \in T$, we have found $N(\mathbf{x}, \delta) \subseteq T$. Therefore, T is open. □

Theorem 4.15 has the practical effect of letting us study open sets **locally**, that is, one neighborhood at a time. This reduction of the region to be investigated is especially useful with a continuous function having an open domain.

Theorem 4.16. *Assume that the domain D of $\mathbf{f}$ is open. Let $\mathbf{b} \in D$ and $\mathbf{c} := \mathbf{f}(\mathbf{b})$. If $\mathbf{f}$ is continuous at $\mathbf{b}$, then for each neighborhood $N(\mathbf{c}, \varepsilon)$ there is a corresponding neighborhood $N(\mathbf{b}, \delta)$ such that $\mathbf{f}(\mathbf{x}) \in N(\mathbf{c}, \varepsilon)$ for each $\mathbf{x}$ in $N(\mathbf{b}, \delta)$.*

Proof. Under the hypothesis, let $N(\mathbf{c}, \varepsilon)$ be specified.

By Theorem 3.9(d), the continuity of $\mathbf{f}$ at $\mathbf{b}$ means that there is $N(\mathbf{b}, r)$ such that $\mathbf{f}(\mathbf{x})$ is in $N(\mathbf{c}, \varepsilon)$ for every $\mathbf{x}$ from $D \cap N(\mathbf{b}, r)$. D being open, some neighborhood of $\mathbf{b}$ is contained in D. That is, there exists s with $N(\mathbf{b}, s) \subseteq D$. Let δ be the smaller of r and s. Then $N(\mathbf{b}, \delta)$ is the intersection of the r- and s-neighborhoods, and so $\mathbf{x} \in N(\mathbf{b}, \delta)$ implies both $\mathbf{x} \in D$ and $\mathbf{f}(\mathbf{x}) \in N(\mathbf{c}, \varepsilon)$. □

Recall our definition of image of a set under a function: $\mathbf{f}(S) := \{\mathbf{f}(\mathbf{x}): \mathbf{x} \in S\}$. In Theorem 4.16, everything from a neighborhood $N(\mathbf{b}, \delta)$ is mapped into $N(\mathbf{c}, \varepsilon)$. We can describe the situation by $\mathbf{f}(N(\mathbf{b}, \delta)) \subseteq N(\mathbf{c}, \varepsilon)$.

We can also view the situation from the range side. Given any set T in the space to which $\mathbf{f}$ maps, write $\mathbf{f}^{-1}(T) := \{\mathbf{x} \in D: \mathbf{f}(\mathbf{x}) \in T\}$ for the **inverse image** (or **preimage**) of T under $\mathbf{f}$.

Example 2. In a number of ways, inverse images are better behaved than images. We illustrate with

$$F(x, y) := \begin{cases} e^{-x} & \text{for } (x, y) \text{ with } x \geq 0, \\ 1 & \text{for } (x, y) \text{ with } x < 0, \end{cases}$$

whose relevant attribute is that its domain is all of $\mathbf{R}^2$. (Compare Exercise 6.)

(a) $F(\text{quadrant II}) = F(\text{quadrant III}) = \{1\}$. Thus the images of two different, in fact disjoint, sets can be the same. Nothing so odd happens with inverse images. For example,

$$F^{-1}([1, \infty)) = \{(x, y): x \leq 0\}$$

and

$$F^{-1}((-\infty, 1)) = \{(x, y): x > 0\},$$

which are inverse images of complements, are themselves complements.

(b) Quadrant II is an open set, but $F(\text{quadrant II})$ is closed. Similarly,

$$S := \{(x, y): x \geq 0\}$$

is closed, but its image contains values from 1 down toward zero, including 1 but not zero. Thus, S is a closed set whose image $(0, 1]$ is neither open nor closed.

Neither situation is possible with inverse images. The interval $\left(\frac{1}{5}, \frac{1}{e}\right) \subseteq \mathbf{R}$ is open, and

$$F^{-1}\left(\left(\frac{1}{5}, \frac{1}{e}\right)\right) = \{(x, y): 1 < x < \ln 5\}$$

is an open strip in $\mathbf{R}^2$. Similarly, the interval $\left(-\infty, \frac{1}{2}\right]$ is a closed set, and

$$F^{-1}\left(\left(-\infty, \frac{1}{2}\right]\right) = \{(x, y): x \geq \ln 2\}$$

is a closed half-plane.

The connection between continuity and neighborhoods is obvious in characterizations like Theorem 4.16 and the last-mentioned Theorem 3.9(d). Using inverse images, we can connect continuity and open sets in general.

Theorem 4.17. *Let* $\mathbf{f}$ *map* $D \subseteq V$ *into another normed space* W*. Then the following are equivalent.*

(a) $\mathbf{f}$ *is continuous on* D.

(b) *For every open subset* O *of* W*, there is a corresponding open subset* P *of* V *such that the inverse image of* O *is precisely the part of* D *in* P*. In symbols,* $\mathbf{f}^{-1}(O) = P \cap D$.

Proof. (a) $\Rightarrow$ (b) Assume that $\mathbf{f}$ is continuous on D. Let O be an open subset of W. We must find an appropriate P.

Let $\mathbf{x}$ be any vector in $\mathbf{f}^{-1}(O)$. By definition, $\mathbf{f}(\mathbf{x})$ is defined and in O. Since O is open, there is a neighborhood $N(\mathbf{f}(\mathbf{x}), \varepsilon(\mathbf{x}))$ contained in O; we use $\varepsilon(\mathbf{x})$ to indicate that the specific ε depends on $\mathbf{x}$. Because $\mathbf{f}$ is continuous at $\mathbf{x}$, there is a neighborhood $N(\mathbf{x}, \delta(\mathbf{x}))$ in which $\mathbf{y} \in D$ implies $\mathbf{f}(\mathbf{y}) \in N(\mathbf{f}(\mathbf{x}), \varepsilon(\mathbf{x}))$. (Note that δ depends on $\mathbf{x}$ both directly, because the continuity may not be uniform, and indirectly, because δ depends on ε, which depends on $\mathbf{x}$.)

The union of all these $N(\mathbf{x}, \delta)$-neighborhoods constitutes an open set, by Theorem 4.15. Call it P; it is the set we need.

For proof, first suppose that $\mathbf{z} \in \mathbf{f}^{-1}(O)$. Then $\mathbf{z}$ is one of the vectors around which we built neighborhoods in the previous paragraph. Therefore, $\mathbf{z} \in N(\mathbf{z}, \delta(\mathbf{z})) \subseteq P$. Having shown that $\mathbf{z} \in \mathbf{f}^{-1}(O)$ implies $\mathbf{z} \in D \cap P$, we conclude that $\mathbf{f}^{-1}(O) \subseteq D \cap P$.

Suppose instead $\mathbf{z} \in D \cap P$. From $\mathbf{z} \in P :=$ union of neighborhoods, we conclude that $\mathbf{z}$ is in one of the $N(\mathbf{x}, \delta(\mathbf{x}))$. By the way we constructed the neighborhoods, the fact that $\mathbf{z} \in D$ implies that $\mathbf{f}(\mathbf{z})$ is in $N(\mathbf{f}(\mathbf{x}), \varepsilon(\mathbf{x})) \subseteq O$. We have arrived at $\mathbf{f}(\mathbf{z}) \in O$, or $\mathbf{z} \in \mathbf{f}^{-1}(O)$. Thus, $\mathbf{z} \in D \cap P$ implies $\mathbf{z} \in \mathbf{f}^{-1}(O)$, establishing $D \cap P \subseteq \mathbf{f}^{-1}(O)$.

We have found P with the property that $D \cap P = \mathbf{f}^{-1}(O)$, as needed.

(b) $\Rightarrow$ (a) Exercise 8.

Theorem 4.18. *Let* $\mathbf{f}$ *map* $D \subseteq V$ *into another normed space* W*, and assume that* D *is open. Then the following are equivalent.*

(a) $\mathbf{f}$ *is continuous on* D.

(b) *For every open subset* O *of* W*, the inverse image* $\mathbf{f}^{-1}(O)$ *is an open subset of* V.

Proof. This equivalence is immediate from Theorem 4.17, because the intersection of the open sets P and D is open (Exercise 9).

Exercises

1. In $\mathbf{R}^2$, which of the following are open sets? Which are closed?

   (a) Quadrant I.

   (b) the graph of $y = 1/x$.

   (c) the polar-coordinates graph $r = e^\theta$, $-\infty < \theta < \infty$.

   (d) the graph of $y = \sin(1/x)$.

2. In $\mathbf{R}^n$, prove that $(\mathbf{a}, \infty) := \{\mathbf{x}: \mathbf{a} < \mathbf{x}\}$ and $(-\infty, \mathbf{b}) := \{\mathbf{x}: \mathbf{x} < \mathbf{b}\}$ are open sets.

3. In $\mathbf{R}^n$, show that if $\mathbf{a} < \mathbf{b}$ is false, then no $\mathbf{x}$ satisfies $\mathbf{a} < \mathbf{x} < \mathbf{b}$.

4. In a normed space V, show that V and $\emptyset$, the empty set, are open subsets of V.

5. In a normed space, which finite sets are open?

6. Assume that $\mathbf{f}$ maps (all of) a normed space V into a normed space W.

   (a) Show that the inverse image of a complement is the complement of the inverse image. In symbols, show that if $T \subseteq W$, then $\mathbf{f}^{-1}(T^*) = [\mathbf{f}^{-1}(T)]^*$ in V.

   (b) Suppose $\mathbf{f}$ is continuous on V. Show that the inverse image of every open subset of W is an open subset of V, and that the inverse image of a closed set is closed.

7. Prove Theorem 4.14.

8. Assume that $\mathbf{f}$ has domain D. Prove that if for each open set O there is an open set P such that $\mathbf{f}^{-1}(O) = D \cap P$, then $\mathbf{f}$ is continuous on D. (Hint: Arrive at the characterization of Theorem 3.9(d).)

9. (a) Prove that the intersection of two open sets is open.

   (b) Prove that the intersection of any finite number of open sets is open.

10. Prove that the union of any finite number of closed sets is closed.

4.5 Continuous Functions on Connected Sets

Returning to Section 4.1, we find one more property of the functions continuous in a box: the intermediate value theorem. In this section we characterize the sets in which the theorem holds.

Definition. A set S in a normed space has the **intermediate value property (IVP)** if every continuous real function on S achieves intermediate values: Whenever

$f(\mathbf{c}) \neq f(\mathbf{d})$ are values of f on S, then so is every real number between $f(\mathbf{c})$ and $f(\mathbf{d})$.

To understand where the IVP rules, we begin with an example where it does not.

Example 1. Consider $F(x, y) := x/|x|$ in $\mathbf{R}^2$.

F is continuous in its domain $D = \{(x, y) \colon x \neq 0\}$. For suppose $(x_i, y_i) \to (x, y) \in D$. The limit is either left or right of the y-axis; let us assume $x > 0$. Then $x_i \to x$ (Reason?), so there exists I such that $i \geq I$ forces $x_i > x/2$. In particular,

$$i \geq I \Rightarrow x_i > 0 \Rightarrow F(x_i, y_i) = 1 = F(x, y).$$

Therefore, $F(x_i, y_i) \to F(x, y)$, and F is continuous.

F is valued 1 in the right half-plane, -1 in the left. Since it never takes the value 0, D does not have the IVP.

The presence of the y-axis between the halves of the domain in Example 1 means that a point where F has one value is surrounded by like vectors. The word "surrounded" is our key. It leads us to the language of open sets.

Theorem 4.19. *Suppose there are two open sets O_1 and O_2 that split the set D in the following way: Every vector in D is in O_1 or O_2, but not both; each of O_1 and O_2 has some vectors from D. Then D does not have the IVP.*

Proof. Define

$$G(\mathbf{x}) := \begin{cases} -1 & \text{if } \mathbf{x} \in D \cap O_1 \\ 1 & \text{if } \mathbf{x} \in D \cap O_2. \end{cases}$$

By assumption, $D \cap O_1$ has some vectors, so $G = -1$ someplace in D. Similarly, $G = 1$ someplace in D. Since the two intersections between them cover all of D, it is clear that $G \neq 0$ throughout D. Certainly, then, G does not achieve intermediate values.

Next, let $\mathbf{b} \in D$. We know $\mathbf{b}$ has to be in O_1 or O_2; for definiteness, say $\mathbf{b} \in O_1$. Because O_1 is open, there is a neighborhood $N(\mathbf{b}, \delta)$ contained in O_1. If $(\mathbf{x}_i)$ is any sequence in D converging to $\mathbf{b}$, then there exists I for which

$$i \geq I \Rightarrow \mathbf{x}_i \in N(\mathbf{b}, \delta) \subseteq O_1 \qquad \text{(Theorem 2.5)},$$

which says that

$$i \geq I \Rightarrow G(\mathbf{x}_i) = -1.$$

Hence

$$\lim_{i \to \infty} G(\mathbf{x}_i) = -1 = G(\mathbf{b}),$$

and G is continuous at $\mathbf{b}$.

Since D admits continuous functions that do not cover intermediate values, D does not have the IVP. □

There is some vocabulary associated with the situation described in Theorem 4.19. Anytime a set S is made up of two nonempty disjoint subsets S_1 and S_2—in symbols, $S = S_1 \cup S_2$ with S_1 and S_2 nonempty but $S_1 \cap S_2$ empty—we call $\{S_1, S_2\}$ a **nontrivial partition** (or just **partition**) of S. If the partition is accomplished with open sets O_1 and O_2 having $S_1 = S \cap O_1$ and $S_2 = S \cap O_2$, then $\{S_1, S_2\}$ is a **disconnection** of S. Notice that O_1 and O_2 do not have to be subsets of S. More important, although they have to be different sets (because each has members of S that the other does not), they do not have to be disjoint; only their shares of S have to be disjoint.

Definition. Let S be a subset of a normed space. If there exists a disconnection of S, then S is a **disconnected set**. If not, then S is **connected**.

Theorem 4.20. *A set S is disconnected iff it is possible to find two closed sets C_1 and C_2 such that $\{C_1 \cap S, C_2 \cap S\}$ is a nontrivial partition of S.*

Proof. Exercise 5.

Example 2. (a) In $\mathbf{R}^n$, every box is connected. After all, by Theorem 4.19 a set with the IVP must be connected, and by Theorem 4.5 every box has the IVP.

(b) In $\mathbf{R}^2$, $\{(x, y)\colon \text{either } y = 0 \text{ or } y = 1/x\}$ is the union of the x-axis, which is closed, and the graph of $y = 1/x$, which is also closed (Exercise 1b in Section 4.4). Theorem 4.20 makes it clear that the union of disjoint nonempty closed sets is always disconnected.

We now show that a set has the IVP iff it is connected, by establishing the inverse of Theorem 4.19.

Theorem 4.21 (The Intermediate Value Theorem). *Every connected set has the IVP.*

Proof. Assume that D lacks the IVP. Thus, there is a continuous f defined on D possessing unequal values $f(\mathbf{a}) < f(\mathbf{b})$ but never taking the intermediate value t.

We know (Exercise 2 in Section 4.4) that $(-\infty, t)$ is an open subset of $\mathbf{R}$. By Theorem 4.17, there is an open set O_1 such that

$$f^{-1}((-\infty, t)) = D \cap O_1.$$

Similarly, there is an open O_2 such that

$$f^{-1}((t, \infty)) = D \cap O_2.$$

These two sets disconnect D. To explain, note that if $\mathbf{x}$ is in D, then $f(\mathbf{x}) = t$ is not allowed. Therefore, either $f(\mathbf{x}) < t$, causing $\mathbf{x} \in D \cap O_1$; or $f(\mathbf{x}) > t$, and

$\mathbf{x} \in D \cap O_2$; and not both. Moreover, $D \cap O_1$ is not empty; $\mathbf{a} \in D \cap O_1$, because $f(\mathbf{a}) < t$. Similarly, $\mathbf{b} \in D \cap O_2$.

We have established that if D does not have the IVP, then D is disconnected.

□

To gain some familiarity with connected sets, we are going to show that two elementary classes are made up of connected sets. These will relate connectedness to our intuitive notion of "joining" points, namely, drawing a line or curve from one point to the other. In Example 1 we exhibited a set that is disconnected because it has two halves separated by a wall. Roughly speaking, if no such barrier exists, so that we can join each member of our set to any of the others, then we are in a connected set.

Theorem 4.5 says that a box in $\mathbf{R}^n$ has the IVP. Unlike the other results in that section, its proof has no relation to the Bolzano–Weierstrass theorem. Instead, the proof depends on the property that the segment joining two points of the box lies entirely within the box. Open boxes and infinite boxes also have this attribute, so the closedness and boundedness of the box are both inessential.

The class of sets with this segment property is interesting enough to merit a name.

Definition. A set S in a normed space is **convex** if for any two points of S, the segment joining them is contained in S.

Example 3. (a) Every ball is convex, and so is every neighborhood. The proof is mainly Theorem 2.1(b).

(b) A sphere is necessarily nonconvex (unless you count a point as a zero-radius sphere). Recall that any line through the center of a sphere has two points in common with the sphere. The segment joining them is a **diameter**, and none of the other points of this segment is on the sphere.

The proof of Theorem 4.5 carries over unchanged with "convex set" in place of "box." Thus, every convex set has the IVP and is therefore connected.

In fact, we can easily go a step further. Nothing in that proof depends on the segment's being a subset of a line. In the proof, the only property we exploited of the segment $\{(1-\alpha)\mathbf{c}+\alpha\mathbf{d}: 0 \le \alpha \le 1\}$ was that it is the range of the continuous mapping $\mathbf{G}(t) := [1-t]\mathbf{c} + t\mathbf{d}$. Any such range would have served the same purpose. For that reason, we make the following definition.

Definition. Given $\mathbf{c}$ and $\mathbf{d}$ in a normed space, a continuous function $\mathbf{g}$ defined on a closed interval $[r, s]$ of $\mathbf{R}$ and having $\mathbf{g}(r) = \mathbf{c}$, $\mathbf{g}(s) = \mathbf{d}$ is called an **arc from c to d**.

Despite our earlier worry about such practice, we will let "arc" serve a dual purpose: It names both the function and the set of points that is the function's range.

Definition. A set S is **arc-connected** if for any two of its points there exists an arc from one to the other (note Exercise 8c) lying (having its range) entirely within S.

Theorem 4.22. *Every arc-connected set is connected.*

Proof. Exercise 7.

Convexity and arc-connectedness are intuitive geometric concepts. (See Exercise 9 for another such idea.) In a convex set, every point can see all the others. In an arc-connected set, the view might be blocked, but you can always get from here to there without leaving the set.

It would be nice if so simple a geometric quality characterized connectedness. We leave until later the demonstration that the converse of Theorem 4.22 is not true. We present instead one last result linking continuous functions and connectedness.

Theorem 4.23. *The continuous image of a connected set is connected. In symbols, let* $\mathbf{f}$ *be a continuous mapping from a domain* D*, contained in a normed space* V*, to a normed space* W*. If* D *is connected, then so is* $\mathbf{f}(D)$*.*

Proof. Assume that $\mathbf{f}$ is continuous on D and $\mathbf{f}(D)$ is disconnected.

By definition, there are open sets O_1 and O_2 of W that partition $\mathbf{f}(D)$ nontrivially. By Theorem 4.17, there are corresponding open sets P_1 and P_2 in V with $\mathbf{f}^{-1}(O_1) = P_1 \cap D, \mathbf{f}^{-1}(O_2) = P_2 \cap D$. We will establish that P_1 and P_2 define a disconnection of D.

If $\mathbf{x} \in D$, then $\mathbf{f}(x) \in \mathbf{f}(D)$. Because O_1 and O_2 partition $\mathbf{f}(D)$, $\mathbf{f}(\mathbf{x})$ is in O_1 or O_2 and not both. If $\mathbf{f}(\mathbf{x}) \in O_1$, then $\mathbf{x} \in \mathbf{f}^{-1}(O_1)$, implying that $\mathbf{x} \in P_1$; if instead $\mathbf{f}(\mathbf{x}) \in O_2$, then similarly we have $\mathbf{x} \in P_2$; and $\mathbf{x} \in P_1$ and $\mathbf{x} \in P_2$ cannot both be true. Also, each of O_1 and O_2 has some images, so each of P_1 and P_2 has some preimages. Therefore, P_1 and P_2 partition D.

From the assumption that $\mathbf{f}(D)$ is disconnected, we arrived at a disconnection of D. Hence if D is connected, then so is $\mathbf{f}(D)$. □

Exercises

1. For each listed subset of $\mathbf{R}^2$: Is it connected? Is it arc-connected?
   (a) $\{(x, y)\colon y = (\sin x)/x\} \cup \{\mathbf{O}\}$
   (b) the polar-coordinates graph $r = e^\theta$, $-\infty < \theta < \infty$, together with $\mathbf{O}$. (Hint: yes.)
2. (a) Prove that $\mathbf{Q}^n$ is not connected.
   (b) Is the complement of $\mathbf{Q}^n$ connected? (Hint: $n = 1$ is wildly different from $n > 1$.)
3. Prove that if S is a connected set in $\mathbf{R}^n$, then the projections $\pi_k(S)$ are connected. Is the converse true?

4. Show that $\emptyset$ and V are connected subsets of any normed space V.

5. Prove Theorem 4.20.

6. Prove that the only subsets of a normed space V that are simultaneously open and closed are V and $\emptyset$.

7. Prove Theorem 4.22. You should need just a small modification of the argument used to prove Theorem 4.5.

8. Let $\mathbf{b}$ and $\mathbf{c}$ be vectors in a set S.

   (a) Show that there is an arc within S from $\mathbf{b}$ to $\mathbf{b}$.

   (b) Suppose S is open. Show that in some neighborhood of $\mathbf{b}$, every vector has an arc from it to $\mathbf{b}$ within S.

   (c) Show that if there is some arc within S from $\mathbf{c}$ to $\mathbf{b}$, then there is one from $\mathbf{b}$ to $\mathbf{c}$.

9. We call a set "star-shaped" if some vector in it can see all the others: S is **star-shaped** if there is $\mathbf{b} \in S$ such that for every $\mathbf{x} \in S$, the segment from $\mathbf{b}$ to $\mathbf{x}$ is contained in S.

   (a) Show that every star-shaped set is arc-connected.

   (b) Is every arc-connected set star-shaped?

   (c) Is every star-shaped set convex?

10. Let $\{\mathbf{x}_1, \ldots, \mathbf{x}_k\}$ be any finite set of vectors. The linear combination $\alpha_1\mathbf{x}_1 + \cdots + \alpha_k\mathbf{x}_k$ is called a **convex combination** if each $\alpha_i \geq 0$ and $\alpha_1 + \cdots + \alpha_k = 1$. For any set S, let $\mathrm{CH}(S)$ be the set of convex combinations of vectors (finitely many at a time) from S.

    (a) Prove that $\mathrm{CH}(S)$ is a convex set.

    (b) *Prove that $\mathrm{CH}(S)$ is the smallest convex set that contains S. (Hint: induction.)

    $\mathrm{CH}(S)$ is called the **convex hull** of S.

    (c) Prove that S is convex iff $S = \mathrm{CH}(S)$.

    (d) Prove that $\mathrm{CH}(S)$ consists of exactly those vectors that are on some line segment with one end in S and the other on some line segment with one end in S and the other on some line segment with one . . . with the other end also in S. (The previous sentence will become intelligible if you sketch $S := \{(-1, 0), (0, 1), (1, 0), (0, 4)\}$, then the edges joining them, then the segments from the vertices to the remote edges.)

    (e) Is the convex hull of S always a closed set? always open? always connected?

4.6 Finite-Dimensional Subspaces of Normed Linear Spaces

We have established in $\mathbf{R}^n$ certain properties of continuous functions that are related to the Bolzano–Weierstrass theorem (BWT). In this section we show that the BWT is related to the finite dimension of $\mathbf{R}^n$ and not to the specific norm that the Pythagorean theorem leads us to use.

Any vector space V of dimension n over $\mathbf{R}$ is isomorphic to $\mathbf{R}^n$. That is, the algebraic properties of V are the same as those of $\mathbf{R}^n$, even if V is a subspace of a bigger space, including an infinite-dimensional one. On the other hand, if we attach metrics to spaces, then differences emerge. Relative to any metric, we can apply the construction of Example 2 in Section 1.5 to produce a metric of (possibly) a very different nature. The differences are not as great among norms; we have seen that normed spaces have a great deal in common. Still, even norms can differ in such fundamental factors as derivability from an inner product (Section 1.4).

To decide what to concentrate on, let us simply remember that we are studying calculus. At this point, we want to know about continuous functions. We therefore focus on convergence.

Definition. Let V be a vector space. Suppose $\| \ \|_a$ and $\| \ \|_b$ denote two norms on V. We say that $\| \ \|_a$ and $\| \ \|_b$ are **equivalent norms** if every sequence that converges as defined by one norm also converges as defined by the other.

Given $\| \ \|_a$, what we mean by convergence of a sequence is this:

$$(\mathbf{x}_i) \text{ converges to } \mathbf{x} \text{ if } \|\mathbf{x} - \mathbf{x}_i\|_a \to 0.$$

The definition of equivalence demands that if some $\mathbf{x}$ satisfies $\|\mathbf{x} - \mathbf{x}_i\|_a \to 0$, then some $\mathbf{y}$ must satisfy $\|\mathbf{y} - \mathbf{x}_i\|_b \to 0$, and conversely. The definition does not state that the $\| \ \|_a$-limit of $(\mathbf{x}_i)$, assuming that it exists, must be the same as the $\| \ \|_b$-limit. But we take care of that immediately.

Theorem 4.24. *Assume that $\| \ \|_a$ and $\| \ \|_b$ are equivalent. Then any sequence that converges under $\| \ \|_a$ has the same limit under $\| \ \|_b$, and vice versa.*

Proof. Assume that the norms are equivalent.

Suppose $(\mathbf{x}_i) \to \mathbf{x}$ under $\| \ \|_a$. Then the sequence $\mathbf{x}_1, \mathbf{x}, \mathbf{x}_2, \mathbf{x}, \mathbf{x}_3, \mathbf{x}, \ldots$ also converges to $\mathbf{x}$ under $\| \ \|_a$. By the assumption of equivalent norms, the latter sequence converges under $\| \ \|_b$. But this sequence has a subsequence $\mathbf{x}, \mathbf{x}, \ldots$. Therefore, its $\| \ \|_b$-limit has to be $\mathbf{x}$. □

Example 1. (a) For any norm $\|\mathbf{x}\|$ on any space, a typical "rescaling" is

$$\text{INCHNORM}(\mathbf{x}) := 12\|\mathbf{x}\|.$$

(Compare the similar statement, related to metrics, preceding Example 2 in Section 1.5).

If $(\mathbf{x}_i)$ is any sequence, then

$$\text{INCHNORM}(\mathbf{x} - \mathbf{x}_i) := 12\|\mathbf{x} - \mathbf{x}_i\|$$

tells us that $(\mathbf{x}_i) \to \mathbf{x}$ is true in both norms or in neither. We conclude that a norm and any of its rescalings are equivalent.

(b) In $\mathbf{R}^n$, we defined (Section 1.4) "maxnorm"by

$$\|\mathbf{x}\|_0 := \max\{|x_1|, \ldots, |x_n|\}.$$

It is easy to see that

$$\|\mathbf{x}\|_0 \leq \|\mathbf{x}\|_2 \leq \sqrt{n}\,\|\mathbf{x}\|_0.$$

It follows that $\|\mathbf{x} - \mathbf{x}_i\|_0 \to 0$ iff $\|\mathbf{x} - \mathbf{x}_i\|_2 \to 0$. Therefore, maxnorm and the Pythagorean norm are equivalent.

(c) In $C[0, 1]$ we know that the two norms are inequivalent. The evidence is in our graph (Figure 3.2) in Section 3.5 of a function g_k with $\|g_k\|_0 = 2^{k/2}$ and $\|g_k\|_2 \leq \sqrt{3}$. For the sequence so defined,

$$\left\|2^{-k/2} g_k\right\|_2 \to 0,$$

so that $\left(2^{-k/2} g_k\right)$ converges to $\mathbf{O}$ under $\|\ \|_2$, but

$$\|2^{-k/2} g_k\|_0 = 1,$$

so $\left(2^{-k/2} g_k\right)$ does not converge to $\mathbf{O}$ under $\|\ \|_0$.

The inequality in (b) above leads to an alternative definition of equivalent norms.

Theorem 4.25. *Two norms $\|\ \|_a$ and $\|\ \|_b$ on a normed space V are equivalent iff there exist positive constants m and M such that for every $\mathbf{x} \in V$,*

$$m\|\mathbf{x}\|_a \leq \|\mathbf{x}\|_b \leq M\|\mathbf{x}\|_a.$$

Proof. $\Rightarrow$ Suppose that no constant M satisfies $\|\mathbf{x}\|_b \leq M\|\mathbf{x}\|_a$ for all $\mathbf{x}$. Corresponding to each i we can find some $\mathbf{x}_i$ for which $\|\mathbf{x}_i\|_b > i\|\mathbf{x}_i\|_a$. Write

$$\mathbf{y}_i := \frac{\mathbf{x}_i}{i\|\mathbf{x}_i\|_a}. \qquad \text{(Is the division legal?)}$$

Then

$$\|\mathbf{y}_i\|_b = \frac{\|\mathbf{x}_i\|_b}{i\|\mathbf{x}_i\|_a} > 1,$$

so $(\mathbf{y}_i)$ does not converge to zero as defined by $\|\ \|_b$. But

$$\|\mathbf{y}_i\|_a = \frac{\|\mathbf{x}_i\|_a}{i\|\mathbf{x}_i\|_a} = \frac{1}{i} \to 0.$$

Hence $(\mathbf{y}_i)$ converges to zero as defined by $\| \ \|_a$. The norms are inequivalent.

By contraposition, therefore, if the norms are equivalent, then there exists M with

$$\|\mathbf{x}\|_b \leq M\|\mathbf{x}\|_a \qquad \text{for all } \mathbf{x}.$$

By similar reasoning, we conclude that there is also K with

$$\|\mathbf{x}\|_a \leq K\|\mathbf{x}\|_b.$$

Taking $m = 1/K$, we obtain the other half of the inequality.

$\Leftarrow$ Exercise 4.

Our basic result in this section is the following important theorem:

Theorem 4.26. *In a finite-dimensional vector space, any two norms are equivalent.*

Proof. Assuming that V has dimension n, let $B := \{\mathbf{v}_1, \ldots, \mathbf{v}_n\}$ be a basis for V. Every $\mathbf{x} \in V$ has a unique expression

$$\mathbf{x} = \alpha_1\mathbf{v}_1 + \cdots + \alpha_n\mathbf{v}_n$$

as a linear combination of $\mathbf{v}_1, \ldots, \mathbf{v}_n$. We define

$$\|\mathbf{x}\|_B := \left(\alpha_1^2 + \cdots + \alpha_n^2\right)^{1/2}. \qquad \text{(Is it actually a norm?)}$$

We will show, via Theorem 4.25, that every norm on V is equivalent to $\| \ \|_B$. It then follows (Exercise 5) that any norm is equivalent to any other.

Let $\|\mathbf{x}\|$ signify another norm defined on V. By the triangle inequality

$$\begin{aligned}\|\mathbf{x}\| &= \|\alpha_1\mathbf{v}_1 + \cdots + \alpha_n\mathbf{v}_n\| \leq |\alpha_1|\,\|\mathbf{v}_1\| + \cdots + |\alpha_n|\,\|\mathbf{v}_n\| \\ &\leq n\max\{|\alpha_1|\,\|\mathbf{v}_1\|, \ldots, |\alpha_n|\,\|\mathbf{v}_n\|\} \\ &\leq n\max\{\|\mathbf{x}\|_B\,\|\mathbf{v}_1\|, \ldots, \|\mathbf{x}\|_B\,\|\mathbf{v}_n\|\}.\end{aligned}$$

Setting $M := n\max\{\|\mathbf{v}_1\|, \ldots, \|\mathbf{v}_n\|\}$, we obtain

$$\|\mathbf{x}\| \leq M\|\mathbf{x}\|_B,$$

where M depends on $\| \ \|$ but not on $\mathbf{x}$. Thus, $\|\mathbf{x}\|_B$ dominates $\|\mathbf{x}\|$.

To show that $\|\mathbf{x}\|$ dominates $\|\mathbf{x}\|_B$, consider first the function

$$f(y_1, \ldots, y_n) := \|y_1\mathbf{v}_1 + \cdots + y_n\mathbf{v}_n\|.$$

It maps $\mathbf{R}^n$ continuously (Why?) to $\mathbf{R}$. The unit sphere is closed and bounded in $\mathbf{R}^n$, so f has a minimum value $m := f(\beta_1, \ldots, \beta_n)$ there. This minimum is nonzero, because $m = 0$ would imply $\|\beta_1\mathbf{v}_1 + \cdots + \beta_n\mathbf{v}_n\| = 0$; this last would force $\beta_1 = \cdots = \beta_n = 0$, because $\mathbf{v}_1, \ldots, \mathbf{v}_n$ is a basis; and the β_i cannot all be zero at a point on the unit sphere.

Next, for $\mathbf{x} = \alpha_1 \mathbf{v}_1 + \cdots + \alpha_n \mathbf{v}_n \neq \mathbf{O}$ in V, examine the ratio

$$\frac{\|\mathbf{x}\|}{\|\mathbf{x}\|_B} = \left\| \left(\frac{\alpha_1}{\|\mathbf{x}\|_B}\right) \mathbf{v}_1 + \cdots + \left(\frac{\alpha_n}{\|\mathbf{x}\|_B}\right) \mathbf{v}_n \right\|.$$

Writing

$$y_j := \frac{\alpha_j}{\|\mathbf{x}\|_B} \qquad \text{for each } j,$$

we see that

$$y_1^2 + \cdots + y_n^2 = \frac{\alpha_1^2}{\|\mathbf{x}\|_B^2} + \cdots + \frac{\alpha_n^2}{\|\mathbf{x}\|_B^2} = 1,$$

so $(y_1, \ldots, y_n)$ is on the unit sphere. Further,

$$\frac{\|\mathbf{x}\|}{\|\mathbf{x}\|_B} = f(y_1, \ldots, y_n).$$

Since m is the minimum of f on the sphere, we know that

$$\frac{\|\mathbf{x}\|}{\|\mathbf{x}\|_B} \geq m.$$

We conclude that

$$m\|\mathbf{x}\|_B \leq \|\mathbf{x}\| \qquad \text{for all } \mathbf{x}, \text{ including } \mathbf{O}.$$

We have found M and m with

$$m\|\mathbf{x}\|_B \leq \|\mathbf{x}\| \leq M\|\mathbf{x}\|_b. \qquad \square$$

Example 2. We know that $\| \ \|_2$ and $\| \ \|_0$ are inequivalent on $C[0, 1]$. Specifically,

$$\|f\|_2 := \left(\int_0^1 f(x)^2\, dx\right)^{1/2} \leq \max |f(x)| = \|f\|_0,$$

so that maxnorm dominates mean-norm, but (Example 1(c)) mean-norm does not dominate maxnorm.

However, their restrictions to any finite-dimensional subspace of $C[0, 1]$ have to be equivalent. Consider the subspace

$$V = \langle\langle \sin 2\pi x, \sin 4\pi x, \sin 6\pi x \rangle\rangle$$

spanned by those three functions. If α, β, and γ are real constants and

$$f(x) := \alpha \sin 2\pi x + \beta \sin 4\pi x + \gamma \sin 6\pi x,$$

then

$$\|f\|_0 \leq |\alpha| + |\beta| + |\gamma| \leq 3\left(\alpha^2 + \beta^2 + \gamma^2\right)^{1/2}.$$

The three sines are orthogonal and of mean-norm $1/\sqrt{2}$ (Example 1 in Section 1.6), so

$$\|f\|_2 = \left(\frac{\alpha^2}{2} + \frac{\beta^2}{2} + \frac{\gamma^2}{2}\right)^{1/2}.$$

Hence on V,

$$\|f\|_2 \leq \|f\|_0 \leq 3\sqrt{2}\|f\|_2.$$

The equivalence of norms means that the theory we have developed for $\mathbf{R}^n$ describes continuous functions with respect to any space-norm combination of finite dimension over $\mathbf{R}$. We will rephrase the important results in three groups. The first group, addressing convergence of sequences, corresponds to the Theorems 2.8–2.10.

Theorem 4.27. *Assume that V is a normed vector space of dimension n over $\mathbf{R}$.*

(a) *Let $B = \{\mathbf{v}_1, \ldots, \mathbf{v}_n\}$ be a basis for V. For any vector*

$$\mathbf{x} = \alpha_1\mathbf{v}_1 + \cdots + \alpha_n\mathbf{v}_n,$$

write $\Pi_j(\mathbf{x}) := \alpha_j$ (the "coordinate of $\mathbf{x}$ along $\mathbf{v}_j$ relative to B"). Then a sequence $(\mathbf{x}_i)$ converges to a vector in V iff each coordinate sequence $(\Pi_j(\mathbf{x}_i))$ converges, and the coordinates of the limit are the limits of the coordinates.

(b) *A sequence in V converges iff it is Cauchy.*

(c) *Every bounded sequence in V has a convergent subsequence.*

Proof. We demonstrate (b) and leave the (analogous) proofs of (a) and (c) to Exercise 6.

$\Rightarrow$ We know that in every normed space, a convergent sequence is Cauchy.

$\Leftarrow$ Let $\|\ \|$ represent the norm in V, and assume that $(\mathbf{x}_i)$ is Cauchy. Thus, $\|\mathbf{x}_i - \mathbf{x}_j\| \to 0$ as $i, j \to \infty$. By Theorem 4.26, B's "Pythagorean norm"—the norm defined by $\|\mathbf{x}\|_B := (\alpha_1^2 + \cdots + \alpha_n^2)^{1/2}$—and $\|\mathbf{x}\|$ are equivalent. By Theorem 4.25, there exists M such that $\|\mathbf{x}_i - \mathbf{x}_j\|_B \leq M\|\mathbf{x}_i - \mathbf{x}_j\|$. Therefore, $(\mathbf{x}_i)$ is a Cauchy sequence relative to this Pythagorean norm. By the argument of Theorem 2.9, $(\mathbf{x}_i)$ has a limit under $\|\ \|_B$. The equivalence of the norms forces $(\mathbf{x}_i)$ to have a limit under $\|\ \|$. □

The second group corresponds to material from Chapter 3. It is essential, because it tells us that a continuous function remains a continuous function, no matter what specific norm we use in the (finite-dimensional) domain to judge continuity.

Theorem 4.28. *Assume that V is a vector space of finite dimension over $\mathbf{R}$.*

(a) *A sequence that converges relative to one norm converges to the same limit relative to any other norm on* V.

(b) *Let* $\mathbf{f}$ *map a subset of* V *to a (fixed) normed space* W. *If* $\mathbf{f}$ *has a limit at* $\mathbf{b}$ *relative to one norm in* V, *then it has the same limit at* $\mathbf{b}$ *relative to any other norm in* V.

(c) *Let* $\mathbf{f}$ *map* $D \subseteq V$ *to a (fixed) normed space* W. *If* $\mathbf{f}$ *is continuous at* $\mathbf{b} \in D$ *relative to one norm in* V, *then it is continuous at* $\mathbf{b}$ *relative to any other norm in* V.

Proof. Exercise 7.

The third group of results extends to any normed space of finite dimension the most important results from Sections 4.2 and 4.3.

Theorem 4.29. *Assume that* V *is a normed linear space of finite dimension over* $\mathbf{R}$.

(a) *If a subset of* V *is closed (or bounded) relative to one norm, then it is closed (respectively, bounded) relative to any other norm.*

(b) *A subset of* V *has the extreme value property iff it is closed and bounded.*

(c) *Every function continuous on a closed bounded set is uniformly continuous.*

Proof. Exercise 8.

One last theorem remains, relating to finite-dimensional subspaces of larger spaces.

Theorem 4.30. *In any normed space, every finite-dimensional subspace is a closed set.*

Proof. Suppose V is a finite-dimensional subset of a normed space W (which need not have finite dimension). Let $B = \{\mathbf{v}_1, \ldots, \mathbf{v}_n\}$ be a basis for V and let $\| \ \|$ denote the norm in W.

Suppose $(\mathbf{x}_i)$ is a sequence from V converging in W to a vector $\mathbf{x}$. Then $(\mathbf{x}_i)$ is a Cauchy sequence in W, so $\|\mathbf{x}_i - \mathbf{x}_j\| \to 0$ as $i, j \to \infty$. Since the restriction $\| \ \|_V$ of $\| \ \|$ to V is a norm in V (How come?) and

$$\|\mathbf{x}_i - \mathbf{x}_j\|_V = \|\mathbf{x}_i - \mathbf{x}_j\| \to 0,$$

$(\mathbf{x}_i)$ is a Cauchy sequence in V. By Theorem 4.27(b), there is a $\| \ \|_V$-limit; that is, there is $\mathbf{y} \in V$ such that $\|\mathbf{y} - \mathbf{x}_i\|_V \to 0$ as $i \to \infty$. From

$$\|\mathbf{y} - \mathbf{x}_i\| = \|\mathbf{y} - \mathbf{x}_i\|_V \to 0,$$

we infer that $\mathbf{y}$ is also the limit in W of $(\mathbf{x}_i)$. Hence $\mathbf{x} = \mathbf{y} \in V$.

We have proved that a closure point of V has to be a member of V. □

Example 3. In $C_0[0, 1]$:

(a) From linear algebra we know that the polynomials of degree less than or equal to n constitute a subspace of dimension $n + 1$. By Theorem 4.30, this subspace is closed. Therefore, the uniform limit of a sequence of polynomials of degree n or less is itself such a polynomial.

(b) The subspace consisting of all polynomials on $[0, 1]$ is of infinite dimension. Theorem 4.30 does not apply, and the subspace is not closed. For evidence, consider

$$p_k(x) := 1 + x + \frac{x^2}{2!} + \cdots + \frac{x^k}{k!}.$$

Our knowledge of power series tells us that (p_k) converges uniformly to a nonzero function that is its own derivative. Such a function cannot be a polynomial. (Why?)

(c) It is another elementary exercise in linear algebra that $W := \{f\colon f\left(\frac{1}{2}\right) = 0\}$ is a subspace of $C_0[0, 1]$. We have our own evidence that its dimension is infinite: $\sin 2\pi x, \sin 4\pi x, \ldots$, being mutually orthogonal, are necessarily independent. So Theorem 4.30 does not apply.

But W is closed anyway. If (f_i) is a sequence converging (uniformly) to f, then

$$f\left(\frac{1}{2}\right) = \lim_{i\to\infty} f_i\left(\frac{1}{2}\right) = 0,$$

and so $f \in W$.

Exercises

1. In $\mathbf{R}^n$:

   (a) Show that

   $$\|\mathbf{x}\|_1 := |x_1| + \cdots + |x_n|$$

   defines a norm.

   (b) Find m and M such that

   $$m\|\mathbf{x}\|_2 \le \|\mathbf{x}\|_1 \le M\|\mathbf{x}\|_2.$$

2. The solutions in $[0, 1]$ of the differential equation

   $$\frac{d^2 f}{dx^2} + 4\pi^2 f = 0$$

   are of the form

   $$f(x) = \alpha \sin 2\pi x + \beta \cos 2\pi x,$$

   α and β real. They form a subspace V of $C[0, 1]$.

(a) Show that

$$\|f\|_2 = \frac{\|f\|_0}{\sqrt{2}} \qquad \text{for every } f \in V.$$

(b) Prove that if (f_i) is a sequence of solutions of the differential equation converging uniformly to a function f, then f is also a solution.

3. Suppose

$$f_i(x) := a_i x^3 + b_i x^2 + c_i x + d_i$$

defines a sequence of cubic functions with

$$\int_0^1 f_i(x)^2\,dx \to 28 \qquad \text{as } i \to \infty.$$

(a) Prove that there exists M such that

$$|f_i(x)| < M \qquad \text{for every } i \text{ and } x.$$

(b) Prove that some subsequence of (f_i) converges uniformly.

4. Prove the second half of Theorem 4.25: If there exist positive constants m and M such that for every $\mathbf{x} \in V$,

$$m\|\mathbf{x}\|_a \leq \|\mathbf{x}\|_b \leq M\|\mathbf{x}\|_a,$$

then the two norms fit our definition of equivalent norms.

5. Prove that equivalence of norms is an equivalence relation. [For equivalence relation, consult Royden, Chapter 1, Section 7.].

6. Prove (a) and (c) of Theorem 4.27.

7. Prove Theorem 4.28. (Hint: The principal job is to translate accurately to symbols. In (c), the hypothesis is that $\|\mathbf{b} - \mathbf{x}_i\|_a \to 0$ implies $\|\mathbf{f}(\mathbf{b}) - \mathbf{f}(\mathbf{x}_i)\|_W \to 0$. Decide what the conclusion says and derive it.)

8. Prove Theorem 4.29.

9. Show that every linear map from a finite-dimensional normed space to any other normed space is continuous. (Hint: Theorems 3.16 and 4.28(c).)

10. In a fixed vector space, show that all these are equivalent:

(a) $\| \ \|_a$ and $\| \ \|_b$ are equivalent norms.

(b) $\|\mathbf{x}\|_a/\|\mathbf{x}\|_b$ and $\|\mathbf{x}\|_b/\|\mathbf{x}\|_a$ are both bounded (functions of $\mathbf{x}$).

(c) $\|\mathbf{x}\|_a/\|\mathbf{x}\|_b$ is bounded and bounded away from zero.

(d) Each norm is a function of bounded magnification relative to the other.

(e) Each norm is a continuous function relative to the other.

5
Topology in Normed Spaces

Introduction

Continuity is the central concept of our study, and we have built up to it by defining numerous ideas in terms of distance. It is natural to describe constructs that depend on distance by the adjective **metric**. Our definitions of bounded set and uniformly continuous function are examples of metric constructs; they are inseparable from distance.

Our definition of open set refers explicitly to distance, but it turns out that openness is not a strictly metric concept. That is, it is possible to define open set without referring to a distance function. Once you do that, others of our concepts follow. For example, closedness (Theorem 4.14), continuity (Theorems 4.17 and 4.18), and connectedness (throughout Section 4.5) are expressible in terms of open sets.

Concepts that can be defined in terms of open sets come under the area of mathematical study called **topology**. Although our business is calculus, we now explore the topological properties of $\mathbf{R}^n$, because knowing the attributes of open sets and some of their relatives will assist our future study of derivatives and integrals of vector functions.

5.1 Connected Sets

We saw that connectedness characterizes the intermediate value property. We also showed that a set whose points can be joined by arcs is necessarily connected, and left open the question of whether *only* such arc-connected sets are connected. In this section we show that the relation it is not that simple, except for an important class of sets.

First the bad news: A set can be connected without being arc-connected.

Example 1. Let $S := \{(x, y): x > 0 \text{ and } y = \sin(10/x), \text{ or } x = 0 \text{ and } -1 \leq y \leq 1\}$ (see Figure 5.1).

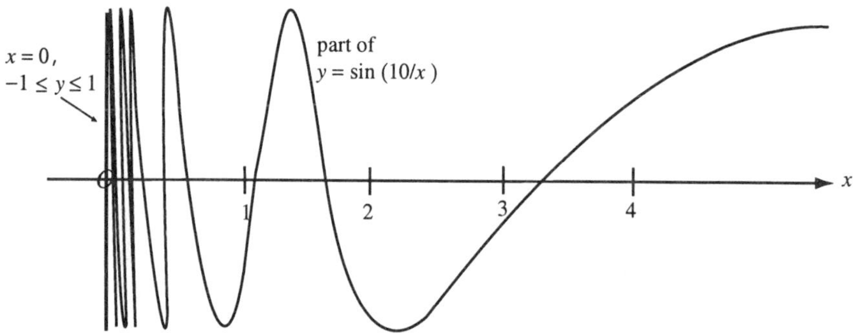

Figure 5.1.

The set S consists of the right-hand piece of the graph of $y = \sin(10/x)$, together with the part of the y-axis between -1 and 1. The figure suggests the difficulty of drawing the portion of S near the y-axis. It is because one piece is "right-hand" that the set is not arc-connected, yet the way that piece squeezes against the y-axis makes the set connected.

Proof that S is not arc-connected is left as Exercise 1. Intuitively, it certainly looks impossible to draw a continuous arc from $(10/\pi, 0)$ to $(0, 0)$ within S; it appears that you would have to jump off the curved part of S to get to the straight part. This idea motivates the demonstration that is guided in the exercise.

To prove the connectedness, we invoke Theorem 4.20. If S were disconnected, there would be closed sets C_1 and C_2 such that $C_1 \cap S$ and $C_2 \cap S$ nontrivially partition S. We will show that such a partition is impossible by establishing that if $C_1 \cap S$ and $C_2 \cap S$ are disjoint and together fill up S, then one of the two is empty.

Assume that C_1 and C_2 are closed sets with $S \subseteq C_1 \cup C_2$ and $C_1 \cap S$ disjoint from $C_2 \cap S$. The curved part of S, the part with $x > 0$, is evidently an arc-connected set. Therefore, it cannot be nontrivially partitioned by C_1 and C_2. Since it is a subset of their union, it must lie entirely in one of them, say C_1.

The points $(10/\pi, 0)$, $(10/2\pi, 0)$, ... are on the curved part, so they are in C_1. Since C_1 is closed, the limit $(0, 0)$ of this sequence must also be in C_1. Hence some of the straight part of S is in C_1. But then all of the straight part must lie in C_1, since this part, like the curved part, is arc-connected, and so cannot be split between C_1 and C_2.

We have shown that all of S is in one of C_1 and C_2. It follows that S cannot be partitioned by closed sets.

Now the good news: In the most basic of the classes of sets we employ, connectedness is equivalent to arc-connectedness.

Theorem 5.1. *An open set is connected iff it is arc-connected.*

Proof. $\Leftarrow$ This is trivial; every arc-connected set is connected (Theorem 4.22).

$\Rightarrow$ Assume that O is open and connected. Suppose $\mathbf{b}$ is any vector in O.

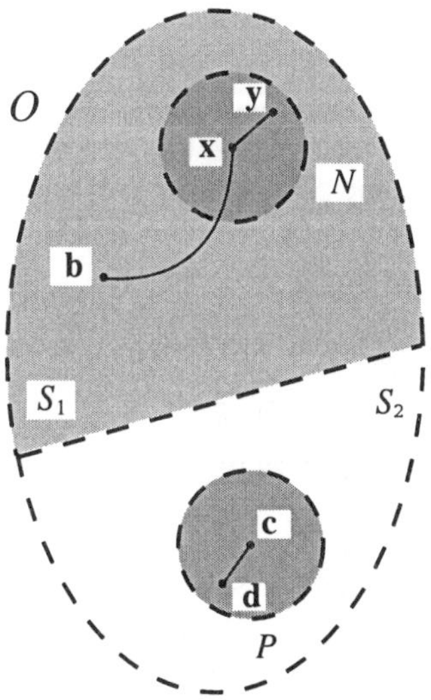

Figure 5.2.

Consider Figure 5.2. Let S_1 represent the set of vectors $\mathbf{x} \in O$ for which there is some arc contained in O from $\mathbf{b}$ to $\mathbf{x}$. (In the figure, O is the large oval, S_1 its shaded upper part. Why is $\mathbf{b}$ necessarily in S_1?) Then S_1 is an open set. For let $\mathbf{x} \in S_1$. First, there is by the nature of S_1 an arc from $\mathbf{b}$ to $\mathbf{x}$ lying within O. Second, because $\mathbf{x}$ is in the open set O, there is a neighborhood N of $\mathbf{x}$ contained in O. Given any $\mathbf{y} \in N$, the segment from $\mathbf{x}$ to $\mathbf{y}$ lies within N, hence within O. Therefore, the arc and segment together make an arc within O from $\mathbf{b}$ to $\mathbf{y}$. (Exercise 3 asks for an analytical proof.) This last says that $\mathbf{y} \in S_1$. Since every $\mathbf{y}$ from N is in S_1, we have $N \subseteq S_1$. Since every $\mathbf{x}$ from S_1 has a neighborhood N contained in S_1, S_1 is open.

The remainder of O, that is, $S_2 := \{\mathbf{c} \in O\colon \text{no arc from } \mathbf{b} \text{ to } \mathbf{c} \text{ lies within } O\}$, is also open. To see this, assume $\mathbf{c} \in S_2$. Because $\mathbf{c} \in O$, some neighborhood P of $\mathbf{c}$ is a subset of O. If one point $\mathbf{d}$ in P were reachable by an arc within O from $\mathbf{b}$, then $\mathbf{c}$ would be reachable by that arc plus the segment from $\mathbf{d}$ to $\mathbf{c}$ (Exercise 3). Hence no point $\mathbf{d}$ in P admits an arc from $\mathbf{b}$ within O. That is, $P \subseteq S_2$. Consequently, S_2 is open.

Now we have $O = S_1 \cup S_2$, where S_1 and S_2 are themselves open and have no points in common. Since S is connected, it cannot be that $\{S_1, S_2\}$ is a partition.

Therefore, one of S_1 or S_2 must be empty. The empty set has to be S_2; as we observed, **b** (in fact, a whole neighborhood of **b**) is in S_1. Therefore, $O = S_1$.

We have shown that the vectors with arcs from **b** fill all of O. Since **b** is arbitrary, we conclude that any two vectors in O can be joined by an arc within O. □

In **R** itself, the connected sets come from a familiar class.

An **interval** in **R** is a set containing the set $\{t\colon A < t < B\}$, where each of A and B can be real or infinite. The interval may include two other points: it may, but does not have to, include A if A is finite; and it similarly may include B if B is finite. If both A and B are included, or one is infinite and the other is included, or both are infinite, then the interval is a box or an infinite box. In that case, we know from examples and exercises that it is a closed set; we may therefore call it a **closed interval**. If both A and B (finite or not) are excluded, then the interval is an open subset of **R**, and we may call it an **open interval**.

Theorem 5.2. *In* **R**, *a set is connected iff it is an interval.*

Proof. $\Leftarrow$ Assume that S is an interval. The intermediate value theorem for functions in **R** [Ross, Theorem 18.2] says that S has the intermediate value property. By Theorem 4.19, S is connected.

$\Rightarrow$ Assume that S is a connected subset of **R**. Write $A := \inf S$, $B := \sup S$, with the understanding that either of these might be infinite. We will show that S is one of the possible intervals from A to B.

(What happens if S is empty? Exercise 2 asks you to take care of that situation. Nevertheless, examine all the steps in the three paragraphs below. You should ascertain that they are still valid, maybe vacuously, in that event.)

Suppose $x \in S$. By definition, $A \le x \le B$, which says that $x \in [A, B]$. We conclude that $S \subseteq [A, B]$.

Suppose instead that $y \in (A, B)$. Because $y < B$ and B is the least upper bound of S, there must be a real $\beta \in S$ with $y < \beta$. Similarly, there must be a real $\alpha < y$ in S. The identity function $\mathrm{ID}(t) := t$ is continuous on the connected set S, so it achieves intermediate values. Since $\mathrm{ID}(t)$ has the value α at one point in S and β at another, there must be some z in S at which $\mathrm{ID}(z) = y$. In other words, $y = z \in S$. We conclude that $(A, B) \subseteq S$.

We now know

$$(A, B) \subseteq S \subseteq [A, B].$$

We cannot determine whether the endpoints are in S, but certainly S is an interval from A to B. □

Exercises

1. Let

$$S := \{(x, y)\colon x > 0 \text{ and } y = \sin(10/x),\ \text{or } x = 0 \text{ and } -1 \le y \le 1\}.$$

Assume that $\mathbf{f}$ is an arc from $(10/\pi, 0)$ to $(0, 0)$ within S. Thus, $\mathbf{f}$ is a continuous mapping of $[a, b]$ into $\mathbf{R}^2$ with $\mathbf{f}(a) = (10/\pi, 0), \mathbf{f}(b) = (0, 0)$, and $\mathbf{f}(t)$ always in S.

(a) Show that there is an interval $[a, a + \delta)$ in which $\mathbf{f}(t)$ is always in the curved (right-hand) part of S.

(b) Show that there is an interval $[a, \varepsilon]$ such that $\mathbf{f}(t)$ is in the curved part for $a \leq t < \varepsilon$ but $\mathbf{f}(\varepsilon)$ is on the straight part. In words: There is a first value $t = \varepsilon$ where $\mathbf{f}(t)$ reaches the y-axis.

(c) Show that for $a \leq t < \varepsilon, \mathbf{f}(t)$ traverses the entire curve from $(10/\pi, 0)$ leftward toward the y-axis.

(d) *Show that for the values $t_5, t_9, t_{13} \ldots$ where $\mathbf{f}(t)$ first reaches $(20/5\pi, 1)$, $(20/9\pi, 1)$, $(20/13\pi, 1), \ldots$ (respectively), necessarily $t_i \to \varepsilon$.

(e) Show that for the values $t_3, t_7, t_{11} \ldots$ where $\mathbf{f}(t)$ first reaches $(20/3\pi, -1)$, $(20/7\pi, -1), \ldots$, necessarily $t_j \to \varepsilon$.

(f) Show that (d) and (e) make a contradiction.

2. Show that in $\mathbf{R}$, $\emptyset$ (which we know to be connected) is the interval from $\inf \emptyset$ to $\sup \emptyset$.

3. (a) Let S be a subset of a normed space. Assume that $\mathbf{f}\colon [r, s] \to S$ is an arc within S from $\mathbf{b}$ to $\mathbf{x}$, and $\mathbf{g}\colon [u, v] \to S$ similarly goes from $\mathbf{x}$ to $\mathbf{y}$. Prove that

$$\mathbf{h}(t) := \begin{cases} \mathbf{f}(t) & \text{if } t \in [r, s], \\ \mathbf{g}(u + (t - s)) & \text{if } t \in [s, s + v - u], \end{cases}$$

defines an arc within S from $\mathbf{b}$ to $\mathbf{y}$.

(b) Why is the segment from $\mathbf{x}$ to $\mathbf{y}$ an arc from $\mathbf{x}$ to $\mathbf{y}$. (Compare Exercise 8b in Section 4.5.)

4. Let S be a subset of a normed space. Define a relation JOINS on S by

$\mathbf{a}$ JOINS $\mathbf{b}$ if there is an arc lying within S from $\mathbf{a}$ to $\mathbf{b}$.

(a) Prove that JOINS is an equivalence relation. (Hint: Exercises 3 here and 8a, c in Section 4.5.)

(b) Show that each equivalence class $J_{\mathbf{a}} := \{\mathbf{x}\colon \mathbf{a} \text{ JOINS } \mathbf{x}\}$ is arc-connected.

(c) Show that each equivalence class is a *maximal* arc-connected subset of S; that is, if $J_{\mathbf{a}}$ is an equivalence class and $J_{\mathbf{a}} \subseteq T \subseteq S$ with $T \neq J_{\mathbf{a}}$, then T is not arc-connected.

5.2 Open Sets

In this section we study the structure of open sets. We need the following lemma.

Theorem 5.3. *In a normed space:*

(a) *The union of any class of open sets is open.*

(b) *The intersection of finitely many open sets is open.*

Proof. (a) Let Σ be a class of open sets and U the union of its members. If $\mathbf{x} \in U$, then by definition of union, there is some $O \in \Sigma$ with $\mathbf{x} \in O$. Since O is an open set, there is a neighborhood $N(\mathbf{x}, \delta) \subseteq O$. Every member of O is a member of U, so $N(\mathbf{x}, \delta) \subseteq U$. Thus, for any $\mathbf{x} \in U$, there is a neighborhood of $\mathbf{x}$ contained in U. Hence U is open.

(b) Exercise 9b in Section 4.4.

Example 1. The intersection of infinitely many open sets need not be open. In $\mathbf{R}^2$:

(a) Let

$$S_i := \{\mathbf{x}\colon \|\mathbf{x}\|_2 < 1/i\} = N(\mathbf{O}, 1/i), \quad i = 1, 2, \ldots .$$

We know that S_i is open, and it is easy to check that $S_1 \cap S_2 \cap \cdots = \{\mathbf{O}\}$, a closed set.

(b) Let T_i be the open box from $(0, 0)$ to $(1 + 1/i, 1 + 1/i)$. Then

$$T_1 \cap T_2 \cap \cdots = \{(x, y)\colon 0 < x \leq 1, 0 < y \leq 1\},$$

sort of a box with no left side or bottom, which is neither open nor closed. (Compare Exercise 1.)

We begin the description of open sets with a definition.

Definition. Let O be an open set and $\mathbf{b} \in O$. The set of elements of O to which there exists an arc from $\mathbf{b}$ lying within O is called the **component of O that holds b.**

Theorem 5.4. *Every open set is the union of its components.*

Proof. Let U be the union of the components of the open set O.

If $\mathbf{x} \in U$, then $\mathbf{x}$ is in some component of O. By definition, a component of O consists of points from O. Therefore $\mathbf{x} \in O$, and $U \subseteq O$.

If instead $\mathbf{x} \in O$, then $\mathbf{x}$ can be joined to $\mathbf{x}$ (Exercise 8a in Section 4.5). Therefore, $\mathbf{x}$ is in the component of O holding $\mathbf{x}$. (In other words, the name "component of O that holds $\mathbf{x}$" is justified!) Since $\mathbf{x}$ is in some component, it is a member of the union of the components. That is, $\mathbf{x} \in U$. We conclude that $O \subseteq U$. $\square$

Example 2. In $\mathbf{R}^2$, let O be the complement $\{(x, y) : x \neq 0\}$ of the y-axis.

The points of O that can be joined to $(-1, 0)$ are exactly those in the left half-plane L. The reason is that if $x < 0$, then the line segment from $(-1, 0)$ to (x, y) is an arc (Exercise 3b in Section 5.1) that stays within L (Exercise 2a), and therefore in O; and if $x > 0$, then any arc from $(-1, 0)$ to (x, y) has to cross the y-axis (Exercise 2b), and therefore leaves O. Hence L is one component. Similarly, the points in the right half R are the ones that can be joined to $(1, 0)$, and R is another component.

Are there other components? If you choose any other point (x, y) of O, then the same argument shows that its component is again L or R, depending on whether $x < 0$ or $x > 0$. Therefore, O has exactly the two components L and R.

Our next three results show that the situation in Example 2 is typical. Specifically, each of the components L and R is an open set, each is connected, each is as big as it could get without losing its connectedness, and the two have no vector in common. The only variation from the example will be how many components there are.

Theorem 5.5. *Each component of an open set is:*

(a) *open,*

(b) *arc-connected,*

(c) *disjoint from every other component.*

Proof. (a) was part of the proof of Theorem 5.1.

(b) Let P be a component of O, and suppose $\mathbf{a}$ and $\mathbf{c}$ are vectors in P. We must find an arc from $\mathbf{a}$ to $\mathbf{c}$. By definition, P has a member $\mathbf{b}$ from which arcs can be drawn to all the other vectors in P. Geometrically, it is obvious that if you can draw an arc from $\mathbf{b}$ to $\mathbf{a}$ and one from $\mathbf{b}$ to $\mathbf{c}$, then the combined paths trace an arc from $\mathbf{a}$ to $\mathbf{c}$. We will demonstrate analytically that the picture is truthful.

An arc from $\mathbf{b}$ to $\mathbf{a}$ is a continuous function

$$\mathbf{f}_1 : [r, s] \to O \qquad \text{with } \mathbf{f}_1(r) = \mathbf{b} \text{ and } \mathbf{f}_1(s) = \mathbf{a}.$$

Given $\mathbf{f}_1$, the function

$$\mathbf{g}(t) := \mathbf{f}_1(s - [t - r])$$

is continuous on $[r, s]$, because it is the composite $\mathbf{f}_1(h)$ for $h(t) := s + r - t$; and it has $\mathbf{g}(r) = \mathbf{a}$, $\mathbf{g}(s) = \mathbf{b}$. Therefore, $\mathbf{g}$ is an arc from $\mathbf{a}$ to $\mathbf{b}$. (This paragraph answers Exercise 8c in Section 4.5).

We now have an arc $\mathbf{g}$ from $\mathbf{a}$ to $\mathbf{b}$ and a second arc $\mathbf{f}_2 : [u, v] \to O$ from $\mathbf{b}$ to $\mathbf{c}$. Then

$$\mathbf{h}(t) := \begin{cases} \mathbf{g}(t) & \text{for } r \leq t \leq s, \\ \mathbf{f}_2(u + t - s) & \text{for } s \leq t \leq s + (v - u), \end{cases}$$

is the arc we need. By the assumptions, $\mathbf{h}$ maps $[r, s+v-u]$ into S, has $\mathbf{h}(r) = \mathbf{a}$, $\mathbf{h}(s+v-u) = \mathbf{c}$, and is continuous on $[r, s)$ and $(s, s+v-u]$. The only question is continuity at $t = s$, which is easily settled: $t \to s^- \Rightarrow \mathbf{h}(t) = \mathbf{g}(t) \to \mathbf{g}(s) = \mathbf{b}$, and $t \to s^+ \Rightarrow \mathbf{h}(t) = \mathbf{f}_2(u+t-s) \to \mathbf{f}_2(u) = \mathbf{b}$. (This answers Exercise 3 in Section 5.1, which enables Exercise 4 in the same section. The latter is frequently employed to prove (b) and (c) here; see Exercise 7.)

(c) Suppose P and Q are two components of O that have a point $\mathbf{c}$ in common. By definition, there are vectors $\mathbf{b}_1$ and $\mathbf{b}_2$ such that P is the set of points reachable by an arc within O from $\mathbf{b}_1$, and Q the set reachable from $\mathbf{b}_2$.

Now let $\mathbf{x} \in P$. Then there is an arc from $\mathbf{b}_1$ to $\mathbf{x}$. Since $\mathbf{c} \in P$, there is also an arc from $\mathbf{b}_1$ to $\mathbf{c}$, which we can reverse to get an arc from $\mathbf{c}$ to $\mathbf{b}_1$. Since $\mathbf{c} \in Q$, there is an arc from $\mathbf{b}_2$ to $\mathbf{c}$. Thus, there are arcs from $\mathbf{b}_2$ to $\mathbf{c}$, $\mathbf{c}$ to $\mathbf{b}_1$, and $\mathbf{b}_1$ to $\mathbf{x}$. By an extension of the argument in part (b), there exists an arc within O from $\mathbf{b}_2$ to $\mathbf{x}$. Hence $\mathbf{x} \in Q$, and we have proved that $P \subseteq Q$.

By symmetry, it must also be that $Q \subseteq P$; that is, $P = Q$. We have established that if P and Q have any points in common, then they are the same. Hence if P and Q are different components, then $P \cap Q$ is empty. □

Theorem 5.6. *Let O be an open set. Then every component of O is a maximal connected subset of O, and vice versa.*

Proof. $\Rightarrow$ Assume that P is a component of the open set O. Then P is a subset of O, by definition, and P is connected, by Theorem 5.5(b). It remains to prove that among connected subsets of O, P is maximal.

First, let $U^\#$ be the union of all the components of O other than P. These other components are open (Theorem 5.5(a)), so their union $U^\#$ is open (Theorem 5.3). By Theorem 5.4, O is the union of all its components, so $O = P \cup U^\#$. Finally, any $x \in U^\#$ is in a component other than P, so (Theorem 5.5(c)) $\mathbf{x}$ is not in P; therefore no vector is in both P and $U^\#$.

Assume now that Q is a subset of O strictly bigger than P. That is, Q contains P and has some member $\mathbf{b}$ of O from outside P. Then P and $U^\#$ are open sets that partition Q. The reason is that $P \cap Q$ is nonempty, because $P \cap Q = P$, and components are never empty; $U^\# \cap Q$ is nonempty, because $\mathbf{b}$ is in it; and everything from Q, being in O, has to be in P or in $U^\#$ and not in both. We conclude that Q is disconnected.

We have shown that if Q is a subset of O strictly bigger that P, then Q is not connected. That is what it means to say that P is a maximal connected subset of O.

$\Leftarrow$ Assume that Q is a maximal connected subset of O.

The argument in ($\Rightarrow$) shows that if Q straddled different components of O, that is, if it had some points in one component P and some in other components, then Q would be disconnected. Therefore, Q must lie entirely within some component P.

If it were the case that $Q \neq P$, then P would be a bigger connected subset of O. Hence if Q is maximal among connected subsets of O, then Q must be all of P. □

Theorem 5.7. *In $\mathbf{R}^n$, the class of components of an open set is countable.*

Proof. Let P be a component of O. If $\mathbf{b}$ is some member of P, then some neighborhood $N(\mathbf{b}, \delta)$ of $\mathbf{b}$ is contained in P. Within any such neighborhood, there exists some member $\mathbf{q}(P) = \mathbf{q}(P, \mathbf{b}, \delta)$ of $\mathbf{Q}^n$.

Consider the association $P \leftrightarrow \mathbf{q}(P)$ between components of O and members of $\mathbf{Q}^n$. If P and Q are different components, then $\mathbf{q}(P) \in P$ and $\mathbf{q}(Q) \in Q$. Since P and Q have to be disjoint, $\mathbf{q}(P)$ and $\mathbf{q}(Q)$ must be different. Hence $P \to \mathbf{q}(P)$ is a one-to-one map from the entire class of components of O to some subset of $\mathbf{Q}^n$. Since $\mathbf{Q}^n$ is countable, so must be the class of components. (See the discussion in Exercise 6 in Section 5.3.) □

Theorems 5.4–5.7 make up a description of the large-scale structure of open sets in normed spaces, especially in $\mathbf{R}^n$. For an open set, components are the basic sub-units, the cells of the set. They are indivisible: If you partition a component into subsets, then some of the subsets will fail to be open (because the component is connected). And they cannot be put together into bigger cells: The union of even two of them will be disconnected.

These theorems do not, on the other hand, promise that the structure is simple.

Example 3. In $\mathbf{R}$, let

$$P_2 := \left(\frac{1}{3}, \frac{2}{3}\right); \quad P_3 := \left(\frac{1}{9}, \frac{2}{9}\right), \quad P_4 := \left(\frac{7}{9}, \frac{8}{9}\right);$$

$$P_5 := \left(\frac{1}{27}, \frac{2}{27}\right), \quad P_6 := \left(\frac{7}{27}, \frac{8}{27}\right),$$

$$P_7 := \left(\frac{19}{27}, \frac{20}{27}\right), \quad P_8 := \left(\frac{25}{27}, \frac{26}{27}\right); \ldots .$$

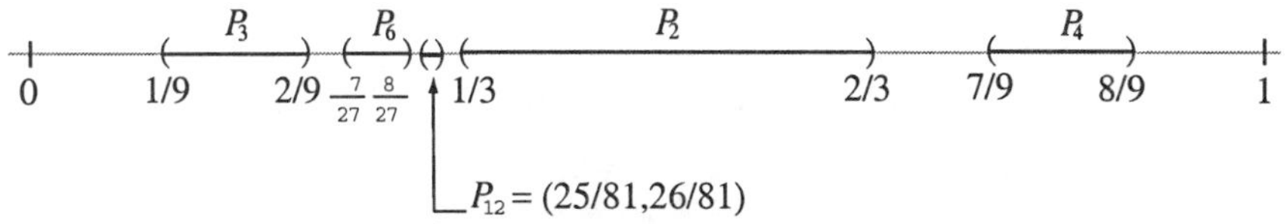

Figure 5.3.

Figure 5.3 offers clear evidence that these intervals are pairwise disjoint. It easily follows that they are the components of their (necessarily open) union O. We make some other observations, suggested by the figure, about this open set.

(a) The point at $x = \frac{1}{3}$, which is not in O, has a component of O immediately to its right, but not to its left. That is, to the right, $(\frac{1}{3}, \frac{2}{3})$ is the first component of O after $\frac{1}{3}$; to the left, there is an infinite family $(\frac{1}{9}, \frac{2}{9})$, $(\frac{7}{27}, \frac{8}{27})$, ... of components of O "converging" to $\frac{1}{3}$, so there is no last component of O before $\frac{1}{3}$.

(b) The latter observation applies to every component of O, and on both sides. That is, there is no first component of O to the right of $(\frac{7}{9}, \frac{8}{9})$, since $(\frac{25}{27}, \frac{26}{27})$, $(\frac{73}{81}, \frac{74}{81})$, ... (among others) are at decreasing distances to the right of $\frac{8}{9}$. Likewise, we may not say "the first component of O to the left of $(\frac{7}{9}, \frac{8}{9})$."

(c) Every nondegenerate (positive length) subinterval of $[0, 1]$ has points of O. In fact, every such subinterval, if not contained in O, is chopped up into an infinity of pieces by an infinity of subintervals of O.

Exercises

1. In $\mathbf{R}$, show that the intersection of the open intervals $(0, 1 + 1/i)$, $i = 1, 2, \ldots$, is the unopen interval $(0, 1]$.

2. In $\mathbf{R}^2$, show that:

   (a) If $x < 0$, then the segment from $(-1, 0)$ to (x, y) lies in the left half-plane.

   (b) If $x > 0$, then any arc from $(-1, 0)$ to (x, y) crosses the y-axis.

3. The formula
$$f(x) := x \sin(\pi/x), \qquad f(0) := 0,$$
defines a continuous function on $\mathbf{R}$. That being so,
$$\{x \colon f(x) > 0\} = f^{-1}((0, \infty))$$
has to be an open set. List its components.

4. For the open set O of Example 3:

   (a) What portion of the interval $[0, 1]$ does O cover? That is, P_2 covers one-third, P_3 and P_4 together cover an additional two-ninths, ...; what is the total length covered?

   (b) Discuss the set of points of $[0, 1]$ that are outside O?

5. Consider this construction in $\mathbf{R}^2$:
$$Q_2 := \left\{(x, y)\colon \frac{1}{3} < x < \frac{2}{3} \text{ or } \frac{1}{3} < y < \frac{2}{3}\right\},$$
$$Q_3 := \left\{(x, y)\colon \frac{1}{9} < x < \frac{2}{9} \text{ or likewise } y\right\},$$
$$Q_4 := \left\{(x, y)\colon \frac{7}{9} < x < \frac{8}{9} \text{ or likewise } y\right\}, \ldots.$$

(a) Draw a picture approximating $O := Q_2 \cup Q_3 \cup \cdots$.

(b) Are $Q_2, Q_3, \ldots$ the components of O?

(c) How much of the unit square $S := [(0, 0), (1, 1)]$ does O occupy?

6. (Sierpińsky's Gasket) Consider this construction in $\mathbf{R}^2$: Divide the unit square into nine congruent squares, and let P_1 be the open middle square. Divide each of the remaining eight squares of size 1/3 by 1/3 into nine congruent squares each, and let $P_2, \ldots, P_{1+8}$ be the eight resulting open middle squares. There remain sixty-four squares of size 1/9 by 1/9, and we subdivide those to produce middle squares labeled $P_{10}, \ldots, P_{1+8+64}$, and so on. Let $O := P_1 \cup P_2 \cup \cdots$.

   (a) Draw O approximately. The remainder $S - O$ of the square is called "Sierpińsky's gasket."

   (b) Are $P_1, P_2, \ldots$ the components of O?

   (c) How much of the unit square $[(0, 0), (1, 1)]$ does O occupy?

   (d) Is any neighborhood within the square free of points of O?

7. Recall (Exercise 4 in Section 5.1) the relation defined in a subset S of a normed space by

$$\mathbf{a} \text{ JOINS } \mathbf{b} \text{ iff there is an arc lying within } S \text{ from } \mathbf{a} \text{ to } \mathbf{b}.$$

   Use the fact that JOINS is an equivalence relation (without using Theorems 5.4–5.6) to give alternative proofs that if S is open, then:

   (a) each equivalence class is a component of S, and vice versa;

   (b) S is the disjoint union of its components;

   (c) each component of S is arc-connected.

8. Show that if S is not open, then the equivalence classes under JOINS may fail to be:

   (a) open;

   (b) maximal connected subsets, even though they have to be maximal arc-connected subsets (Exercise 4c in Section 5.1);

   (c) countably many.

5.3 Closed Sets

The organization of this section is similar to that of the previous one: We have a theorem about unions and intersections, a look at the structure of closed sets (revealing mostly that we cannot hope to find the simplicity of structure that exists in open sets), and introduction of an important example.

Theorem 5.8. *In a normed space:*

(a) *The intersection of any class of closed sets is closed.*

(b) *The union of finitely many closed sets is closed.*

Proof. The proof can easily be done by reference to Theorems 4.14 and 5.3 and DeMorgan's laws. We leave that method as Exercise 1. We exhibit here a separate proof, because it gives us added insight into properties of closed sets.

Suppose Σ is a family of closed sets and T is the intersection of its members. To prove that T is closed, we must show that if $(\mathbf{x}_i)$ is a sequence from T converging to $\mathbf{x}$, then $\mathbf{x}$ is in T.

Assume that $(\mathbf{x}_i)$ is such a sequence. Let S be any of the sets in Σ. Each $\mathbf{x}_i$, belonging to the intersection, must belong to all of the sets in Σ. In particular, each $\mathbf{x}_i$ is in S. Because $(\mathbf{x}_i) \to \mathbf{x}$ and S is closed, it must be that $\mathbf{x} \in S$.

We have proved that $\mathbf{x}$ belongs to every member S of Σ. Therefore, $\mathbf{x}$ belongs to the intersection: $\mathbf{x} \in T$.

(b) Suppose $S_1, \ldots, S_k$ are closed sets and $U = S_1 \cup \cdots \cup S_k$. Assume that $(\mathbf{x}_i)$ is a sequence from U converging to $\mathbf{x}$. Each $\mathbf{x}_i$, being in the union, must belong to one or more of $S_1, \ldots, S_k$. We cannot tell which $\mathbf{x}_i$ belongs to which S_j, but one thing is for sure: It cannot be true that each S_j has just finitely many terms $\mathbf{x}_i$, because then there would be only a finite number of terms. Therefore, some subsequence $(\mathbf{x}_{j(i)})$ comes from a single set; call it S_m.

Now, $(\mathbf{x}_{j(i)})$ still converges to $\mathbf{x}$ (Theorem 2.4). Since $(\mathbf{x}_{j(i)})$ comes from the closed set S_m, we conclude that $\mathbf{x} \in S_m$. Since we have found one set to which $\mathbf{x}$ belongs, we have $\mathbf{x} \in U$. It follows that U is closed. □

Note that in Theorem 5.8(a) we deal with the intersection of any collection of closed sets, no matter how many. If you could do the same with the *union* of any collection, then every set would be closed. After all, every set is the union of its singletons—any S is the union of the class $\Sigma := \{\{x\} \colon \mathbf{x} \in S\}$—and singletons are closed sets.

In Theorem 5.8(b) we exclude not only arbitrary collections of closed sets, but even countably infinite ones. The union of a sequence of closed sets may be closed, open, neither, or both (Exercise 2).

Next we want to see whether a closed set has something like the component structure of an open set. According to Exercise 4 in Section 5.1, every set can be viewed as the union of disjoint arc-connected sets that are maximal with respect to being arc-connected. For an open set, the maximal arc-connected subsets ("MACS") are open and (in $\mathbf{R}^n$) only countably numerous. We might therefore hope, for example, to characterize closed sets in terms of closed MACS. The next example indicates that no such description is possible.

Example 1. (a) Recall the set $S := \{(x, y) \colon x > 0 \text{ and } y = \sin(10/x), \text{ or } x = 0 \text{ and } -1 \leq y \leq 1\}$. In this set, the straight $(x = 0)$ and curved $(x > 0)$ parts are MACS. Hence S is a closed set with just two MACS, but only the straight MACS is closed. (Why is S closed and the curved part not?)

(b) Let $T := \{(x, 0) : x \text{ is rational}\}$. It is clear that you cannot join (by an arc within T) any member of T to any other. Hence each member of T is a MACS; that is, every singleton is a MACS. Therefore, T has a countable family of MACS, all of them closed, but T is not a closed set. (See also Exercise 3.)

In $\mathbf{R}$, the set of rationals is full of holes: Given two rationals $q_1 \neq q_2$, there exists an irrational real s between them. So, in $\mathbf{R}$, $\{\mathbf{Q} \cap (-\infty, s), \mathbf{Q} \cap (s, \infty)\}$ is a disconnection of $\mathbf{Q}$. Similarly, for the set T in Example 1(b), the line $x = s$ separates the plane into open halves that disconnect T. A set with the property that any two of its members $\mathbf{b} \neq \mathbf{c}$ give rise to a disconnection $\{S_1, S_2\}$ with $\mathbf{b} \in S_1$ and $\mathbf{c} \in S_2$ is said to be **totally disconnected**. In a totally disconnected set, each member fills up an arc-connected component.

In $\mathbf{R}$, any strictly monotonic sequence gives a totally disconnected set; so does the set of irrationals. It is hard to think up an example of a totally disconnected set that is uncountable, bounded, and closed. Our next example displays the most famous of oddball sets of reals, a set that provides an amazing number of counterexamples to the properties we would view as normal. (See Exercises 4–6 for some of these anomalies.)

Example 2 (The Cantor Set). In Example 3 of Section 5.2 we constructed an open subset O of $[0, 1]$ by gathering open subintervals $P_2, P_3, \ldots$. Let C represent the points left over in the unit interval. Clearly, $C \subseteq [0, 1]$ is bounded. Also, $C = [0, 1] - O = [0, 1] \cap O^*$ is the intersection of closed sets, so C is closed.

(a) C consists of those reals in $[0, 1]$ that have a ternary (base 3) expansion consisting of 0's and 2's.

[Analytical arguments related to the Cantor set demand some familiarity with the representation of real numbers in terms of powers of a fixed integer (the base). The integer 10 gives the familiar **decimal** numeration; 2 gives **binary**, 3 gives **ternary**. Ross, Section 16, covers the decimal case. The general case is best stated in, oddly enough, an exercise: Royden, Chapter 2, Section 4, Exercise 22.]

For evidence, represent the ternary expansion $a_1/3 + a_2/3^2 + a_3/3^3 + \cdots$, in which each a_i is 0, 1, or 2, by $\#a_1a_2a_3\ldots$. By analogy with "decimal," we will call this last expression a **ternamal**. Thus,

$$\#1 := \frac{1}{3} \quad \text{and} \quad \#2020\ldots := \frac{2}{3} + \frac{0}{9} + \frac{2}{27} + \frac{0}{81} + \cdots = \frac{3}{4}.$$

Any real number between $\frac{1}{3}$ and $\frac{2}{3}$ has a ternamal of the form $\#1abc\ldots$. For such a ternamal, there is no hope of losing that first 1, except that

$$\frac{1}{3} = \#1 = \#0222\ldots \quad \text{and} \quad \frac{2}{3} = \#1222\ldots = \#2000\ldots = \#2.$$

Conversely, any real number with an irreplaceable 1 in the first ternamal place, like

$$\#111\ldots = \frac{1}{3} + \frac{1}{9} + \cdots = \frac{1}{2},$$

belongs to the interval $\left(\frac{1}{3}, \frac{2}{3}\right) = P_2$. Therefore, P_2 removes from $[0, 1]$ precisely the reals that have an irreplaceable 1 in the first ternamal column.

A ternamal that does not have a 1 first but does have a 1 second looks like either $\#01abc\ldots$ or $\#21abc\ldots$. Hence such a real number is either between

$$\#01 = \#00222\ldots = \frac{1}{9} \quad \text{and} \quad \#01222\ldots = \#02 = \frac{2}{9},$$

so that it belongs to P_3; or between

$$\#21 = \#20222\ldots = \frac{7}{9} \quad \text{and} \quad \#21222\ldots = \#22 = \frac{8}{9},$$

so that it belongs to P_4. Hence removing P_3 and P_4 from $[0, 1]$ pulls out the reals whose ternamals have their first irreplaceable 1 in the second ternamal column.

By induction, we can show that the ternamals with an irreplaceable 1 occurring for the first time in column number k belong to one of $P_{(2^{k-1}+1)}, \ldots, P_{2^k}$. It follows that $O = P_2 \cup P_3 \cup \cdots$ consists of those reals in $[0, 1]$ for which the one (or the two) ternary expansion(s) has (both have) at least one 1, and C is as we indicated.

(b) C does not contain any nondegenerate subintervals of $[0, 1]$. (This was our claim in Example 3(c) in Section 5.2).

The geometric evidence is clear (indeed, clearer than the analytic, but we will provide some of the latter, too). After the removal of the middle third P_2, the remaining set has two intervals that are each only $\frac{1}{3}$ long. After removal of P_3 and P_4, there are four intervals left, length $= \frac{1}{9}$. In general,

$$C(k) := [0, 1] - P_2 - P_3 - \cdots - P_{2^k}$$

has 2^k pieces, each only 3^{-k} long. Since C is a subset of every $C(k)$, it follows that any interval contained in C has length $\leq 3^{-k}$ for arbitrary k. Therefore, only singleton subintervals like $\left[\frac{1}{3}, \frac{1}{3}\right]$ are contained in C.

A different way of saying the same thing is this: If x and y are unequal members of C, then there is a member z of O between them. This statement has an easy analytic proof. If x and y are unequal members of the Cantor set, then their ternamals must look like $x = \#ab\ldots c0de\ldots$ and $y = \#ab\ldots c2DE\ldots$; in words, at the first place where the ternamals differ, one has a 0 and the other a 2. In that case, $z := \#ab\ldots c111\ldots$ is strictly between x and y and cannot belong to C.

Exercises

1. Use the theorems mentioned plus DeMorgan's laws to prove Theorem 5.8.

2. In $\mathbf{R}^2$, find sequences of unequal closed sets whose unions are:

   (a) closed and not open;

   (b) open and not closed;

   (c) neither open nor closed;

(d) both open and closed.

3. (a) Show that $T := \{(x, 0): x \text{ is rational}\}$ (whose MACS are closed sets) is neither open nor closed.

   (b) What can be said about a set U with a *finite* number of closed MACS?

   (c) Can a subset of $\mathbf{R}^n$ have uncountably many closed MACS?

   (d) Can a subset of $\mathbf{R}^n$ have uncountably many open MACS?

4. Show that the Cantor set is totally disconnected.

5. Show that in every neighborhood of a Cantor point, there exist:

   (a) a member of O (the complement of the Cantor set);

   (b) a complete subinterval of O (Hint: Prove instead that in every neighborhood of a member of the Cantor set, there are other elements of the Cantor set.);

   (c) an infinity of subintervals of O.

6. To deal with countability of the Cantor set C, let us agree to some definitions. We call a set S **infinite** if there is a one-to-one mapping of the natural numbers $\mathbf{N}$ into S (compare Exercise 9 of Section 4.2). For an infinite set S, if there is also some one-to-one mapping of S into $\mathbf{N}$, then we say that S is **countable** or **countably infinite**; if not, then S is **uncountable** (**uncountably infinite**).

   [Georg Cantor made many contributions to mathematics. Some of the most far-reaching had to do with the nature of infinity, including the result that [0, 1] is uncountable. See Kline, Chapter 41, Section 7.].

   (a) Prove that the set of endpoints of the components of $O := [0, 1] - C$ is countable.

   (b) Prove that there is a one-to-one mapping from (all of) [0, 1] into C. (Hint: Every real number in the unit interval has a binary expansion; thus, in base 2,

   $$\&010101\ldots := \frac{0}{2} + \frac{1}{4} + \frac{0}{8} + \frac{1}{16} + \cdots = \frac{1}{3}.)$$

   (c) Given that [0, 1] is uncountable, prove that C is uncountable. (Contrast this statement plus part (a) with Exercises 5b and 5c.)

5.4 Interior, Boundary, and Closure

Most of the sets we work with are not as peculiar as the Cantor set and its complement. Typically they have a main part that is open, together with a "thin" skin bordering the main part. In this section we attach names to these parts and discuss the general structure of point sets.

Definition. Given a subset S of a normed space, the **interior of** S is

$$\text{int}(S) := \{\mathbf{x} \in S: \text{ some neighborhood } N(\mathbf{x}, \delta) \subseteq S\}.$$

Example 1. The interior of a ball $B(\mathbf{b}, \varepsilon)$ is the neighborhood $N(\mathbf{b}, \varepsilon)$.

Any vector in the ball is either in the neighborhood or on the sphere. First, let $\mathbf{x} \in N(\mathbf{b}, \varepsilon)$. Then we know that $\mathbf{x}$ has a neighborhood $N(\mathbf{x}, \delta) \subseteq N(\mathbf{b}, \varepsilon)$, because $N(\mathbf{b}, \varepsilon)$ is open (Example 1(a) of Section 4.4). Therefore, $\mathbf{x} \in N(\mathbf{x}, \delta) \subseteq B(\mathbf{b}, \varepsilon)$. This last says that $\mathbf{x}$ belongs to the interior of $B(\mathbf{b}, \varepsilon)$. Second, let $\mathbf{y}$ be on the spherical surface $S(\mathbf{b}, \varepsilon)$. Every neighborhood $N(\mathbf{y}, s)$ includes points like

$$\mathbf{y} + \left(\frac{s}{2}\right)(\mathbf{y} - \mathbf{b})/\varepsilon = \mathbf{b} + \frac{(\varepsilon + s/2)}{\varepsilon}(\mathbf{y} - \mathbf{b})$$

that are not in $B(\mathbf{b}, \varepsilon)$. (Verify!) Therefore, $\mathbf{y}$ is not in the interior of $B(\mathbf{b}, \varepsilon)$.

Hence the points of $\text{int}(B(\mathbf{b}, \varepsilon))$ are precisely those of $N(\mathbf{b}, \varepsilon)$.

Theorem 5.9. *For every set* S:

(a) $\text{int}(S) \subseteq S$.

(b) *S is open iff S equals its interior.*

(c) *The interior of S is an open set. In fact,*

(d) *The interior of S is the largest subset of S that is open.*

Proof. (a) is immediate from the definition.

(b) By definition of open set, S is open iff each point of S has a neighborhood contained in S. The latter condition is equivalent to $S \subseteq \text{int}(S)$ (Reason?). In view of (a), $S \subseteq \text{int}(S)$ is equivalent to $S = \text{int}(S)$.

(c) Let $T = \text{int}(S)$. To prove that T is open, we have to show that each $\mathbf{b} \in T$ has a neighborhood $N(\mathbf{b}, \varepsilon) \subseteq T$. Accordingly, assume $\mathbf{b} \in T$.

By definition of interior, there exists $N(\mathbf{b}, \varepsilon) \subseteq S$. Suppose $\mathbf{x}$ is any member of $N(\mathbf{b}, \varepsilon)$ (possibly $\mathbf{b}$ itself). By familiar properties of neighborhoods, there is a neighborhood $N(\mathbf{x}, \delta) \subseteq N(\mathbf{b}, \varepsilon) \subseteq S$. Because this neighborhood of $\mathbf{x}$ is contained in S, we conclude that $\mathbf{x} \in \text{int}(S) = T$. Thus, the assumption $\mathbf{x} \in N(\mathbf{b}, \varepsilon)$ leads to $\mathbf{x} \in T$. We have shown that $N(\mathbf{b}, \varepsilon) \subseteq T$.

(d) To prove that T is the biggest open subset of S, we need to prove that every subset U of S that is open has $U \subseteq T$.

Assume that $U \subseteq S$ is open. If $\mathbf{x} \in U$, then because U is open there is a neighborhood $N(\mathbf{x}, \delta) \subseteq U$. Since $N(\mathbf{x}, \delta) \subseteq U \subseteq S$, this means that some neighborhood of $\mathbf{x}$ is contained in S. Therefore, $\mathbf{x}$ is an interior point of S; $\mathbf{x} \in T$. We have proved that $U \subseteq T$. □

The members of $\text{int}(S)$ are surrounded by friends (other elements of S). If $\mathbf{b} \in S$ is not surrounded by friends—if there is no neighborhood of $\mathbf{b}$ filled entirely with members of S—then every neighborhood of $\mathbf{b}$ has both members of S (including $\mathbf{b}$) and nonmembers of S.

Definition. The **boundary** $\mathrm{bd}(S)$ of S consists of those vectors $\mathbf{b}$ whose every neighborhood contains some points from S and some from the complement S^*.

Notice that the definition does not require $\mathbf{b}$ to belong to S. By virtue of the definition's symmetry with respect to S and S^*, it is clear that *the boundary of S is also the boundary of S^**; this is certainly a desirable trait for something named "boundary."

Example 2. Our earlier description, "thin," of boundary applies to familiar sets, but by no means universally.

(a) The boundary of a ball is its surface.

Return to Example 1. We said there that if $\mathbf{y} \in S(\mathbf{b}, \varepsilon)$, then any neighborhood $N(\mathbf{y}, s)$ has points like $\mathbf{b} + (\varepsilon + s/2)(\mathbf{y} - \mathbf{b})/\varepsilon$ from outside the ball $B(\mathbf{b}, \varepsilon)$. The same neighborhood has $\mathbf{y}$ itself from inside the ball. Hence $\mathbf{y} \in \mathrm{bd}(B(\mathbf{b}, \varepsilon))$.

If $\mathbf{z}$ is outside the ball (if $\|\mathbf{z} - \mathbf{b}\| > \varepsilon$) then $N(\mathbf{z}, t)$ has no points from the ball as long as $t < \|\mathbf{z} - \mathbf{b}\| - \varepsilon$ (Theorem 2.2(a)). Hence $\mathbf{z}$ is not a boundary point.

We conclude that the boundary of $B(\mathbf{b}, \varepsilon)$ coincides with $S(\mathbf{b}, \varepsilon)$.

(b) In $\mathbf{R}^2$, let

$$T := \left\{\mathbf{x}\colon \text{ either } \|\mathbf{x}\|_2 < 1, \text{ or } \mathbf{x} \in \mathbf{Q}^2 \text{ and } 1 \le \|\mathbf{x}\|_2 < 2\right\}.$$

The boundary of T is an annular region (see Figure 5.4).

Since the open unit ball is an open subset of T, Theorem 5.9(d) tells us that $N(\mathbf{O}, 1) \subseteq \mathrm{int}(T)$. In fact, it is clear that $N(\mathbf{O}, 1)$ is the entire interior, since every $\mathbf{x}$ with $\|\mathbf{x}\|_2 \ge 1$ is itself outside T or has neighbors (with irrational coordinates) outside T.

Each point in the ring $\{\mathbf{x}\colon 1 \le \|\mathbf{x}\|_2 \le 2\}$ has neighbors with rational coordinates, and these points are in T if they have norm < 2, and neighbors with (some) irrational coordinates, which are in T^* if they have norm ≥ 1. Therefore, this ring is part of the boundary of T. Notice that the outer circle $\{\mathbf{x}\colon \|\mathbf{x}\|_2 = 2\}$ is entirely in T^*.

Finally, $\{\mathbf{x}\colon \|\mathbf{x}\|_2 > 2\}$ is an open set contained in T^*. Therefore, this set is contained in the interior of T^*. Each of its points is therefore surrounded by T^*, and cannot be part of the boundary of T. We have shown that the (closed) ring is $\mathrm{bd}(T)$.

Figure 5.4 is a good Venn rendering of how a set T breaks up the space. Any point of the space falls into exactly one of three regions: It is surrounded by T, and so is in $\mathrm{int}(T)$; or is surrounded by T^*, and is in $\mathrm{int}(T^*)$; or has nearby points from each of T and T^*, and is in $\mathrm{bd}(T)$. Thus, these three are disjoint sets whose union is the whole space. In Example 2(b), the boundary is actually split up between T and T^*, but in a given situation it could have points only from T or only from T^*

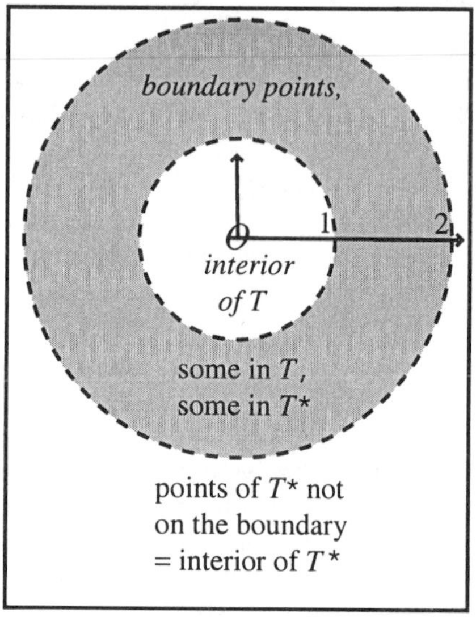

Figure 5.4.

(Exercise 9). The remainder of T^*, the points from T^* that are not on the boundary, make up $\text{int}(T^*)$. We will call the interior of T^* the **exterior** of T; we discuss it in the exercises.

In the example, both boundaries turned out to be closed. That always happens.

Theorem 5.10. *The boundary of any set is closed.*

Proof. Suppose $(\mathbf{x}_i)$ is a sequence of boundary points (members of the boundary) of the set S converging to $\mathbf{x}$. We must show that $\mathbf{x}$ is a boundary point of S.

Choose an arbitrary neighborhood N of $\mathbf{x}$. From $(\mathbf{x}_i) \to \mathbf{x}$, we deduce (Theorem 2.5) that there is some term $\mathbf{x}_I \in N$. Because N is an open set, a neighborhood $N(\mathbf{x}_I, \delta)$ is contained in N. Because $\mathbf{x}_I$ is a boundary point of S, there are members $\mathbf{y} \in S$ and $\mathbf{z} \in S^*$ in $N(\mathbf{x}_I, \delta)$. Hence we have $\mathbf{y}$ from S and $\mathbf{z}$ from S^* in N, and $\mathbf{x} \in \text{bd}(S)$. □

The next notion is a kind of dual—a mirror image—to the concept of interior.

Definition. For a set S, the set $\text{cl}(S)$ of closure points of S is called the **closure** of S.

Theorem 5.11. *For every set S:*

(a) *S is a subset of* $\text{cl}(S)$.

(b) *S is closed iff S equals its closure.*

(c) *The closure is a closed set. In fact,*

(d) *The closure is the smallest closed set that contains S.*

Proof. (a) is immediate from, among others, Theorem 3.1.
(b) is immediate from part (a) and Theorem 4.7.
(c), (d) Exercise 7.

The name "closure" is standard; in fact, it is the reason we chose the name "closure points." The mirror-reversed similarity between Theorems 5.9 and 5.11 is why we refer to closure and interior as "dual" ideas. We have one more duality result.

Theorem 5.12. *For every set, the closure of the set and the interior of the complement are complements*:

$$(\mathrm{cl}(S))^* = \mathrm{int}(S^*) = \text{exterior of } S.$$

Proof. Exercise 8.

Example 3. With the definitions we have, we can add to our earlier picture of set anatomy.

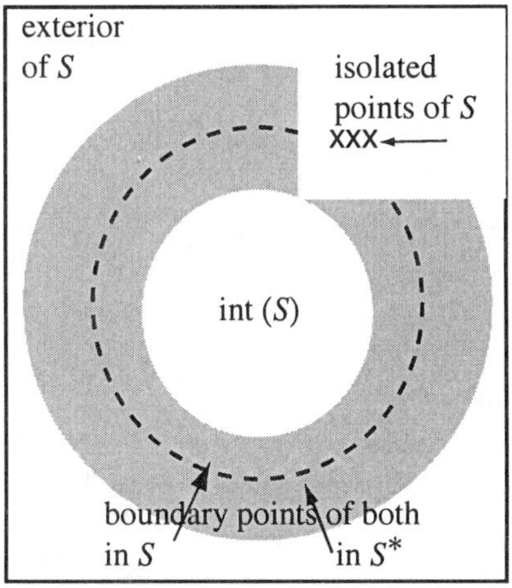

Figure 5.5.

We have already indicated that a space containing a set S partitions into $\mathrm{int}(S)$, $\mathrm{int}(S^*)$ = exterior of S, and $\mathrm{bd}(S) = \mathrm{bd}(S^*)$. The boundary may have parts from S and S^*, shown as the inner and outer halves of the shaded ring in Figure 5.5. In turn, these parts divide into two kinds of points.

A point of S that is on the boundary may be surrounded by points from S^*. We have described those as "isolated points" of S (illustrated). The only other possibility for a point of $S \cap \mathrm{bd}(S)$ is that every neighborhood of it contains both other points of S and some points of S^*. We have called those "accumulation points" of S, and clearly they are also accumulation points of S^*. (On the dashed circle, any dash that happens to lie in S consists of such points.) The S^* part of the boundary splits up symmetrically, into isolated points of S^* and accumulation points of both S and S^* belonging to S^*.

We finish the section with a theorem that provides an alternative definition of connected set. Our definition and the dual definition from Theorem 4.20 both make reference to sets unrelated to S. The characterization in the next theorem has the advantage of being internal to S.

Theorem 5.13. *For a subset S of a normed linear space, the following are equivalent*:

(a) *S is connected.*

(b) *For every (nontrivial) partition $\{S_1, S_2\}$ of S, one of S_1 and S_2 has closure points of the other.*

Proof. (a) $\Rightarrow$ (b) Assume that S is connected.

Let $\{S_1, S_2\}$ be a partition of S. Write $C_1 := \mathrm{cl}(S_1)$ and $C_2 := \mathrm{cl}(S_2)$. Then C_1 and C_2 are closed sets. Since C_1 contains S_1, it follows that $S_1 \subseteq C_1 \cap S$. Similarly, $S_2 \subseteq C_2 \cap S$. Therefore, $C_1 \cap S$ and $C_2 \cap S$ are nonempty and their union is all of S. Since these two cannot disconnect S, it must be that they have some vector $\mathbf{x}$ in common.

Because $\mathbf{x}$ is in S, we infer that it is in either S_1 or S_2. Because $\mathbf{x}$ is in both $\mathrm{cl}(S_1)$ and $\mathrm{cl}(S_2)$, it is a closure point of both. Evidently, then, either S_1 has a closure point of S_2, or vice versa.

(b) $\Rightarrow$ (a) Assume that S is disconnected.

By Theorem 4.20, there exist closed sets D_1 and D_2 such that $T_1 := D_1 \cap S$, $T_2 := D_2 \cap S$ define a partition of S. Now, D_1 is a closed superset of T_1, so $\mathrm{cl}(T_1) \subseteq D_1$. (Reason?) That is, every closure point $\mathbf{y}$ of T_1 is in D_1. If $\mathbf{y}$ is out of S, then $\mathbf{y}$ is not in T_2. If instead $\mathbf{y}$ is in S, then it is in $D_1 \cap S$, which by assumption is disjoint from $D_2 \cap S = T_2$; so again $\mathbf{y}$ is not in T_2. Therefore, every closure point of T_1 is outside T_2.

By symmetry, every closure point of T_2 is outside T_1. We have found a partition $\{T_1, T_2\}$ of S with neither subset having a closure point of the other. □

Exercises

1. In $\mathbf{R}^2$, what are the interior and boundary of:

   (a) the y-axis?

(b) quadrant I?

(c) $\mathbf{Q}^2$?

2. In $\mathbf{R}^2$, what are the closure and the exterior of:

   (a) a ball?

   (b) a neighborhood?

   (c) a line?

   (d) quadrant I?

   (e) $\mathbf{Q}^2$?

3. For an arbitrary set S, characterize:

   (a) the interior of the interior.

   (b) the interior of the exterior.

   (c) the exterior of the boundary.

4. For a neighborhood $N(\mathbf{b}, \delta)$ in a normed space, what is:

   (a) the interior of the boundary?

   (b) the boundary of the interior?

   (c) the boundary of the boundary?

   (d) the boundary of the exterior?

   (e) the exterior of the interior?

   (f) the exterior of the exterior?

5. Answer the questions in Exercise 4 for

$$T := \left\{ \mathbf{x} \in \mathbf{R}^2 \colon \text{either } \|\mathbf{x}\|_2 < 1, \text{ or } \mathbf{x} \in \mathbf{Q}^2 \text{ and } 1 \le \|\mathbf{x}\|_2 < 2 \right\}.$$

   (Consult Example 2(b).)

6. For the Cantor set, what are:

   (a) the interior?

   (b) the boundary?

   (c) the closure?

   (d) the set of isolated points?

   (e) the set of accumulation points?

7. Prove parts (c) and (d) of Theorem 5.11.

8. Prove Theorem 5.12: The complement of the closure of any set is the interior of the complement.

9. Show that for a subset of a normed space:
   (a) the closure is the union of the set's interior and boundary;
   (b) the set is closed iff it contains all of its boundary;
   (c) the set is open iff it contains none of its boundary.

10. Let S be any set in a normed space.
    (a) Show that the accumulation points of S form a closed set.
    (b) Which points are accumulation points of S but not of S^*? Which are accumulation points of S^* but not S?
    (c) Need the isolated points of S form a closed set?

11. (a) Can the closure of the interior of a set be equal to the set? Can it be smaller (a proper subset)? Can it be bigger? Can it be neither subset nor superset?
    (b) Answer the four questions in (a) for the interior of the closure.
    (c) What is the exterior of the closure of a set? What is the closure of the exterior?

12. Show that the closure of a:
    (a) bounded set is bounded;
    (b) convex set is convex; (Hint: Draw the picture in $\mathbf{R}^2$.)
    (c) connected set is connected. (Hint: Theorem 5.13.)

5.5 Compact Sets

Recall our statement (originally Theorem 3.10(a)) that a continuous function is locally bounded. Specifically, if $\mathbf{f}$ is continuous on S, then for each $\mathbf{x} \in S$ there exist a neighborhood $N(\mathbf{x}, \delta(\mathbf{x}))$ and a bound $M(\mathbf{x})$ such that

$$\|\mathbf{f}(\mathbf{y})\| < M(\mathbf{x}) \qquad \text{for } \mathbf{y} \in N(\mathbf{x}, \delta(\mathbf{x})) \cap S.$$

If we could be sure that one of the $M(\mathbf{x})$ were largest, call it M, then $\|\mathbf{f}\|$ would be bounded by M throughout S. We have here a family $\{N(\mathbf{x}, \delta(\mathbf{x}))\}$ of neighborhoods whose union encompasses all of S and for which we need certain associated numbers to include a biggest one. One way to guarantee the existence of such a maximum is to find a *finite* family of neighborhoods that does the same job. This possibility leads us to the following definition.

Definition. A set S in a normed space is **compact** if it has the **Heine–Borel property**: Whenever Ω is a class of open sets whose union contains S, it is possible to find a finite subclass of Ω whose union also contains S.

The definition requires something involving classes (sets of sets). Let Ω be a class of sets, all of them open, and let U be the union of those sets. If $S \subseteq U$, then Ω is called an **open cover(ing) of** (or **for**) S. Notice that an open covering is not an open set; it is a set of sets, not of points. Its union U is a set of points, and by Theorem 5.3(a) it happens to be open, but that is not what we focus on.

A family of sets picked from Ω (**subclass** or **subfamily** or **subcollection**) whose union still contains S is a **subcover(ing) from** Ω. If that family is finite—if $O_1, \ldots, O_k$ all belong to Ω and $S \subseteq O_1 \cup \cdots \cup O_k$—then $\{O_1, \ldots, O_k\}$ is a **finite subcover(ing) from** Ω. For S to be compact, the definition demands this: From every open cover of S we can extract a finite subcover.

Example 1. In $\mathbf{R}$, let O_r be the open interval $(r, r+2)$, and let $\Omega := \{O_r : r \in \mathbf{R}\}$. (A family consisting of sets S_ρ corresponding to members ρ of some set is called an **indexed family** of sets. The sets do not have to be open, and the indexing set need not be $\mathbf{R}$. "Indexed family" is a generalization of "sequence." For example, the sequence $O_1, O_2, \ldots$ can be viewed as the family $\{O_i : i \in \mathbf{N}\}$ indexed by $\mathbf{N}$.)

(a) The family Ω is an open covering for $\mathbf{R}$. The reason is that any $s \in \mathbf{R}$ belongs to some member of Ω; $s \in O_{s-1}$, for example. Therefore, $\mathbf{R} \subseteq \bigcup_{r \in \mathbf{R}} O_r$. (The symbols $\bigcup_{r \in \mathbf{R}}$ and $\bigcap_{r \in \mathbf{R}}$ are standard notations for the union and intersection of the members of an indexed family.)

(b) If $\Omega^\#$ is the subcollection $\{O_0, O_{-1}, O_1, O_{-2}, O_2, \ldots\}$, then $\Omega^\#$ is also an open covering for $\mathbf{R}$, because a real numbers s is in the set O_k with k the biggest integer for which $k < s$. Therefore, $\Omega^\#$ is a subcovering from Ω.

(c) If Γ is a finite subcollection $\{O_\alpha, O_\beta, \ldots, O_\omega\}$, then Γ is not a cover for $\mathbf{R}$. For let $r := \max\{\alpha, \beta, \ldots, \omega\}$. Then

$$s \in O_\alpha \cup \cdots \cup O_\omega \Rightarrow s < r + 2,$$

so the union of the sets in Γ is bounded above.

(d) We have shown that Ω is an open covering of $\mathbf{R}$, and that no finite subcollection from Ω is a subcovering of $\mathbf{R}$. We conclude that $\mathbf{R}$ is not compact.

Example 2. Every bounded closed interval in $\mathbf{R}$ is compact.

We choose this example because it allows an elementary argument [from Royden, Chapter 2, Section 5, half the proof of Theorem 15] based on the defining property of $\mathbf{R}$. The argument is adaptable; see, for instance, Exercise 1b in Section 5.1.

Let Ω be an open cover for the closed interval $[a, b]$. Some member O of Ω contains a, and since O is open, there is a neighborhood $(a - \delta, a + \delta)$ contained in O. Thus, there is $\delta > 0$ such that the interval $[a, a + .9\delta]$ can be covered by a single set O from Ω.

If $a + \delta \leq b$, then we can go past $a + \delta$. That is, $a + \delta$ belongs to some $P \in \Omega$, and so some neighborhood $(a + \delta - \varepsilon, a + \delta + \varepsilon)$ is contained in P. Hence $[a, a + \delta + .9\varepsilon]$ is covered by $O \cup P$.

How far can we continue this way? Let c be the end. More precisely, let

$$T := \{t : [a, t] \text{ can be covered by finitely many sets from } \Omega\} \text{ and } c := \sup T.$$

We already know that $c > a$. Suppose c were less than or equal to b. Then c would belong to some $Q \in \Omega$, and there would be an interval $(c - u, c + u)$ within Q. Since $c - u < c$, there would have to be $t > c - u$ in T. Thus, $[a, t]$ could be covered by a finite subcollection $O_1, \ldots, O_k$. But then

$$\begin{aligned} [a, c + .9u] &= [a, t] \cup [t, c + .9u] \\ &\subseteq (O_1 \cup \cdots \cup O_k) \cup (c - u, c + u) \\ &\subseteq O_1 \cup \cdots \cup O_k \cup Q. \end{aligned}$$

This would be a contradiction, because by definition of c, you cannot cover past c with finitely many sets from Ω.

Therefore, we abandon the possibility that $c \le b$; necessarily $b < c$. Hence there is t between b and c such that $[a, t]$ can be covered by finitely many sets from Ω. Those same sets cover $[a, b]$.

We introduced compactness as a condition that might help us to prove that continuous functions are bounded. We now employ it exactly that way.

Theorem 5.14. *Every compact set has the extreme value property.*

Proof. Assume that S is a compact set, and suppose f is real and continuous on S.

As we observed, local boundedness tells us that for each $\mathbf{x} \in S$ there exist a neighborhood $N(\mathbf{x})$ and a bound $M(\mathbf{x})$ such that

$$|f(\mathbf{y})| < M(\mathbf{x}) \qquad \text{for } \mathbf{y} \in S \text{ in } N(\mathbf{x}).$$

Let $\Omega := \{N(\mathbf{x}) : \mathbf{x} \in S\}$. Since each member of S belongs to at least one of the sets in Ω, Ω is an open covering of S. By assumption, there exists a finite subcover $\{N(\mathbf{x}_1), \ldots, N(\mathbf{x}_k)\}$. Each of these neighborhoods has an associated $M(\mathbf{x}_j)$. Write

$$M := \max\{M(\mathbf{x}_1), \ldots, M(\mathbf{x}_k)\}.$$

Now let $\mathbf{y} \in S$. By definition of subcovering, $\mathbf{y} \in N(\mathbf{x}_1) \cup \cdots \cup N(\mathbf{x}_k)$, so there exists some j with $\mathbf{y} \in N(\mathbf{x}_j)$. From the way we defined $N(\mathbf{x}_j)$, it follows that

$$|f(\mathbf{y})| < M(\mathbf{x}_j) \le M.$$

Since M is independent of y, we have proved that f is bounded. □

Theorem 5.15. *Let S be a compact subset of any normed vector space.*

(a) *S is closed and bounded.*

(b) *S is sequentially compact.*

(c) *Every continuous function mapping S to a normed space is uniformly continuous.*

Proof. (a) and (b) are immediate from Theorem 5.14, using Theorems 4.10 and 4.11.

(c) follows from (b) and Exercise 2 in Section 4.3.

The connection (c) with uniform continuity has historical significance. Compactness is one of the most important topological concepts. Its origin traces back to Heine's proof, using finite subcoverings from infinite coverings, that a continuous function on a closed interval in $\mathbf{R}$ is uniformly continuous. Emile Borel extended the idea of finite subcoverings to more general settings [Kline, Chapter 40, Section 2].

With these historical remarks in mind, we give the converse of Theorem 5.14 the following form.

Theorem 5.16. *In a normed space, every set with the extreme value property is compact. In particular:*
(The Heine–Borel Theorem) *In* $\mathbf{R}^n$, *if* S *is a closed, bounded set, then every open covering of* S *yields a finite subcovering.*

Proof. Assume that S has the extreme value property, and let Ω be an open covering of S. We will produce a finite subcovering from Ω.

By Theorem 4.10, S has to be bounded; there is M such that $\|\mathbf{y}\| < M$ for every $\mathbf{y} \in S$.

First, fix any $\mathbf{x} \in S$ and $O \in \Omega$. Look at the distance from x to the complement O^*,

$$d(\mathbf{x}, O^*) := \inf\{\|\mathbf{x} - \mathbf{z}\| : \mathbf{z} \in O^*\}.$$

(Compare Exercises 8 in Section 3.4, 9b in Section 4.3.) If $d(\mathbf{x}, O^*)$ is as big as $2M$, then all of S will be inside O. That is, $d(\mathbf{x}, O^*) \geq 2M$ means that every point within $2M$ of $\mathbf{x}$ is in O; all of S is within $2M$ of $\mathbf{x}$, because $\mathbf{y} \in S \Rightarrow \|\mathbf{y}\| < M \Rightarrow \|\mathbf{x} - \mathbf{y}\| \leq \|\mathbf{x}\| + \|\mathbf{y}\| < 2M$. That will make O by itself a finite subcover; we are done. Therefore, we may assume that $d(\mathbf{x}, O^*) < 2M$ for every $\mathbf{x}$ and O.

Now continue with just $\mathbf{x}$ fixed. Since Ω is an open cover of S, there is some $P \in \Omega$ that contains $\mathbf{x}$. The set P is open, so there is a neighborhood $N(\mathbf{x}, \delta) \subseteq P$. Therefore, every $\mathbf{y} \in P^*$ is outside $N(\mathbf{x}, \delta)$; that is,

$$\mathbf{y} \in P^* \Rightarrow \|\mathbf{x} - \mathbf{y}\| \geq \delta.$$

This shows that

$$d(\mathbf{x}, P^*) \geq \delta > 0.$$

It follows that

$$0 < \sup\{d(\mathbf{x}, O^*) \colon O \in \Omega\} \leq 2M,$$

and

$$f(\mathbf{x}) := \sup\{d(\mathbf{x}, O^*) \colon O \in \Omega\}$$

is a positive real-valued function on S.

For $\mathbf{x}$ and $\mathbf{y}$ in S and $O \in \Omega$, we have

$$d(\mathbf{x}, O^*) \leq d(\mathbf{y}, O^*) + \|\mathbf{x} - \mathbf{y}\| \qquad \text{(Exercise 5 in Section 4.3).}$$

Taking the supremum with respect to O^* on both sides, we obtain

$$f(\mathbf{x}) \leq f(\mathbf{y}) + \|\mathbf{x} - \mathbf{y}\|.$$

By symmetry,

$$f(\mathbf{y}) \leq f(\mathbf{x}) + \|\mathbf{x} - \mathbf{y}\|,$$

and we conclude that f is contractive:

$$|f(\mathbf{x}) - f(\mathbf{y})| \leq \|\mathbf{x} - \mathbf{y}\|.$$

Hence f is continuous. Since S has the extreme value property, f must achieve its infimum. Thus, f has a minimum value $2m := f(\mathbf{b})$, which by the nature of f is positive.

What we have so far is that Ω affords a certain breathing room to all members of S. Specifically, if $\mathbf{x} \in S$, then $\sup\{d(\mathbf{x}, O^*)\} \geq 2m$, so there must be in Ω an open O (necessarily containing $\mathbf{x}$) such that $d(\mathbf{x}, O^*) > m$.

Let now $\mathbf{x}_1$ be a single member of S. There is an $O_1 \in \Omega$ that has $\mathbf{x}_1$ in it and satisfies $d(\mathbf{x}_1, O_1^*) > m$. If O_1 does not cover S, then there is $\mathbf{x}_2 \in S$ outside O_1. Since $\mathbf{x}_2 \in O_1^*$, we have $d(\mathbf{x}_1, \mathbf{x}_2) > m$. Further, there must exist O_2 including $\mathbf{x}_2$ with $d(\mathbf{x}_2, O_2^*) > m$. If O_1 and O_2 do not together cover S, then there is $\mathbf{x}_3 \in S$ outside both. Since $\mathbf{x}_3 \in O_1^*$, we have $d(\mathbf{x}_3, \mathbf{x}_1) > m$; since $\mathbf{x}_3 \in O_2^*$, $d(\mathbf{x}_3, \mathbf{x}_2) > m$. If O_1, O_2, and O_3 do not together cover S, then ..., except that this process cannot go on indefinitely. If it did, then we would produce a sequence $(\mathbf{x}_i)$ with terms all further than m apart. Such a sequence would possess no Cauchy, and therefore no convergent, subsequences. Such a sequence is impossible, because S, having the extreme value property, must be sequentially compact (Theorem 4.11).

Therefore, the process eventually runs out of candidates. At some k, there are no members of S outside all of $O_1, \ldots, O_k$. At that point, $\{O_1, \ldots, O_k\}$ is a finite subcover from Ω. We have proved that S is compact. (The Heine–Borel theorem follows from the remarks below Theorem 4.10.) □

Most treatments of the completeness properties of continuous functions, like the extreme value theorem, use compactness as a starting point. It is easy to show that a compact set is sequentially compact, and that a sequentially compact set has the extreme value property. Then Theorem 5.16 closes the circle. The labor expended in the proof of Theorem 5.16 is unavoidable. Every treatment seems to require a long proof, a series of lemmas, or additional concepts to prove that extreme value property or sequential compactness implies compactness. Our choice of organization—specifically, extreme value property first—was based on a desire for a consistently bottom-up approach, with our needs for the study of calculus driving the creation of concepts and results.

Compact sets have interesting properties on their own (without reference to continuous functions). We show some in the next two theorems and in the exercises.

Theorem 5.17. *Suppose S is compact. Then a subset of S is compact iff it is closed.*

Proof. $\Rightarrow$ This is automatic: Every compact set is closed (Theorem 5.15(a)).

$\Leftarrow$ Assume that T is closed, and let Σ be an open cover for T. Then $\Sigma \cup \{T^*\}$ is an open cover for S. After all, T^* is open, and any vector from S is either in T, in which case it belongs to one of the open sets in Σ; or not in T, in which case it belongs to T^*. Since S is compact, it is possible to cover S with just a finite subcollection $U_1, \dots, U_k$, plus possibly T^*. Therefore,

$$T \subseteq S \subseteq U_1 \cup \cdots \cup U_k \cup T^*.$$

But T^* does not contribute to covering T. It must be that $T \subseteq U_1 \cup \cdots \cup U_k$.

We have found within Σ a finite subcover. Hence T is compact. □

Theorem 5.18 (The Nested-Sets Theorem). *Suppose S is a compact set in the normed space W, and (T_i) is a decreasing sequence ($T_1 \supseteq T_2 \supseteq \cdots$) of nonempty closed subsets of S. Then their intersection $T_1 \cap T_2 \cap \cdots$ is not empty.*

Proof. Suppose that (T_i) is a decreasing sequence of closed subsets of S with empty intersection. By DeMorgan's laws,

$$\emptyset = T_1 \cap T_2 \cap \cdots$$

becomes

$$W = T_1{}^* \cup T_2{}^* \cup \cdots .$$

That is, $\{T_i{}^*\}$ covers the entire normed space. In particular, $\{T_i{}^*\}$ covers S. Since S is compact, there is a finite subcover $T_1{}^*, \dots, T_k{}^*$ of S. Since (T_i) is decreasing, $(T_i{}^*)$ must be increasing, so

$$T_1{}^* \cup \cdots \cup T_k{}^* = T_k{}^*.$$

Therefore, $S \subseteq T_k{}^*$, and

$$T_k = T_k \cap S \subseteq T_k \cap T_k{}^* = \emptyset.$$

We have proved that if the intersection is empty, then so is one of the T_i. □

Example 3. The conclusion in Theorem 5.18 is by no means trivial. The hypotheses that S is compact and the T_i closed are essential.

(a) In $\mathbf{R}$, which is not compact, define

$$T_i := (-i, i)^* = (-\infty, -i] \cup [i, \infty).$$

Then each T_i is closed, each contains less of $\mathbf{R}$ than the previous one, and none is empty. But every real number is eventually left out, so $T_1 \cap T_2 \cap \cdots$ is empty.

(b) Within the compact set $S := [0, 1]$, let $T_i := (0, 1/i)$. Then clearly, T_i is nonempty and contains T_{i+1}. But $T_1 \cap T_2 \cap \cdots$ is again empty.

See also Exercise 5.

We end the section with two more results related to continuous functions.

Theorem 5.19. *The continuous image of a compact set is compact.*

Proof. Since compactness is equivalent to (extreme value property and therefore to) sequential compactness, the result follows from Theorem 4.13. Nevertheless, in the spirit of this chapter, we exhibit a topological proof.

Let $\mathbf{f}$ be continuous on the compact set S. We must prove that $\mathbf{f}(S)$ is compact. Thus, assume that Ω is an open cover for $\mathbf{f}(S)$. We need to produce a finite subcover.

Let $\mathbf{x} \in S$. Since $\mathbf{f}(\mathbf{x}) \in \mathbf{f}(S)$ and Ω covers $\mathbf{f}(S)$, there exists $O(\mathbf{x}) \in \Omega$ with $\mathbf{f}(\mathbf{x}) \in O(\mathbf{x})$. Hence $\mathbf{x} \in \mathbf{f}^{-1}(O(\mathbf{x}))$. By Theorem 4.17(b), this inverse image is the intersection with S of an open set $P(\mathbf{x})$, so that we have $\mathbf{x} \in P(\mathbf{x}) \cap S$.

This being true for arbitrary $\mathbf{x}$, we see that $\{P(\mathbf{x})\}$ is an open covering for S. Since S is compact, there is a finite subcovering $\{P(\mathbf{x}_1), \ldots, P(\mathbf{x}_k)\}$.

Look at the corresponding $O(\mathbf{x}_1), \ldots, O(\mathbf{x}_k)$. If $\mathbf{y} \in \mathbf{f}(S)$, then there is $\mathbf{x} \in S$ with $\mathbf{y} = \mathbf{f}(\mathbf{x})$. This $\mathbf{x}$ is in some $P(\mathbf{x}_j)$, so that

$$\mathbf{x} \in P(\mathbf{x}_j) \cap S = \mathbf{f}^{-1}(O(\mathbf{x}_j)),$$

forcing

$$\mathbf{y} = \mathbf{f}(\mathbf{x}) \in O(\mathbf{x}_j).$$

We have shown that $O(\mathbf{x}_1), \ldots, O(\mathbf{x}_k)$ cover $\mathbf{f}(S)$. □

Theorem 5.20. *Let* $\mathbf{f}$ *map* T *from a normed space into a complete normed space* W.

(a) *If* $\mathbf{f}$ *is uniformly continuous on* T, *then* $\mathbf{f}$ *can be extended to a continuous function on the closure* $\mathrm{cl}(T)$.

(b) *Suppose* T *is a subset of a compact set* S. *Then the converse of (a) is true.*

Proof. (a) Assume that $\mathbf{f}$ is uniformly continuous on T.

To begin, we prove that $\lim_{\mathbf{x}\to\mathbf{b}} \mathbf{f}(\mathbf{x})$ exists and is finite for every $\mathbf{b} \in \mathrm{cl}(T)$. Assume that $\mathbf{b}$ is in the closure. By definition, there are sequences from T converging to $\mathbf{b}$. Let $(\mathbf{x}_i)$ be such a sequence. Then $(\mathbf{x}_i)$ is Cauchy. By the general principle noted in the proof of Theorem 4.12, $(\mathbf{f}(\mathbf{x}_i))$ (which is defined because $\mathbf{x}_i$ is from T) is Cauchy. Since W is complete, $(\mathbf{f}(\mathbf{x}_i))$ must converge to a vector in W. This proves that the finite limit exists.

Accordingly, we may define

$$\mathbf{F}(\mathbf{y}) := \lim_{\mathbf{x}\to\mathbf{y}} \mathbf{f}(\mathbf{x}) \qquad \text{for all } \mathbf{y} \in \mathrm{cl}(T).$$

This is certainly an extension of $\mathbf{f}$: If $\mathbf{y} \in T$, then $\mathbf{F}(\mathbf{y}) := \lim_{\mathbf{x}\to\mathbf{y}} \mathbf{f}(\mathbf{x}) = \mathbf{f}(\mathbf{y})$ because $\mathbf{f}$ is continuous on T. The remaining question is whether $\mathbf{F}$ is continuous on $\text{cl}(T)$.

Choose any $\mathbf{b} \in \text{cl}(T)$ and $\varepsilon > 0$. Because $\mathbf{F}(\mathbf{b})$ is a limit of $\mathbf{f}(\mathbf{x})$, there is a neighborhood $N(\mathbf{b}, \delta_1)$ in which

$$\mathbf{x} \in T \Rightarrow \|\mathbf{f}(\mathbf{x}) - \mathbf{F}(\mathbf{b})\| < \varepsilon/2.$$

Let $\mathbf{y}$ be any member of $\text{cl}(T)$ in $N(\mathbf{b}, \delta_1)$. Some neighborhood $N(\mathbf{y}, \delta_2)$ is contained in $N(\mathbf{b}, \delta_1)$. Because $\mathbf{F}(\mathbf{y})$ is a limit of $\mathbf{f}(\mathbf{x})$, within $N(\mathbf{y}, \delta_2)$ there must be a vector $\mathbf{z} \in T$ for which

$$\|\mathbf{F}(\mathbf{y}) - \mathbf{f}(\mathbf{z})\| < \varepsilon/2.$$

Because $\mathbf{z} \in N(\mathbf{y}, \delta_2) \subseteq N(\mathbf{b}, \delta_1)$, we have

$$\|\mathbf{f}(\mathbf{z}) - \mathbf{F}(\mathbf{b})\| < \varepsilon/2.$$

Therefore,

$$\|\mathbf{F}(\mathbf{b}) - \mathbf{F}(\mathbf{y})\| < \varepsilon.$$

We have shown that a given tolerance (requirement)

$$\|\mathbf{F}(\mathbf{b}) - \mathbf{F}(\mathbf{y})\| < \varepsilon$$

can be met by confining $\mathbf{y} \in \text{cl}(T)$ to a neighborhood of $\mathbf{b}$. Thus $\mathbf{F}$ is continuous at $\mathbf{b}$.

(b) Given that $T \subseteq S$ and S is compact, assume that $\mathbf{f}$ can be extended to a function $\mathbf{F}$. That is, $\mathbf{F}$ is continuous on $\text{cl}(T)$, and for each $\mathbf{x} \in T$ we have $\mathbf{F}(\mathbf{x}) = \mathbf{f}(\mathbf{x})$.

Since S is a compact set, S is closed (Theorem 5.15(a)). Since S is a closed superset of T, S must contain $\text{cl}(T)$ (Theorem 5.11(d)). Thus, $\text{cl}(T)$ is a closed subset of the compact set S. By Theorem 5.17, $\text{cl}(T)$ is compact.

Now, $\mathbf{F}$ is a continuous function on the compact set $\text{cl}(T)$. By Theorem 5.15(c), $\mathbf{F}$ is uniformly continuous on $\text{cl}(T)$. That is, $\varepsilon > 0$ gives rise to $\delta > 0$ such that

$$\mathbf{x}, \mathbf{y} \in \text{cl}(T), \ \|\mathbf{x} - \mathbf{y}\| < \delta \Rightarrow \|\mathbf{F}(\mathbf{x}) - \mathbf{F}(\mathbf{y})\| < \varepsilon.$$

If $\mathbf{x}$ and $\mathbf{y}$ are in T itself, then they are in $\text{cl}(T)$, and $\mathbf{F}(\mathbf{x}) = \mathbf{f}(\mathbf{x})$, $\mathbf{F}(\mathbf{y}) = \mathbf{f}(\mathbf{y})$. Hence

$$\mathbf{x}, \mathbf{y} \in T, \ \|\mathbf{x} - \mathbf{y}\| < \delta \Rightarrow \|\mathbf{f}(\mathbf{x}) - \mathbf{f}(\mathbf{y})\| < \varepsilon.$$

We conclude that $\mathbf{f}$ is uniformly continuous on T. □

The setting in Theorem 5.20 is not exotic. For functions between Euclidean spaces—the case of greatest interest to us—the space to which the function maps is complete, and every bounded set is a subset of a compact set. Specifically, if T is bounded, then $\text{cl}(T)$ is closed and bounded (Exercise 12a in Section 5.4), so $\text{cl}(T)$ is a compact set containing T. What Theorem 5.20 then says is that you can extend a continuous function continuously to the boundary of T iff the function is uniformly continuous.

Exercises

1. Let $\mathbf{f}$ map the unit ball in $\mathbf{R}^n$ continuously into a normed space W.

   (a) Assume that $S \subseteq B(\mathbf{O}, 1)$ is closed. Show that $\mathbf{f}(S)$ is closed.

   (b) Suppose $O \subseteq B(\mathbf{O}, 1)$ is open. Need $\mathbf{f}(O)$ be open?

   (c) Suppose $T \subseteq W$ is compact. Show that $\mathbf{f}^{-1}(T)$ is compact. (Hint: Theorem 4.17.)

2. Assume that $\mathbf{f}$ is uniformly continuous on a set $S \subseteq \mathbf{R}^n$.

   (a) Show that if S is bounded, then $\mathbf{f}(S)$ is bounded.

   (b) If S is closed, need $\mathbf{f}(S)$ be closed?

   (c) If S is closed and bounded, need $\mathbf{f}(S)$ be closed? Need it be bounded?

   (d) If $O \subseteq S$ is open and bounded, need $\mathbf{f}(O)$ be open? Need it be bounded?

   (e) Can (a) fail if $\mathbf{f}$ is continuous but not uniformly?

3. Give another proof of the second part of Theorem 5.17 by showing that a closed subset of a compact set is sequentially compact.

4. In Theorem 5.18, suppose S is compact, (T_i) is a decreasing sequence of nonempty closed subsets of S, and the diameter (Exercise 7 of Section 4.3) of T_i approaches 0 as $i \to \infty$. Prove that $T_1 \cap T_2 \cap \cdots$ possesses exactly one vector. (In $\mathbf{R}$, it is usual to take $T_i := [a_i, b_i]$, with $b_i - a_i \to 0$. This case is called the **nested-intervals theorem**, and sometimes so is our Theorem 5.18. We avoid the name "interval" because none of what we have written here depends on connectedness.)

5. A family Ω of sets has the **finite-intersection property** if every finite subfamily from Ω has nonempty intersection, in other words if no matter how you pick $T_1, \ldots, T_k$ from Ω, the intersection $T_1 \cap \cdots \cap T_k$ has something in it.

   (a) Prove that if S is a compact set, then every family Ω of *closed* subsets of S that possesses the finite-intersection property also has nonempty intersection.

   (b) Prove that the converse of (a) is false.

6. Assume that S is a compact set in a normed space.

   (a) Let $\mathbf{x}$ be outside S. Show that it is possible to "separate $\mathbf{x}$ and S" with disjoint open sets O_1 and O_2 such that $\mathbf{x} \in O_1$ and $S \subseteq O_2$.

   (b) Suppose T is compact and disjoint from S. Show that it is possible to "separate S and T" with disjoint open sets O_1 and O_2 such that $S \subseteq O_1$ and $T \subseteq O_2$.

(c) Show that (a) and (b) may fail if S is not compact.

7. Suppose S is compact and T is closed and disjoint from S. (Contrast Exercise 6(b).) Show that there exist disjoint open sets O_1 and O_2 such that $S \subseteq O_1$ and $T \subseteq O_2$.

8. *Prove that if S and T are disjoint subsets of a normed space, with S compact and T closed, then there exists a real function f, continuous on the whole space, with $f(\mathbf{x}) = 0$ throughout S and $f(\mathbf{x}) = 1$ throughout T.

5.6 Compactness in Infinite Dimensions

In this section we end our study of continuity by separating the completeness properties of continuous functions from the closed-boundedness of the domain, which is the property on which we based our original results.

We have seen (Theorems 4.11, 5.14, and 5.16) that in every normed linear space, the extreme value property, sequential compactness, and compactness are equivalent. In the special case of $\mathbf{R}^n$, a subset has those properties iff it is closed and bounded (Theorem 5.15 and Heine–Borel). This last is always true in finite dimensions.

Theorem 5.21. *Let V be a finite-dimensional normed linear space. A subset of V is compact iff it is closed and bounded.*

Proof. $\Rightarrow$ This is automatic (Theorem 5.15).

$\Leftarrow$ Assume that $S \subseteq V$ is closed and bounded. By Theorem 4.29, S has the extreme value property. Therefore (Theorem 5.16), S is compact. □

We have seen also (Sections 1.6, 2.4) that in $C_0[0, 1]$ and $C_2[0, 1]$ there are sequences from the unit ball with no convergent subsequences. Thus, in each space the unit ball is a closed bounded set that is not sequentially compact. Consequently, in those spaces, compactness and closed-boundedness are not equivalent. We next show that this situation obtains in every space of infinite dimension.

It will be helpful first to recall the **Gram–Schmidt process**. In an inner product space of finite dimension, the process is normally applied to produce an orthonormal basis. Its basic idea is as follows. Suppose that $\mathbf{u}_1, \mathbf{u}_2, \ldots, \mathbf{u}_k$ are orthonormal vectors spanning a subspace V and $\mathbf{x}$ is not in V. The vector

$$\mathbf{proj}(\mathbf{x}, V) := (\mathbf{x} \bullet \mathbf{u}_1)\mathbf{u}_1 + \cdots + (\mathbf{x} \bullet \mathbf{u}_k)\mathbf{u}_k$$

is called the **projection of x onto** V. (See Figure 5.6.) It has at least two important properties (compare Theorems 1.2(a) and (c)): First, among the vectors of V, it is the closest to $\mathbf{x}$; second, $\mathbf{x} - \mathbf{proj}(\mathbf{x}, V)$ is orthogonal to (every vector in) V. The process takes

$$\mathbf{u}_{k+1} := \frac{(\mathbf{x} - \mathbf{proj}(\mathbf{x}, V))}{\|\mathbf{x} - \mathbf{proj}(\mathbf{x}, V)\|_2}$$

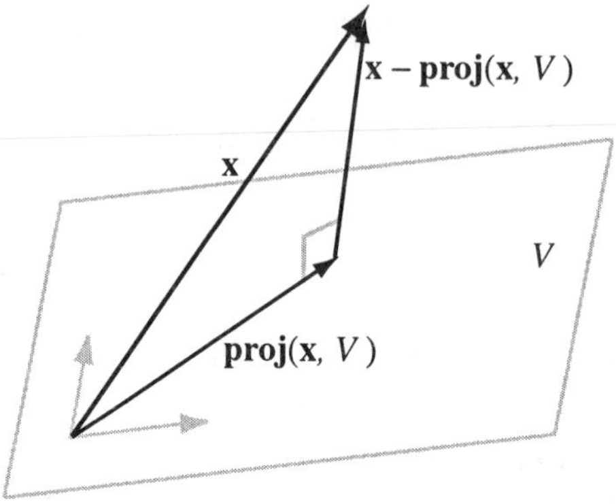

Figure 5.6.

to produce a new orthonormal basis $\{\mathbf{u}_1, \ldots, \mathbf{u}_{k+1}\}$ for the space of dimension $n+1$ that contains V and includes $\mathbf{x}$.

In an arbitrary normed space, we can mimic the Gram–Schmidt process.

Theorem 5.22. *Suppose V is a finite-dimensional subspace of a normed space W. If $\mathbf{x} \in W$ is outside V, then:*

(a) *There exists $\mathbf{v} \in V$ that is as close to $\mathbf{x}$ as a vector from V can get, that is,*

$$\|\mathbf{x} - \mathbf{v}\| = d(\mathbf{x}, V).$$

(b) *The unit vector $\mathbf{u}$ in the direction of $\mathbf{x} - \mathbf{v}$ has $d(\mathbf{u}, V) = 1$.*

Proof. (a) By definition of set distance and infimum, there is a sequence $(\mathbf{v}_i)$ from V with

$$d(\mathbf{x}, V) := \inf\{\|\mathbf{x} - \mathbf{y}\| : \mathbf{y} \in V\} = \lim_{i\to\infty} \|\mathbf{x} - \mathbf{v}_i\|.$$

The convergent real sequence $(\|\mathbf{x} - \mathbf{v}_i\|)$ must be bounded, and therefore so is

$$\|\mathbf{v}_i\| \le \|\mathbf{x}\| + \|\mathbf{v}_i - \mathbf{x}\|.$$

Thus, $(\mathbf{v}_i)$ is a bounded sequence in V. V being finite-dimensional, we get to use the Bolzano–Weierstrass theorem (Theorem 4.27(c)). We conclude that some subsequence $(\mathbf{v}_{j(i)})$ converges to a vector $\mathbf{v}$ in V. Then

$$\|\mathbf{x} - \mathbf{v}\| = \lim_{i\to\infty} \|\mathbf{x} - \mathbf{v}_{j(i)}\| = d(\mathbf{x}, V). \qquad \text{(Reasons?)}$$

(b) Let

$$\alpha := \|\mathbf{x} - \mathbf{v}\| \quad \text{and} \quad \mathbf{u} := (\mathbf{x} - \mathbf{v})/\alpha.$$

(Why is $\|\mathbf{x} - \mathbf{v}\| \ne 0$?) The origin is at distance 1 from $\mathbf{u}$, so $d(\mathbf{u}, V)$ is 1 or less.

Can it be less? If so, there would exist $\mathbf{y} \in V$ with $\|\mathbf{u}-\mathbf{y}\| < 1$. We would have

$$\|(\mathbf{x}-\mathbf{v})-\alpha\mathbf{y}\| = \|\alpha\mathbf{u}-\alpha\mathbf{y}\| < \alpha = \|\mathbf{x}-\mathbf{v}\|.$$

The last inequality says that $\mathbf{v}+\alpha\mathbf{y}$, which belongs to V, is closer to $\mathbf{x}$ than $\mathbf{v}$ is. That is not possible, because $\mathbf{v}$ is as close as a vector can get. Hence $d(\mathbf{u}, V) = 1$.
□

Notice that in Theorem 5.22, as with the Gram–Schmidt process, we end up with a vector $\mathbf{v}$ that is at minimal distance from $\mathbf{x}$. Still, there are some differences.

Example 1. (a) In any inner product space, the "closest vector" is unique; this is not so in general.

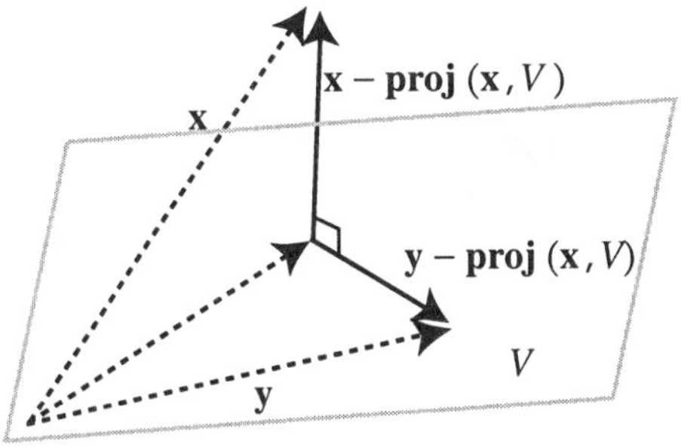

Figure 5.7.

Figure 5.7 suggests the situation in $\mathbf{R}^3$. Let us grant (Exercise 3) that if $\mathbf{x} \notin V$, then $\mathbf{x} - \mathbf{proj}(\mathbf{x}, V)$ is orthogonal to every vector lying in V. Then $\mathbf{x}$, $\mathbf{proj}(\mathbf{x}, V)$, and any $\mathbf{y} \in V$ are vertices of a right triangle. By the Pythagorean theorem,

$$\|\mathbf{x}-\mathbf{y}\|_2^2 = \|\mathbf{x}-\mathbf{proj}(\mathbf{x}, V)\|_2^2 + \|\mathbf{y}-\mathbf{proj}(\mathbf{x}, V)\|_2^2.$$

Therefore,

$$\|\mathbf{x}-\mathbf{y}\|_2 > \|\mathbf{x}-\mathbf{proj}(\mathbf{x}, V)\|_2,$$

except when $\mathbf{y} = \mathbf{proj}(\mathbf{x}, V)$; in other words, the projection is strictly closest to $\mathbf{x}$.

By contrast, consider the Cartesian plane under maxnorm:

$$\|(x, y)\|_0 := \max\{|x|, |y|\}.$$

The distance from $(1, 1)$ to the x-axis—which is the subspace spanned by $(1, 0)$—is

$$\inf\{\|(1,1)-(x,0)\|_0\} = \inf\{\max(|1-x|, 1)\}.$$

This infimum is clearly 1, but it is achieved for every $x \in [0, 2]$. Hence all the points from $(0, 0)$ to $(2, 0)$ are at the minimal distance from $(1, 1)$. (Compare Exercise 3 in Section 1.4.)

(b) In the latter example above, $\mathbf{u} := (1, 1) - (1, 0)$ is at distance 1 from $(\pm 1, 0)$. In an inner product space, the unit vector along $\mathbf{x} - \mathbf{proj}(\mathbf{x}, V)$ would be at distance $\sqrt{2}$ from the unit vectors in V.

In general, then, we do not have the neat arrangement characteristic of perpendicular unit vectors in $\mathbf{R}^n$. Nevertheless, the following information suffices.

Theorem 5.23. *In any infinite-dimensional normed space W, we may define a sequence $(\mathbf{u}_i)$ of unit vectors such that each $\mathbf{u}_i$ is at distance 1 or more from any of the others.*

Proof. We define the sequence recursively. To start, let $\mathbf{u}_1$ be any unit vector in W. Once $\mathbf{u}_1, \ldots, \mathbf{u}_k$ are defined, we know they do not span W, so there is $\mathbf{x}$ outside $V := \langle\langle \mathbf{u}_1, \ldots, \mathbf{u}_k \rangle\rangle$. Finding $\mathbf{u}_{k+1}$ as specified in Theorem 5.22(b), we have

$$d(\mathbf{u}_{k+1}, \mathbf{u}_i) \geq d(\mathbf{u}_{k+1}, V) = 1, \qquad i = 1, \ldots, k. \qquad \square$$

The existence of such a sequence in a space means that the finite-dimensional results related to the Bolzano–Weierstrass theorem fail in the space as a whole.

Theorem 5.24. *In a normed space of infinite dimension, it is always possible to find:*

(a) *bounded sequences with no convergent subsequences;*

(b) *closed, bounded sets that are not compact;*

(c) *closed, bounded sets admitting continuous functions that are unbounded, bounded continuous functions that do not reach extremes, and continuous functions that are not uniformly continuous.*

Proof. Refer to the sequence $(\mathbf{u}_i)$ from Theorem 5.23.

(a) The separation among its terms means that $(\mathbf{u}_i)$ has no Cauchy subsequence. Therefore, $(\mathbf{u}_i)$ has no convergent subsequence, even though $(\mathbf{u}_i)$ is bounded.

(b) See the paragraph following Theorem 5.21.

(c) If we define

$$f(\mathbf{u}_i) := (-1)^i i,$$

then f is continuous on $\{\mathbf{u}_1, \mathbf{u}_2, \ldots\}$, indeed, uniformly continuous (compare Example 2 in Section 4.3). But f is unbounded above and below. On the same set,

$$g(\mathbf{u}_i) := (-1)^i \left[1 + e^{-1/i}\right]$$

has supremum 2 and infimum -2, but does not take on either value.

We leave the last statement as Exercise 5. $\square$

Exercises

1. Prove that in a normed space, the unit ball is compact iff the dimension is finite.

2. (a) Consider the set P of quadratic polynomials $p(x) = bx^2 + cx + d$ that satisfy

$$-1 \leq p(x) \leq 1 \qquad \text{for all } x \in [0, 1].$$

Show that there is a positive M such that $|b| \leq M$ for every $p \in P$. (In words: For a fixed degree, bounded sets of polynomials have bounded coefficients.)

(b) Give an example of a sequence

$$p_i(x) = a_{k(i)}x^{k(i)} + \cdots + b_i x^2 + c_i x + d_i$$

of polynomials that satisfy $-1 \leq p_i(x) \leq 1$ for $x \in [0, 1]$ but for which the sequence (b_i) of quadratic coefficients is unbounded.

3. In an inner product space, suppose $\{\mathbf{u}_1, \ldots, \mathbf{u}_k\}$ is an orthonormal set and $\mathbf{x}$ is a vector. Show that $\mathbf{x} - [(\mathbf{x} \bullet \mathbf{u}_1)\mathbf{u}_1 + \cdots + (\mathbf{x} \bullet \mathbf{u}_k)\mathbf{u}_k]$ is orthogonal to every linear combination $\alpha_1\mathbf{u}_1 + \cdots + \alpha_k\mathbf{u}_k$.

4. If f is continuous on $[0, 1]$, the number

$$\alpha_j := \sqrt{2}\int_0^1 f(x)\sin 2j\pi x\, dx$$

is called a **Fourier coefficient** of f.

(a) Show that

$$f_k(x) := \alpha_1\sqrt{2}\sin 2\pi x + \cdots + \alpha_k\sqrt{2}\sin 2k\pi x$$

defines the closest linear combination of $\sin 2\pi x, \ldots, \sin 2k\pi x$ to f in $C_2[0, 1]$.

(b) Show that

$$\alpha_1^2 + \cdots + \alpha_k^2 \leq \|f\|_2^2 \qquad \text{for each } k.$$

(c) Show that the inequality in (b) is strict, unless f actually is a linear combination of $\sin 2\pi x, \ldots, \sin 2k\pi x$.

5. Find an example of a function that is continuous but not uniformly continuous on a closed, bounded set. (Hint: Show first that for the sequence $(\mathbf{u}_i)$ of Theorem 5.23, $B\left(\mathbf{u}_2, \frac{1}{2}\right) \cup B\left(\mathbf{u}_3, \frac{1}{3}\right) \cup \ldots$ is closed.)

6. Find an example of disjoint nonempty sets, S sequentially compact, T closed, such that no point of T is closest to S. (In other words, show that in Exercise 9e of Section 4.3, the finite dimension of $\mathbf{R}^n$ is essential. Contrast with Exercise 9a in that section.)

Solutions to Exercises

The solutions here are intended to be instructive. They are therefore sprinkled with comments, alternative approaches, and questions. Except for leaving such questions to be pondered by the reader, these answers are complete.

Many of them were worked out by Sera Cremonini, *summa cum laude* graduate of City College, whose knowledge of mathematics is all the more wonderful because it was not her major.

Chapter 1

Section 1.1

1. (a) The graph is a rectangular hyperbola—one whose asymptotes $y = \pm x$ are at right angles—crossing the x-axis at $(\pm 1, 0)$. See the figure for this solution.

   (b) No. $\left(2, \sqrt{3}\right)$ and $\left(2, -\sqrt{3}\right)$ both satisfy the equation, but $\sqrt{3}$ and $-\sqrt{3}$ cannot both be $g(2)$. (Note that an example is essential, and that many exist.)

2. This is familiar from the definition of function. Keep Exercise 1 in mind.

   (a) It must pass the "vertical-line test": No vertical line can have as many as two points from the set. Observe that the graph in (1) fails.

   (b) For any value of x, if (x, y_1) and (x, y_2) are both in the set, then $y_1 = y_2$.

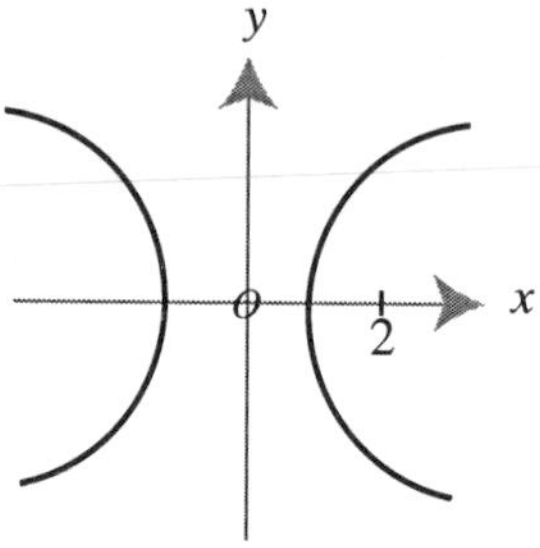

Exercise 1a

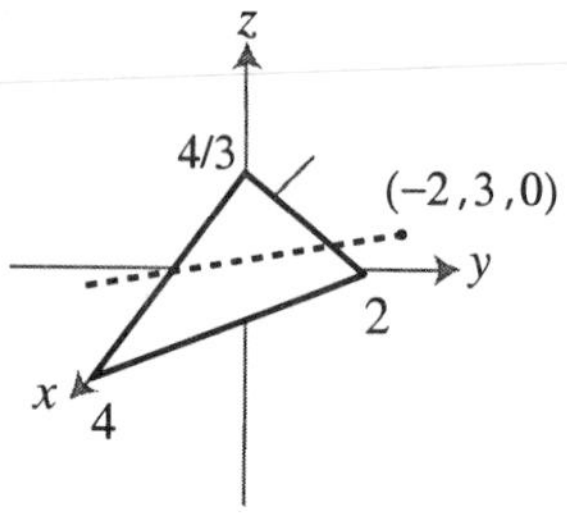

Exercises 3a, c

3. (a) Algebraic experience tells us that the graph is a plane. Without worrying too much about proof (one is suggested by Exercise 8 in Section 1.2), we can give a good idea of the graph. First, the intercepts $x = 4$, $y = 2$, and $z = \frac{4}{3}$ are evident. Second, the lines joining the intercepts are on the graph. For example, points with $x + 2y = 4$ in the $z = 0$ plane necessarily satisfy the equation. Hence the edges of the triangle in the figure are in the solution set. The plane of that triangle is the graph of the equation.

   (b) Since (a) is a plane, this is the intersection of two planes. The planes are evidently unequal, and setting $z = 0$ gives xy-plane lines of unequal slope. Therefore, the planes have nontrivial intersection; that is, they intersect in a line.

   (c) The system's augmented matrix reduces to

$$\begin{matrix} 1 & 0 & -1 & -2 \\ 0 & 1 & 2 & 3 \end{matrix}$$

   giving $z =$ arbitrary t, $y = 3 - 2t$, $x = -2 + t$. Those equations are the parametric description of a line, shown dashed in the figure.

4. If a, b, and c are all 0, then $\mathbf{R}^2$ is the solution set. If $a = b = 0$ and $c \neq 0$, then the set is empty. Anything else—if either a or b is nonzero—allows us to solve for either x or y and recognize the equation of a line. To put it differently, if a or b is nonzero, then the set is a one-parameter family that we can recognize as a Cartesian line.

5. (a) Same idea as in (4): $a = b = c = d = 0 \Rightarrow$ all solutions; $a = b = c = 0 \neq d \Rightarrow$ none; otherwise, solution has two free variables, parametrizes a plane.

   (b) If a through h are all zero, then $\mathbf{R}^3$ is the set. If either a, b, and c are zero with $d \neq 0$, or if e, f, g are zero with $h \neq 0$, then one of the equations has no solutions; the system cannot have any. Otherwise, three possibilities remain. Case 1: One equation is a multiple of the other; either they both represent the same plane, and that plane is the set, or the "one" says $0x+0y+0z=0$, and the plane of the "other" is the solution. Case 2: One of (a, b, c) and (e, f, g) is a multiple of the other, but d and h are not in the same ratio; in this case, the planes are parallel, and the system has no solution. Case 3: Neither of (a, b, c) and (e, f, g) is a multiple of the other, in which case ("one-parameter solution") the planes intersect in a line, irrespective of d and h.

Section 1.2

1. (a) Slope 4 tells us that the line is in the direction of the vector $\mathbf{c} := (1, 4)$, with $\mathbf{b} := (3, 5)$.

   (b) We calculate slope $= \frac{1}{2}$. Then $\mathbf{c}$ has to be $(2, 1)$ or some multiple. We can take $\mathbf{b} := (-2, -1)$, then check that $(4, 2) = \mathbf{b}+3\mathbf{c}$. Alternatively, set $\mathbf{b}_2 := (4, 2)$; then $(-2, -1) = \mathbf{b}_2 + (-3)\mathbf{c}$.

   (c) The slope points us along $\mathbf{c} := (1, m)$, and the intercept locates $\mathbf{b} := (0, b)$.

2. (a) Point–slope form, where the slope has to be $-\frac{4}{2}$: $y - 1 = -2(x + 2)$. Alternatively, parametrically: $x = -2 + 2t$, $y = 1 - 4t$.

   (b) $y - 1 = \left(-\frac{2}{4}\right)(x + 2)$, which goes through the origin.

3. (a) Yes and yes. The line joining $(1, 2)$ and $(3, 4)$ consists of vectors of the form $(1-\alpha)(1, 2)+\alpha(3, 4)$. Is $(5, 6)$ one such? Solve $(1-\alpha)(1, 2)+\alpha(3, 4) = (5, 6)$: $\alpha = 2$, answering the first question. Then we need not answer the second separately: Since $(5, 6) = -1(1, 2) + 2(3, 4)$, we have $(1, 2) = 2(3, 4)+(-1)(5, 6)$, putting $(1, 2)$ on the line joining the other two.

   (b) The argument is within (a): $\mathbf{c}$ is on the line joining $\mathbf{a}$ and $\mathbf{b}$ $\Leftrightarrow$ $\mathbf{c} = (1-\alpha)\mathbf{a} + \alpha\mathbf{b}$ for some α $\Leftrightarrow$ $\mathbf{a} = -\alpha/(1-\alpha)\mathbf{b} + 1/(1-\alpha)\mathbf{b}$. (How come $\alpha \neq 1$?) The last says that $\mathbf{a}$ is on the line joining $\mathbf{b}$ and $\mathbf{c}$, because the coefficients add up to 1.

4. For comparison, calculate $\mathbf{b} - \mathbf{a}$ and $\mathbf{c} - \mathbf{a}$ for the vectors in Exercises 3a, 5, and 6.

   $\Rightarrow$ Assume that $\mathbf{a}$, $\mathbf{b}$, and $\mathbf{c}$ are on the line $\mathbf{d} + \langle\langle \mathbf{e} \rangle\rangle$; that is, $\mathbf{a} = \mathbf{d} + \alpha\mathbf{e}$, $\mathbf{b} = \mathbf{d} + \beta\mathbf{e}$, $\mathbf{c} = \mathbf{d} + \gamma\mathbf{e}$. Then $\mathbf{b} - \mathbf{a}$ and $\mathbf{c} - \mathbf{a}$ are multiples of $\mathbf{e}$; they are dependent.

   $\Leftarrow$ Assume $\mathbf{b} - \mathbf{a} = \alpha\mathbf{e}$ and $\mathbf{c} - \mathbf{a} = \beta\mathbf{e}$. Then $\mathbf{a} = \mathbf{a} + 0\mathbf{e}$, $\mathbf{b} = \mathbf{a} + \alpha\mathbf{e}$, and $\mathbf{c} = \mathbf{a} + \beta\mathbf{e}$ are all on the line $\mathbf{a} + \langle\langle \mathbf{e} \rangle\rangle$.

5. Yes and no. They have to be dependent; they are three vectors in $\mathbf{R}^2$. For collinearity, Exercise 4 tells us to look at $\mathbf{b} - \mathbf{a} = (2, 2)$ and $\mathbf{c} - \mathbf{a} = (5, 3)$, which are independent.

6. Yes and yes. Turn the vectors into rows in a matrix and do row reduction: first row times -6 added to second, first row times -11 added to third, resulting second row times -2 added to third, third row is now all zeros. This means that $\mathbf{c}-11\mathbf{a}-2(\mathbf{b}-6\mathbf{a}) = 1\mathbf{a}-2\mathbf{b}+1\mathbf{c} = \mathbf{O}$, and the vectors are dependent. Rewrite the last as $\mathbf{c} = 2\mathbf{b} - 1\mathbf{a}$, and we see that $\mathbf{c}$ is on the line joining the other two, because the coefficients sum to 1.

7. Algebraically: If three or more vectors are on the line $\mathbf{d} + \langle\!\langle \mathbf{e} \rangle\!\rangle$, then they are linear combinations of $\mathbf{d}$ and $\mathbf{e}$; three combinations of two vectors are necessarily dependent. Geometrically: If vectors lie along a line, then they are on the plane (or line) determined by that line and the origin; that plane being a two-dimensional subspace, the span of the vectors has at most dimension 2; they are dependent. For the converse, see Exercise 5.

8. Say $a \neq 0$. The solutions are given by $y =$ arbitrary $= t$, $x = c/a - (b/a)t$, which can be written $(x, y) = (c/a, 0) + t(-b/a, 1)$. (Check that the right side is what the hint refers to.) The right side describes the line $(c/a, 0) + \langle\!\langle (-b/a, 1) \rangle\!\rangle$.

9. Assume $\mathbf{a} \neq \mathbf{b}$ and that they are both on the line $\mathbf{c} + \langle\!\langle \mathbf{d} \rangle\!\rangle$. Thus, $\mathbf{a} = \mathbf{c} + \alpha\mathbf{d}$, $\mathbf{b} = \mathbf{c} + \beta\mathbf{d}$. We solve: $\mathbf{c} = (\beta\mathbf{a} - \alpha\mathbf{b})/(\beta - \alpha)$, $\mathbf{d} = (\mathbf{b} - \mathbf{a})/(\beta - \alpha)$. (Why is $\beta \neq \alpha$?)

   Now, suppose $\mathbf{p} := k\mathbf{a} + l\mathbf{b}$ is on the line joining $\mathbf{a}$ and $\mathbf{b}$. We check that $\mathbf{p} = (k+l)\mathbf{c} + (k\alpha + l\beta)\mathbf{d} = \mathbf{c} + m\mathbf{d}$, which is on the line $\mathbf{c} + \langle\!\langle \mathbf{d} \rangle\!\rangle$. (Why is $k + l = 1$?) That shows that "the line joining $\mathbf{a}$ and $\mathbf{b}$" is a subset of $\mathbf{c} + \langle\!\langle \mathbf{d} \rangle\!\rangle$. Conversely, suppose $\mathbf{q} := \mathbf{c} + r\mathbf{d}$ is on the line $\mathbf{c} + \langle\!\langle \mathbf{d} \rangle\!\rangle$. We check that $\mathbf{q} = (\beta - r)/(\beta - \alpha)\mathbf{a} + (-\alpha + r)/(\beta - \alpha)\mathbf{b}$, a combination that is on the line joining $\mathbf{a}$ and $\mathbf{b}$. (Why?) Hence $\mathbf{c} + \langle\!\langle \mathbf{d} \rangle\!\rangle$ is a subset of the line joining $\mathbf{a}$ and $\mathbf{b}$. This proves equality of the lines.

Section 1.3

1. To minimize the distance between (6, 8) and $(x, x/2)$, minimize instead its square, $d^2 := (x-6)^2 + (x/2-8)^2 = 5x^2/4 - 20x + 100$. By calculus: Its derivative $5x/2 - 20$ is zero when $x = 8$, where its second derivative is positive. By algebra: $d^2 = \frac{5([x-8]^2+16)}{4}$ is least when $x = 8$.

2. (a) By definition, the angle is $\cos^{-1}(\mathbf{a} \bullet \mathbf{b}/\|\mathbf{a}\|_2\|\mathbf{b}\|_2) = \cos^{-1}\left(\left[\sqrt{3}\sqrt{3} + 1(-1) + 0\right]/\left[\sqrt{4}\sqrt{4}\right]\right) = \pi/3$.

   (b) The (*your*) sketch should show that the angle between $\mathbf{a}$ and $-\mathbf{b}$ is the supplement of the angle between $\mathbf{a}$ and $\mathbf{b}$. This is borne out by the definition: because $\mathbf{a} \bullet (-\mathbf{b}) = -(\mathbf{a} \bullet \mathbf{b})$, we have (using the values here) $\cos^{-1}\left(-\frac{1}{2}\right) = \pi - \cos^{-1}\left(\frac{1}{2}\right) = 2\pi/3$.

   (c) From $\mathbf{a} + \mathbf{b} = \left(2\sqrt{3}, 0, 0\right)$, the calculation is $\cos^{-1}(6/4\sqrt{3}) = \cos^{-1}\left(\sqrt{3}/2\right) = \pi/6$. That is predictable. Vectors $\mathbf{a}$ and $\mathbf{b}$, being equally long, are sides of a rhombus. In the sketch, $\mathbf{a}+\mathbf{b}$ is the diagonal of the rhombus. Therefore, it bisects the angle between $\mathbf{a}$ and $\mathbf{b}$.

3. This is vector-space material.

   (a) $\mathbf{x} \bullet \mathbf{O} = \mathbf{x} \bullet (\mathbf{O} + \mathbf{O})$ (vector identity) $= \mathbf{x} \bullet \mathbf{O} + \mathbf{x} \bullet \mathbf{O}$ (Theorem 1.1(b)). Subtract $\mathbf{x} \bullet \mathbf{O}$ from both sides and you get $\mathbf{x} \bullet \mathbf{O} = 0$.

   (b) Yes. $(1, 2) \bullet (-2, 1) = 0$ is an example in $\mathbf{R}^2$.

   (c) No. If $\mathbf{x}$ is orthogonal to all, then in particular, $\mathbf{x} \bullet \mathbf{x} = 0$, forcing $\mathbf{x} = \mathbf{O}$. (Reason?)

   (d) In $\mathbf{R}^1$, yes vacuously; there are no "vectors that are not . . . ." In dimension > 1, no: given $\mathbf{x} \neq \mathbf{O}$ and another vector $\mathbf{y}$ that is not a multiple of $\mathbf{x}$, it is impossible for $\mathbf{x}$ to be orthogonal to both $\mathbf{y}$ and to $\mathbf{x} + \mathbf{y}$, because then $\mathbf{x} \bullet \mathbf{x} = \mathbf{x} \bullet (\mathbf{x} + \mathbf{y}) - \mathbf{x} \bullet \mathbf{y} = 0$.

4. Angle is $-\pi \Leftrightarrow \mathbf{x} \bullet \mathbf{y}/\|\mathbf{x}\|_2\|\mathbf{y}\|_2 = -1 \Rightarrow \mathbf{x} \bullet \mathbf{y} = -\|\mathbf{x}\|_2\|\mathbf{y}\|_2 \Rightarrow$ one of $\mathbf{x}$ and $\mathbf{y}$ is a multiple of the other (Cauchy's inequality). Since they are nonzero, we write $\mathbf{x} = \alpha\mathbf{y}$. Then $\alpha(\mathbf{y} \bullet \mathbf{y}) = (\alpha\mathbf{y}) \bullet \mathbf{y} = -\|\alpha\mathbf{y}\|_2\|\mathbf{y}\|_2 = -|\alpha|\ \|\mathbf{y}\|_2\|\mathbf{y}\|_2$ implies $\alpha = -|\alpha|$, and α is negative. Conversely, $\mathbf{x} = \alpha\mathbf{y}$ with α negative gives $\mathbf{x} \bullet \mathbf{y}/\|\mathbf{x}\|_2\|\mathbf{y}\|_2 = \alpha/|\alpha| = -1$, making the angle π.

5. (a) By definition, $\|\mathbf{x} \pm \mathbf{y}\|_2^2 = (\mathbf{x} \pm \mathbf{y}) \bullet (\mathbf{x} \pm \mathbf{y})$, which by numerous uses of Theorem 1.1 match $\mathbf{x} \bullet \mathbf{x} \pm 2\mathbf{x} \bullet \mathbf{y} + \mathbf{y} \bullet \mathbf{y}$ (all respectively). If $\mathbf{x}$ and $\mathbf{y}$ are orthogonal, then we have $\|\mathbf{x} + \mathbf{y}\|_2^2 = \mathbf{x} \bullet \mathbf{x} + \mathbf{y} \bullet \mathbf{y} = \|\mathbf{x} - \mathbf{y}\|_2^2$.

   (b) In the figure, we see that the hypothesis says that $\mathbf{x}$ and $\mathbf{y}$ are the sides of a rectangle. The equation says that in a rectangle, the diagonals are equally long.

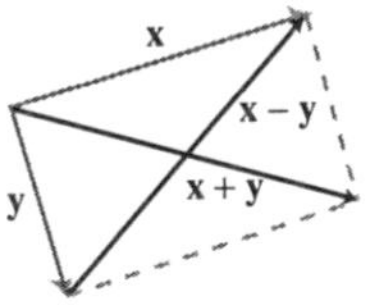

Exercise 5b

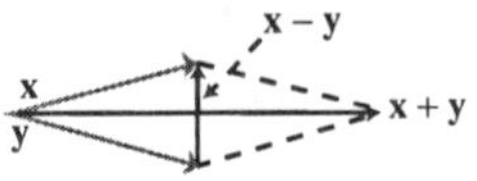

Exercise 6b

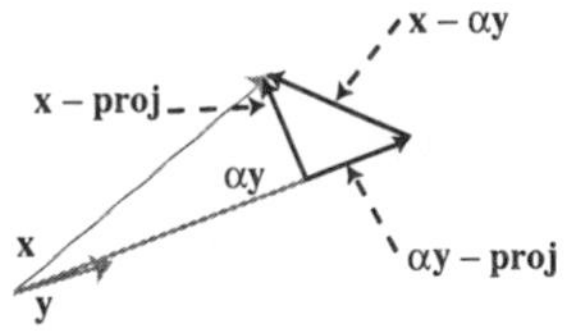

Exercise 8c

6. (a) By multiple uses of Theorem 1.1, we show that $(\mathbf{x} + \mathbf{y}) \bullet (\mathbf{x} - \mathbf{y}) = \mathbf{x} \bullet \mathbf{x} - \mathbf{y} \bullet \mathbf{y} = \|\mathbf{x}\|_2^2 - \|\mathbf{y}\|_2^2$. If those last two norms match, then $\mathbf{x} + \mathbf{y}$ and $\mathbf{x} - \mathbf{y}$ are orthogonal.

   (b) In the figure, $\|\mathbf{x}\|_2 = \|\mathbf{y}\|_2$ says that $\mathbf{x}$ and $\mathbf{y}$ are the sides of a rhombus. The equation says that in a rhombus, the diagonals are perpendicular. (Compare Exercise 2c.)

7. (a) Symmetry: $\mathbf{x} \bullet \mathbf{y} := x_1y_1 + \cdots + x_ny_n = y_1x_1 + \cdots + y_nx_n = \mathbf{y} \bullet \mathbf{x}$ by commutativity of real multiplication.

(b) Distributivity: $(\mathbf{x} + \mathbf{y}) \bullet \mathbf{z} := (x_1 + y_1)z_1 + \cdots + (x_n + y_n)z_n = (x_1 z_1 + x_n z_n) + \cdots + (y_1 z_1 + y_n z_n)$ by several uses of the real-field axioms; the last is $\mathbf{x} \bullet \mathbf{z} + \mathbf{y} \bullet \mathbf{z}$. Similarly for the other half.

(c) Homogeneity: $(\alpha\mathbf{x}) \bullet \mathbf{y} := (\alpha x_1)y_1 + \cdots + (\alpha x_n)y_n = \alpha(x_1 y_1 + \cdots + x_n y_n) = \alpha(\mathbf{x} \bullet \mathbf{y})$. The next-to-last quantity also matches $x_1(\alpha y_1) + \cdots + x_n(\alpha y_n) = \mathbf{x} \bullet (\alpha\mathbf{y})$.

(d) Positive definiteness: $\mathbf{x} \bullet \mathbf{x} := x_1 x_1 + \cdots + x_n x_n = x_1{}^2 + \cdots + x_n{}^2$. Squares being nonnegative, the last is a sum of nonnegative terms, so $\mathbf{x} \bullet \mathbf{x} \geq 0$. Moreover, if any of the squares is positive, then the sum is at least that positive. Hence if $\mathbf{x} \bullet \mathbf{x} = 0$, then all the x_i have to be zero, and $\mathbf{x} = \mathbf{O}$.

8. (a) $[\mathbf{x} - \mathbf{proj}(\mathbf{x}, \mathbf{y})] \bullet \mathbf{y} = \mathbf{x} \bullet \mathbf{y} - \mathbf{proj}(\mathbf{x}, \mathbf{y}) \bullet \mathbf{y} := \mathbf{x} \bullet \mathbf{y} - [\mathbf{x} \bullet \mathbf{y} / \mathbf{y} \bullet \mathbf{y}]\mathbf{y} \bullet \mathbf{y} = 0$ by parts (b) and (c) of Theorem 1.1.

(b) (See the answer to (5).) $\|\mathbf{x} - \mathbf{y}\|_2^2 := (\mathbf{x} - \mathbf{y}) \bullet (\mathbf{x} - \mathbf{y}) = \mathbf{x} \bullet \mathbf{x} - 2\mathbf{x} \bullet \mathbf{y} + \mathbf{y} \bullet \mathbf{y}$ (symmetry, homogeneity). If $\mathbf{x}$ is orthogonal to $\mathbf{y}$, then $\|\mathbf{x} - \mathbf{y}\|_2^2 = \mathbf{x} \bullet \mathbf{x} + \mathbf{y} \bullet \mathbf{y} = \|\mathbf{x}\|_2^2 + \|\mathbf{y}\|_2^2$.

(c) See the figure for this solution. Let $\alpha\mathbf{y}$ be a multiple of $\mathbf{y}$. Because $\mathbf{x} - \mathbf{proj}(\mathbf{x}, \mathbf{y})$ is perpendicular to $\mathbf{y}$ (part (a)), it is perpendicular to all multiples of $\mathbf{y}$, and therefore to $\alpha\mathbf{y} - \mathbf{proj}(\mathbf{x}, \mathbf{y})$. (Why is $\mathbf{x} - \mathbf{proj}(\mathbf{x}, \mathbf{y})$ perpendicular to multiples of $\mathbf{y}$? Why is $\alpha\mathbf{y} - \mathbf{proj}(\mathbf{x}, \mathbf{y})$ a multiple of $\mathbf{y}$?) By the Pythagorean theorem, $\|\mathbf{x} - \alpha\mathbf{y}\|_2^2 = \|\mathbf{x} - \mathbf{proj}(\mathbf{x}, \mathbf{y})\|_2^2 + \|\alpha\mathbf{y} - \mathbf{proj}(\mathbf{x}, \mathbf{y})\|_2^2$. Clearly, the shortest length $\|\mathbf{x} - \alpha\mathbf{y}\|_2$ occurs precisely for $\alpha\mathbf{y} = \mathbf{proj}(\mathbf{x}, \mathbf{y})$.

9. We have seen that $\|\mathbf{x} - \mathbf{y}\|_2^2 = \mathbf{x} \bullet \mathbf{x} - 2\mathbf{x} \bullet \mathbf{y} + \mathbf{y} \bullet \mathbf{y}$. On the right, $\mathbf{x} \bullet \mathbf{x}$ and $\mathbf{y} \bullet \mathbf{y}$ are $\|\mathbf{x}\|_2^2$ and $\|\mathbf{y}\|_2^2$ by definition of norm, $\mathbf{x} \bullet \mathbf{y}$ is $\|\mathbf{x}\|_2 \|\mathbf{y}\|_2 \cos\theta$ by definition of angle.

Section 1.4

1. (a) $\|\mathbf{a}\|_2 := \sqrt{(1^2 + 2^2 + 4^2)} = \sqrt{21}$, $\|\mathbf{b}\|_2 := \sqrt{((-3)^2 + 4^2 + 5^2)} = 5\sqrt{2}$, $\|\mathbf{a} + \mathbf{b}\|_2 := \sqrt{((-2)^2 + 6^2 + 9^2)} = 11$.

(b) It suffices to check that the smaller two add up to more than the third: $\sqrt{21} + 5\sqrt{2} > 4.5 + 7 > 11$.

(c) By properties of norm, $\|3\mathbf{a}\|_2 = 3\|\mathbf{a}\|_2 = 3\sqrt{21}$ and $\|3\mathbf{a} + 3\mathbf{b}\|_2 = 3\|\mathbf{a} + \mathbf{b}\|_2 = 33$.

2. (a) The calculations are direct: $\|\mathbf{c}\|_2 = \sqrt{84} = 2\sqrt{21}$, $\|\mathbf{d}\|_2 = 3\sqrt{21}$, $\|\mathbf{c} + \mathbf{d}\|_2 = \sqrt{21}$.

(b) Clearly, two of them sum to the third, so the vectors form a degenerate triangle. More specifically: $\|\mathbf{c}\|_2 + \|\mathbf{c} + \mathbf{d}\|_2 = \|\mathbf{d}\|_2 = \|-\mathbf{d}\|_2 = \|\mathbf{c} - (\mathbf{c} + \mathbf{d})\|_2$ forces (Theorem 1.6(a)) $\mathbf{c} + \mathbf{d} = -\beta\mathbf{c}$, and $\mathbf{d} = (-1 - \beta)\mathbf{c}$. Indeed, we see $\mathbf{d} = -1.5\mathbf{c}$.

(c) In view of (b), we have $\|30\mathbf{c}+40\mathbf{d}\|_2 = \|30\mathbf{c}-60\mathbf{c}\|_2 = \|-30\mathbf{c}\|_2 = 30\|\mathbf{c}\|_2 = 30\sqrt{84}$.

3. (a) (x, y) has unit maxnorm iff $|x| \leq 1$, $|y| \leq 1$, and one of them equals 1. Hence the "unit circle" is $\{(x, y)\colon x = \pm 1 \text{ and } -1 \leq y \leq 1, \text{ or } y = \pm 1 \text{ and } -1 \leq x \leq 1\}$. In our usual language, that set is the square with vertices at the four points $(\pm 1, \pm 1)$.

   (b) If (x, y) is on the "circle," then $\|(2, 0) - (x, y)\|_0 := \max\{|2 - x|, |y|\} = 2 - x$. This length is 1 iff $x = 1$; all the points on the right side of the square fit the description.

4. Clearly, $\|(1, 0) - (-1, 0)\|_0 = 2$, so this is the length we are seeking. We noted that for (x, y) on the "circle," $|x| \leq 1$ and $|y| \leq 1$. Hence $\|(1, 0) - (x, y)\|_0 := \max\{|1 - x|, |y|\}$ can be 2 iff $x = -1$. Thus the chord from $(1, 0)$ to any point on the left side of the square is of length 2; no uniqueness.

5. (a) By Theorem 1.1(c), $\mathbf{x} \bullet \mathbf{x} \geq 0$, with equality iff $\mathbf{x} = \mathbf{O}$. The same holds for $(\mathbf{x} \bullet \mathbf{x})^{1/2}$.

   (b) $\|\alpha\mathbf{x}\|_2 := (\alpha\mathbf{x} \bullet \alpha\mathbf{x})^{1/2} = (\alpha\alpha)^{1/2}(\mathbf{x} \bullet \mathbf{x})^{1/2}$ (Theorem 1.1(c)) $= |\alpha|\, \|\mathbf{x}\|_2$ (definitions).

6. The law of cosines says that $\|\mathbf{x}-\mathbf{y}\|_2^2 = \|\mathbf{x}\|_2^2 + \|\mathbf{y}\|_2^2 - 2\|\mathbf{x}\|_2\|\mathbf{y}\|_2 \cos\theta$. Since $\cos\theta \geq -1$, we have $\|\mathbf{x} - \mathbf{y}\|_2^2 \leq \|\mathbf{x}\|_2^2 + \|\mathbf{y}\|_2^2 + 2\|\mathbf{x}\|_2\|\mathbf{y}\|_2 = (\|\mathbf{x}\|_2 + \|\mathbf{y}\|_2)^2$. The inequality follows.

7. Set $\mathbf{x} := \mathbf{a}$, $\mathbf{y} := -\mathbf{b}$.

   (a) $\Rightarrow$ (b): From $\|\mathbf{a} - \mathbf{b}\|_2 \leq \|\mathbf{a}\|_2 + \|\mathbf{b}\|_2$, we get $\|\mathbf{x} + \mathbf{y}\|_2 \leq \|\mathbf{x}\|_2 + \|-\mathbf{y}\|_2 = \|\mathbf{x}\|_2 + \|\mathbf{y}\|_2$, with equality iff $\mathbf{y} := -\mathbf{b} = -(\alpha\mathbf{a}) = (-\alpha)\mathbf{x}$ for an $\alpha \leq 0$.

   (b) $\Rightarrow$ (a): From $\|\mathbf{x} + \mathbf{y}\|_2 \leq \|\mathbf{x}\|_2 + \|\mathbf{y}\|_2$, we get $\|\mathbf{a} - \mathbf{b}\|_2 \leq \|\mathbf{a}\|_2 + \|-\mathbf{b}\|_2 = \|\mathbf{a}\|_2 + \|\mathbf{b}\|_2$, with equality iff $\mathbf{a} := \mathbf{x} = \beta\mathbf{y} = (-\beta)\mathbf{b}$ with $\beta \geq 0$.

8. (a) Positive definiteness: If $\mathbf{x} = (0, \ldots, 0)$, then $\|\mathbf{x}\|_0 := \max\{0, \ldots, 0\} = 0$. If instead $\mathbf{x} \neq \mathbf{O}$, then some $x_i \neq 0$, so $|x_i| > 0$; in this case, $\|\mathbf{x}\|_0 := \max\{|x_1|, \ldots, |x_n|\} \geq |x_i| > 0$.

   (b) Radial homogeneity: $\|\alpha\mathbf{x}\|_0 := \max\{|\alpha x_1|, \ldots, |\alpha x_n|\} = \max\{|\alpha|\,|x_1|, \ldots, |\alpha|\,|x_n|\}$. Because the common factor $|\alpha|$ is nonnegative, the biggest $|x_i|$ gives the biggest $|\alpha|\,|x_i|$. Hence

$$\|\alpha\mathbf{x}\|_0 = |\alpha| \max\{|x_1|, \ldots, |x_n|\} = |\alpha|\, \|\mathbf{x}\|_0.$$

   (c) Triangle inequality: $\|\mathbf{x} + \mathbf{y}\|_0 := \max\{|x_1 + y_1|, \ldots, |x_n + y_n|\}$. For each i, we have $|x_i + y_i| \leq |x_i| + |y_i| \leq \|\mathbf{x}\|_0 + \|\mathbf{y}\|_0$. (Why?) Hence $\max\{|x_i + y_i|\} \leq \|\mathbf{x}\|_0 + \|\mathbf{y}\|_0$.

Section 1.5

1. Clearly, $\mathbf{y} - \mathbf{x} = 52(1, 1, 1)$ and $\mathbf{z} - \mathbf{x} = 16(1, 1, 1)$ are dependent, so $\mathbf{x}$, $\mathbf{y}$, and $\mathbf{z}$ are on a line (Exercise 4 in Section 1.2). In fact, the sizes make clear that $\mathbf{z}$ is between $\mathbf{x}$ and $\mathbf{y}$. Therefore, it should work out that $d(\mathbf{x}, \mathbf{z}) + d(\mathbf{z}, \mathbf{y}) = d(\mathbf{x}, \mathbf{y})$. Verify!

2. (a) It is the perpendicular bisector of their segment: the y-axis. The distances from (x, y) to $(\pm 1, 0)$ are $\|(x, y) - (\pm 1, 0)\|_2 = \left((x \pm 1)^2 + y^2\right)^{1/2}$. Setting their squares equal, we obtain $2x = -2x$, equivalent to $x = 0$.

   (b) Not just a line: the y-axis is part of it, but it also contains the vee-shaped region above $y = 1$ between the lines $y = x + 1$ and $y = -x + 1$, plus its mirror image below the x-axis.

   To see that, view only quadrant I and its edges. There, the two distances are $d_0((x, y), (\pm 1, 0)) := \max\{|x \mp 1|, |y|\} = \max\{|x \mp 1|, y\}$. Clearly, $x = 0$ makes the two distances match, $\max\{1, y\}$. For $x > 0$, $|x + 1| = x + 1$ exceeds $|x - 1| = \pm(x - 1)$; therefore, if $y \geq x + 1$, then the two maxes are both y, and if $y < x + 1$, then one max is $x + 1$ and the other is less. We conclude that in this quarter-plane, the required points satisfy $y \geq x + 1$. By symmetry, the entire set satisfies $|y| \geq |x| + 1$.

3. The three parts of Theorem 1.9 are named after three properties of norms, and those norm properties are the explanations:

   (a) $d(\mathbf{x}, \mathbf{y}) := \|\mathbf{x} - \mathbf{y}\| = d(\mathbf{y}, \mathbf{x})$ because norms are symmetric;

   (b) $d(\mathbf{x}, \mathbf{y}) := \|\mathbf{x} - \mathbf{y}\| > 0$, except when $\mathbf{x} = \mathbf{y}$, because norms are positive definite;

   (c) $d(\mathbf{x}, \mathbf{z}) := \|\mathbf{x} - \mathbf{z}\| \leq d(\mathbf{x}, \mathbf{y}) + d(\mathbf{y}, \mathbf{z})$ because norms satisfy the triangle inequality.

4. (a) $d(\alpha\mathbf{x}, \alpha\mathbf{y}) := \|\alpha\mathbf{x} - \alpha\mathbf{y}\| = |\alpha|\,\|\mathbf{x} - \mathbf{y}\|$ (radial homogeneity of norms) $= |\alpha| d(\mathbf{x}, \mathbf{y})$.

   (b) $d(\mathbf{x}, \mathbf{y}) := \|\mathbf{x} - \mathbf{y}\| = \|(\mathbf{x} + \mathbf{z}) - (\mathbf{y} + \mathbf{z})\|$ (vector algebra) $= d(\mathbf{x} + \mathbf{z}, \mathbf{y} + \mathbf{z})$.

Section 1.6

1. As x goes from 0 to 1, f decreases from 1 to -1, g increases from 1 to 2, and $f - g$ decreases from 0 to -3. Hence $\max|f| = 1$, $\max|g| = 2$, and $\max|f - g| = 3$; we have $\|f - g\|_0 = \|f\|_0 + \|g\|_0$. But $\|f - g\|_2 = \|f\|_2 + \|g\|_2$ is impossible. It would violate Theorem 1.6, because f and g are not multiples: $f(\sqrt{2}/2) = 0 \neq g(\sqrt{2}/2)$.

2. The cosines c_2 and c_1 are 2 apart: At $x = \frac{1}{2}$, $\cos 2\pi x = -1$ and $\cos 4\pi x = 1$; therefore, $2 \le \sup|c_2 - c_1| \le \sup|c_2| + \sup|c_1| = 2$, and $\|c_2 - c_1\|_0 = 2$. Cosine c_1 and sine s_1 are $\sqrt{2}$ distant: $\cos 2\pi x - \sin 2\pi x = \sqrt{2}\sin(\pi/4 - 2\pi x)$, so that $\|c_1 - s_1\|_0 := \sup|c_1 - s_1| = \sqrt{2}$.

3. (a) From trigonometry, $H = F - 2G$.

   (b) We cannot have $f = \alpha g$. Algebraic evidence: f has three zeros, but g, being quadratic, cannot have that many; in fact, $g \ge \frac{23}{16}$. Calculus evidence: If f were αg, we would have $f' = \alpha g'$, $f'' = \alpha g''$, ...; but g''' is already 0, f''' not.

4. Positive definite: $\|f\|_0 := \sup|f| \ge 0$, and $\|f\|_0 = 0 \Leftrightarrow |f(x)| = 0$ for every $x \Leftrightarrow f = \mathbf{O}$.

   Radially homogeneous: one way to proceed is to recall that $\sup|f(x)|$ is some value $|f(b)|$. Then $|\alpha f(x)| = |\alpha|\,|f(x)| \le |\alpha|\,|f(b)|$ for all x, with equality at $x = b$. Hence $\sup|\alpha f(x)| = |\alpha|\,|f(b)|$, which says that $\|\alpha f\|_0 = |\alpha|\,\|f\|_0$.

5. Both parts follow easily from properties of inner products. We have seen that $\|\mathbf{x}+\mathbf{y}\|_2^2 \pm \|\mathbf{x}-\mathbf{y}\|_2^2 := (\mathbf{x}+\mathbf{y}) \bullet (\mathbf{x}+\mathbf{y}) \pm (\mathbf{x}-\mathbf{y}) \bullet (\mathbf{x}-\mathbf{y}) = \|\mathbf{x}\|_2^2 + 2\mathbf{x} \bullet \mathbf{y} + \|\mathbf{y}\|_2^2 \pm \|\mathbf{x}\|_2^2 \mp 2\mathbf{x} \bullet \mathbf{y} \pm \|\mathbf{y}\|_2^2$. The upper signs give (a), the lower (b).

6. (a) Yes. The product of integrable functions is integrable.

   (b) Yes, yes, no. It is symmetric, because $fg = gf$. Linear, by properties of integrals. Not positive definite: $F(x) := 0$ for $0 \le x < 1$, $:= 1$ for $x = 1$, defines an integrable function that is not the zero of the space $I[0, 1]$, but has $F \bullet \bullet F = 0$.

7. (a) Induction: $f(1) = 1f(1)$; and $f(k) = kf(1)$ forces $f(k+1) = f(k) + f(1)$ (additivity) $= kf(1) + f(1) = (k+1)f(1)$.

   (b) $f(0) = f(0+0) = f(0) + f(0)$ forces $f(0) = 0f(1)$. Likewise, $f(k) + f(-k) = f(k + -k) = 0$ (just proved) forces $f(-k) = -f(k) = -(kf(1)) = -kf(1)$.

   (c) We may assume $m > 0$. For any s, $mf(s) = f(s) + \cdots + f(s) = f(s + \cdots + s) = f(ms)$. Hence $mf(k/m) = f(k) = kf(1)$. Divide by m to get $f(k/m) = k/mf(1)$.

   (d) (ε–δ argument) Assume that f is continuous at $x = b$. Choose any $\varepsilon > 0$. There is a corresponding δ such that $|x - b| < \delta$ implies $|f(x) - f(b)| < \varepsilon$.

   Now suppose $|y - 0| < \delta$. The point $y + b$ is within δ of b, so $|f(y + b) - f(b)| < \varepsilon$. On the other hand, $f(y+b) - f(b) = f(y)$. We conclude that $|f(y) - f(0)| = |f(y)| < \varepsilon$. We have proved that f is continuous at $y = 0$.

(e) (sequence argument) Assume that f is continuous at $x = b$. For any sequence (x_i) converging to b, we have $f(x_i) \to f(b)$.

Now let $y_i \to c$. We see that $f(y_i) = f(y_i + [b - c]) + f(c - b) \to f(b) + f(c - b)$, because $y_i + [b - c] \to b$. Since the sum $f(b) + f(c - b)$ matches $f(c)$, we have shown that $f(y_i) \to f(c)$, proving that f is continuous at $y = c$.

(f) Let z be irrational. By density of the rationals, there is a sequence (q_i) of rationals converging to z. By part (e), f is continuous at $x = z$. Therefore, $f(z) = \lim f(q_i) = \lim(q_i f(1))$ (by part (c)) $= f(1)\lim(q_i) = f(1)z$.

Chapter 2

Section 2.1

1. Write $\mathbf{z} := \beta\mathbf{x} + \alpha\mathbf{y}$. We discussed in Section 1.2 that $\mathbf{z}$ is on the segment from $\mathbf{x}$ to $\mathbf{y}$. The lengths are $\|\mathbf{x}-\mathbf{z}\| = \|(1-\beta)\mathbf{x}-\alpha\mathbf{y}\| = \alpha\|\mathbf{x}-\mathbf{y}\|$ (because $\alpha \geq 0$) and $\|\mathbf{z} - \mathbf{y}\| = \|\beta\mathbf{x} + (\alpha - 1)\mathbf{y}\| = \|\beta\mathbf{x} - (1 - \alpha)\mathbf{y}\| = \beta\|\mathbf{x} - \mathbf{y}\|$. The ratio is clear.

2. Write $d := \|\mathbf{x} - \mathbf{y}\|$. The hypothesis $y \in N$ says that $d < r$.

   (a) Pick a fixed $s \leq r - d$. Then $r \geq d + s$, and we are in the situation of Theorem 2.2(c). Hence $N(\mathbf{y}, s) \subseteq N$.

   (b) It suffices to show that $N \subseteq N(\mathbf{y}, 2r)$. That $z \in N$ implies $\|\mathbf{z} - \mathbf{y}\| \leq \|\mathbf{z} - \mathbf{x}\| + \|\mathbf{x} - \mathbf{y}\| < r + d < 2r$, which in turn implies $z \in N(\mathbf{y}, 2r)$. The inclusion follows.

3. (a) If $r + t$ exceeded d, then the neighborhoods would intersect (Theorem 2.2(b)), and so would the balls. If $r + t$ equaled d, then the point $t/(r+t)\mathbf{x} + r/(r+t)\mathbf{y}$ would be on both spheres (Verify!), and would therefore be common to the balls. Hence $r + t < d$.

   (b) and (c) Using the triangle inequality, we get $d := \|\mathbf{x}-\mathbf{y}\| \leq \|\mathbf{x}-\mathbf{z}\| + \|\mathbf{z} - \mathbf{w}\| + \|\mathbf{w} - \mathbf{y}\| \leq r + \|\mathbf{z} - \mathbf{w}\| + t$. Hence $\|\mathbf{z} - \mathbf{w}\| \geq d - r - t$, and the inequality is strict if either $\|\mathbf{x} - \mathbf{z}\| < r$ or $\|\mathbf{w} - \mathbf{y}\| < t$.

   These parts prove that the least $\|\mathbf{z} - \mathbf{w}\|$ *might* be is $d - r - t$, and that the only way this minimum can be achieved is if $\mathbf{z}$ and $\mathbf{w}$ are on the surfaces of the respective balls.

   (d) For any point $(1-\alpha)\mathbf{x}+\alpha\mathbf{y}$ on the line joining $\mathbf{x}$ and $\mathbf{y}$, the distance to $\mathbf{x}$ $\|\mathbf{x}-((1-\alpha)\mathbf{x}+\alpha\mathbf{y})\| = |\alpha|\,\|\mathbf{x}-\mathbf{y}\|$ is r iff $\alpha = \pm r/d$. That is, the line meets the sphere in just two points, of which $\mathbf{u} := (1 - r/d)\mathbf{x} + r/d\,\mathbf{y}$ is the one on the segment ($\alpha > 0$).

(e) Either reasoning as in (d), or observing that the point we want breaks the segment from $\mathbf{x}$ to $\mathbf{y}$ into the ratio $(d-t):t$ (compare Exercise 1), we get the same answer: $\mathbf{v} := t/d\,\mathbf{x} + (1-t/d)\mathbf{y}$.

(f) $\|\mathbf{u}-\mathbf{v}\| = \|(1-r/d)\mathbf{x}+r/d\,\mathbf{y}-t/d\,\mathbf{x}-(1-t/d)\mathbf{y}\| = \|(1-r/d-t/d)(\mathbf{x}-\mathbf{y})\| = (1-r/d-t/d)d$.

Parts (b) and (c) established the smallest conceivable distance $d-r-t$. Parts (d)–(f) show that this minimum is actually achieved along the segment, and maybe other places.

(g) Assume that $\mathbf{z}$ and $\mathbf{w}$ are on the two spheres. In (b), we saw that $d := \|\mathbf{x}-\mathbf{y}\| \le \|\mathbf{x}-\mathbf{z}\| + \|\mathbf{z}-\mathbf{w}\| + \|\mathbf{w}-\mathbf{y}\| = r + \|\mathbf{z}-\mathbf{w}\| + t$. In an inner product space, the inequality is strict, unless $\mathbf{z}$ is on the segment from $\mathbf{x}$ to $\mathbf{w}$ and $\mathbf{w}$ is on the segment from $\mathbf{z}$ to $\mathbf{y}$ (Theorem 1.10). In that case, $\mathbf{z} = (1-\alpha)\mathbf{x} + \alpha\mathbf{w}$ and $\mathbf{w} = (1-\beta)\mathbf{z} + \beta\mathbf{y}$. Solving those equations for $\mathbf{z}$ and $\mathbf{w}$ in terms of $\mathbf{x}$ and $\mathbf{y}$, we deduce that $\mathbf{z}$ and $\mathbf{w}$ are on the segment from $\mathbf{x}$ to $\mathbf{y}$.

(h) By (g), this situation guarantees that $\mathbf{z}$ and $\mathbf{w}$ are on the segment from $\mathbf{x}$ to $\mathbf{y}$. By (d) and (e), the only such points on the spheres are $\mathbf{z} = \mathbf{u}$ and $\mathbf{w} = \mathbf{v}$.

4. By Exercise 2, some neighborhood $N(\mathbf{u}, p)$ is contained in $N(\mathbf{x}, r)$ and some $N(\mathbf{u}, q) \subseteq N(\mathbf{y}, t)$. Then $N(\mathbf{u}, \min\{p, q\})$ fits within both.

Section 2.2

1. (a) Yes. Because $i\sin(1/i) = \sin(1/i)/(1/i) \to 1$ and $|\cos i/i| \le 1/i \to 0$ as $i \to \infty$, we should expect $(\mathbf{x}_i)$ to converge to $(1, 0)$. We verify: $\|\mathbf{x}_i - (1,0)\|_2^2 = (i\sin(1/i)-1)^2 + \cos^2 i/i^2 \to 0$.

(b) No. The terms are $(0, 1), (-1, 0), (0, -1), (1, 0), (0, 1), \ldots$. Clearly, $\mathbf{y}_1, \mathbf{y}_5, \mathbf{y}_9, \ldots$ converges to $(0, 1)$, $\mathbf{y}_2, \mathbf{y}_6, \mathbf{y}_{10}, \ldots$ to $(-1, 0)$. With unlike subsequences, $(\mathbf{y}_i)$ diverges (Theorem 2.4(a)).

(c) No. The terms begin $(0, 1), (-2, 0), (0, -3), \ldots$. Verify that $\|\mathbf{z}_i\|_2 = i$. The sequence is unbounded, so it cannot have a finite limit (Theorem 2.7(c)).

(d) No limit. Terms are $(0, 0), (8, 8), (0, 0), (32, 32), \ldots$; the subsequence of odd-numbered terms tends to $\mathbf{O}$, the evens to infinity.

2. Only (c). In the (c) answer, we said that $\|\mathbf{z}_i\| = i$, which does approach ∞. In the others: (a) is bounded (it converges), so cannot approach ∞; (b) is bounded because each norm is 1; (d) has a subsequence not going to ∞.

3. (a) Assume $(\mathbf{x}_i) \to \mathbf{x}$ and $(\mathbf{y}_i) \to \mathbf{y}$. Then $\|(\alpha\mathbf{x}_i + \beta\mathbf{y}_i) - (\alpha\mathbf{x} + \beta\mathbf{y})\| = \|\alpha(\mathbf{x}_i - \mathbf{x}) + \beta(\mathbf{y}_i - \mathbf{y})\| \le \alpha\|\mathbf{x}_i - \mathbf{x}\| + \beta\|\mathbf{y}_i - \mathbf{y}\| \to 0 + 0$, which says that $\alpha\mathbf{x}_i + \beta\mathbf{y}_i \to \alpha\mathbf{x} + \beta\mathbf{y}$.

(b) No. Take $(\mathbf{x}_i)$ to be $(1, 1), (2, 2), \ldots$ and $\mathbf{y}_i := (-1)^i \mathbf{x}_i$. They both converge to ∞, but $(\mathbf{x}_i + \mathbf{y}_i)$ behaves like 1(d).

4. (a) Since the difference of norms is at most the norm of the difference (Theorem 1.8), we have $| \, \|\mathbf{x}_i\| - \|\mathbf{x}\| \, | \leq \|\mathbf{x}_i - \mathbf{x}\|$. If $\mathbf{x}_i \to \mathbf{x}$, then the right side tends to 0; so must the left.

(b) $\mathbf{x}_i := (-1)^i(1, 0)$ has $\|\mathbf{x}_i\| \to \|(1, 0)\|$, but not $\mathbf{x}_i \to (1, 0)$. (Why?)

(c) $\mathbf{x}_i \to \mathbf{O} \Leftrightarrow \|\mathbf{x}_i - \mathbf{O}\| \to 0 \Leftrightarrow \|\mathbf{x}_i\| \to 0$.

5. By definition, $\mathbf{x}_i \to \mathbf{x}$ means that for any $\varepsilon > 0$, there exists $I(\varepsilon)$ such that $i \geq I(\varepsilon)$ implies $\|\mathbf{x}_i - \mathbf{x}\| < \varepsilon$. Just substitute r for ε, then note that $\|\mathbf{x}_i - \mathbf{x}\| < r \Leftrightarrow \mathbf{x}_i \in N(\mathbf{x}, r)$.

6. (a) $(\mathbf{x}_i)$ is bounded $\Leftrightarrow$ there is M with $\|\mathbf{x}_i\| < M$ for all $i \Leftrightarrow \|\mathbf{x}_i - \mathbf{O}\| < M$ for all $i \Leftrightarrow \mathbf{x}_i \in N(\mathbf{O}, M)$ for all $i \Leftrightarrow \{\mathbf{x}_i\} \subseteq N(\mathbf{O}, M)$.

(b) $\mathbf{x}_i \to \infty$ means $\|\mathbf{x}_i\| \to \infty$. The latter is defined as follows: For any $M > 0$, there exists I such that $i \geq I$ implies $\|\mathbf{x}_i\| \geq M$. Thus, the numbers $\|\mathbf{x}_i\|$ have no upper bound. Hence $(\mathbf{x}_i)$ is unbounded.

7. (a) Exercise 1b.

(b) Exercise 1d. See also the remarks in answer to Exercise 2.

Section 2.3

1. (a) Converges to $\left(\frac{1}{2}, 1, \frac{\pi}{2}\right)$; verify that the coordinate sequences converge accordingly.

(b) Diverges, because the third coordinates diverge: $\tan \pi/4 = 1$, $\tan 3\pi/4 = -1, \ldots$.

(c) Converges to infinity, because the norm exceeds the second coordinate e^i.

2. (a) $(1, 0), (2, 0), (3, 0), \ldots$. The distance between any two terms is ≥ 1. Therefore, no subsequence $(\mathbf{x}_{j(i)})$ can have $\|\mathbf{x}_{j(i)} - \mathbf{x}_{j(k)}\|_2 \to 0$; no subsequence can be Cauchy.

(b) A sequence has the property iff it converges to ∞. If $(\mathbf{x}_i) \to \infty$, then $\|\mathbf{x}_i - \mathbf{x}_j\|_2 \to \infty$ as $j \to \infty$, no matter how you choose i. Conversely, if $(\mathbf{x}_i)$ does not tend to ∞, then for some M, there is an infinity of terms with norms below M (by definition). Those constitute a bounded subsequence. By BWT, the last must have in turn a subsequence converging to a finite limit. The latter subsequence must be Cauchy.

3. (a) No. The x-coordinates are all 1, 0, or -1. A convergent subsequence cannot have an infinity of two of those, because then it would in turn have unlike subsequences. Hence, eventually the x-coordinates become constantly 1, forcing sublimit $(1, 0)$; or they become -1, forcing

sublimit $= (-1, 0)$; or they become 0, and we apply similar reasoning to deduce that y stabilizes at either 1 or -1.

(b) Similar to (a): a convergent subsequence cannot have an infinity of odd-numbered terms and an infinity of evens, so eventually it takes the road to $\mathbf{O}$ or to ∞.

4. (a, c) They converge, so all subsequences converge to the parent limit (Theorem 2.4(a)*, on p. 41). That is the only sublimit. (Compare Exercise 8a.)

(b) The coordinates look like $x = \sin s/s \to 1$ (as $s \to 0$), $y = (1 - \cos s)/s^2 \to 1/2$, and $z = \pm 1$. Hence (reasoning as in Exercise 3a) a convergent subsequence must approach $(1, 1/2, 1)$ or $(1, 1/2, -1)$.

5. Yes, under our convention that $\mathbf{R}^1$ is not the same as $\mathbf{R}$. In $\mathbf{R}^1$, $\mathbf{x}_i := [-1]^i i$ has norm (= absolute value) i. This norm tends to infinity, so $(\mathbf{x}_i) \to \infty$, the only sublimit.

6. (a) Suppose $|\pi_j(\mathbf{x}_i)| \to \infty$. Then $\|\mathbf{x}_i\|_2^2 = \pi_1(\mathbf{x}_i)^2 + \cdots + \pi_n(\mathbf{x}_i)^2 \geq \pi_j(\mathbf{x}_i)^2 \to \infty$, and $(\mathbf{x}_i) \to \infty$.

(b) $(1, 0), (0, 2), (3, 0), (0, 4), \ldots$ converges to ∞, because $\|\mathbf{x}_i\|_2 = i$; but the two coordinate sequences oscillate.

7. (a) Assume $(\mathbf{x}_i) \to \mathbf{x}$. Then for any $\varepsilon > 0$, there exists I such that $i \geq I$ implies $\|\mathbf{x}_i - \mathbf{x}\| < \varepsilon/2$. By the triangle inequality, $\|\mathbf{x}_i - \mathbf{x}_j\| = \|\mathbf{x}_i - \mathbf{x} + \mathbf{x} - \mathbf{x}_j\| \leq \|\mathbf{x}_i - \mathbf{x}\| + \|\mathbf{x} - \mathbf{x}_j\| < \varepsilon/2 + \varepsilon/2$ for every $i, j \geq I$. Hence $(\mathbf{x}_i)$ is Cauchy.

(b) Use $\varepsilon := 1$ in the definition of Cauchy. There exists a corresponding I such that $i, j \geq I \Rightarrow \|\mathbf{x}_i - \mathbf{x}_j\| < 1$. Set $M := \|\mathbf{x}_1\| + \cdots + \|\mathbf{x}_I\| + 1$. Then clearly, $\|\mathbf{x}_j\| < M$ for $j \leq I$. For $j > I$, $\|\mathbf{x}_j\| \leq \|\mathbf{x}_j - \mathbf{x}_I\| + \|\mathbf{x}_I\| < 1 + \|\mathbf{x}_I\| \leq M$. This says that $(\mathbf{x}_i)$ is bounded.

8. (a) A sublimit is the limit of some subsequence. If the parent sequence is convergent, then (Theorem 2.4(a)* on p. 41) the limit of any subsequence is the parent limit. Hence the parent limit is the only sublimit.

(b) Assume that $\mathbf{B}$ is the unique sublimit of $(\mathbf{x}_i)$.

The argument is the same whether $\mathbf{B}$ is a vector or ∞. Let N be any neighborhood of $\mathbf{B}$. There cannot be infinitely many $\mathbf{x}_i$ outside N: Such a multitude would constitute a subsequence $(\mathbf{x}_{j(i)})$; by the last paragraph of this section, $(\mathbf{x}_{j(i)})$ would in turn have a subsequence $(\mathbf{x}_{k(j(i))})$ converging to some (possibly infinite) $\mathbf{C}$; $\mathbf{C}$ could not be $\mathbf{B}$, because then the sequence $(\mathbf{x}_{k(j(i))})$ would have to enter N (Theorem 2.5*); $\mathbf{C}$ would be a second sublimit of $(\mathbf{x}_i)$. Hence only finitely many terms are outside N. Thus, starting with some index I, we have $\mathbf{x}_I$, $\mathbf{x}_{I+1}, \ldots \in N$. This being true for arbitrary N, Theorem 2.5* tells us that $\mathbf{B}$ is the limit of $(\mathbf{x}_i)$. The sequence converges. (Why does this argument require $\mathbf{R}^n$?)

9. $\Rightarrow$ Assume that $\mathbf{x}$ is a sublimit of $(\mathbf{x}_i)$ and N a neighborhood of $\mathbf{x}$. By definition of sublimit, there exists a subsequence $(\mathbf{x}_{j(i)})$ converging to $\mathbf{x}$. By Theorem 2.5, there is I such that $i \geq I \Rightarrow \mathbf{x}_{j(i)} \in N$. Thus, N contains $\mathbf{x}_{j(I)}, \mathbf{x}_{j(I+1)}, \ldots$.

   $\Leftarrow$ Assume that every neighborhood of $\mathbf{x}$ has infinitely many terms. Then $N(\mathbf{x}, 1)$ has some term $\mathbf{x}_{j(1)}$. Also, $N\left(\mathbf{x}, \frac{1}{2}\right)$ has terms $\mathbf{x}_i$ with $i > j(1)$, because only finitely many terms are numbered $j(1)$ or less; let $\mathbf{x}_{j(2)}$ be such a term. Also $N\left(\mathbf{x}, \frac{1}{3}\right)$ must have $\mathbf{x}_{j(3)}$ with $j(3) > j(2)$, etc. Thus, we recursively define an increasing sequence $j(i)$ with $\|\mathbf{x}_{j(i)} - \mathbf{x}\| < 1/i$. That is, $(\mathbf{x}_{j(i)}) \to \mathbf{x}$, and $\mathbf{x}$ is a sublimit.

10. $\Rightarrow$ Assume that $\mathbf{x}$ (or ∞) is a sublimit of $(\mathbf{x}_i)$, so that some subsequence $(\mathbf{x}_{j(i)})$ goes to that value. Let ε and I be chosen. By Theorem 2.5, there is a starting point J for which $\mathbf{x}_{j(J)}, \mathbf{x}_{j(J+1)}, \ldots$, are all in $N(\mathbf{x}, \varepsilon)$ (respectively, outside $N(\mathbf{O}, 2/\varepsilon)$). Then $i := J + I$ satisfies $i > I$ and $\|\mathbf{x}_{j(i)} - \mathbf{x}\| < \varepsilon$ (respectively, $\|\mathbf{x}_{j(i)}\| > 1/\varepsilon$).

    $\Leftarrow$ Mimic the argument in Exercise 9. For finite $\mathbf{x}$, set $\varepsilon := 1$, $I := 1$, and pick the corresponding index $j(1)$; set $\varepsilon := \frac{1}{2}$, $I := 2 + j(1)$, and pick the corresponding index $j(2)$; set $\varepsilon := \frac{1}{3}$, $I := 3 + j(2)$, and pick the corresponding index $j(3)$; etc. Then $(\mathbf{x}_{j(i)})$ is a subsequence with $\|\mathbf{x}_{j(i)} - \mathbf{x}\| < 1/i$, making $\mathbf{x}$ a sublimit.

Section 2.4

1. (a) For each graph, the upper horizontal segment is at $y = 1$, the lower at $y = 0$.

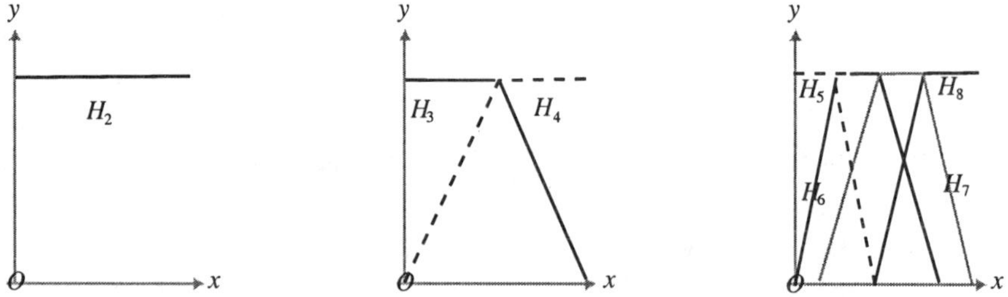

   (b) Let $2^k + 1 \leq i \leq 2^{k+1}$. We know that $H_i \geq 0$, and H_i is constantly 1 on an interval of length $1/2^k$. Hence $\int_0^1 {H_i}^2 \geq 1/2^k$. On the intervals to either side, the rise/fall of H_i guarantees that $\int {H_i}^2 < \int H_i = .5(1/2^k)$, and there are no more than two such intervals. Since $H_i = 0$ the rest of the time, we conclude that $(1 + .5 + .5)(1/2^k) > \int_0^1 H_i^2$.

   (c) Part (b) says that $\|H_i\|_2 \to 0$, so $H_i \to \mathbf{O}$. (Recall Exercise 4c in Section 2.2.)

2. (a) Yes. $0 \le H_i \le 1$, and 1 is a value of H_i, so $\|H_i\|_0 := \sup |H_i| = 1$.

   (b) No. It does not have Cauchy subsequences, because any two terms are 1 apart: If $i < j$, then $-1 \le H_i - H_j \le 1$ always; also, in part of the subinterval in which $H_i = 1$, H_j is either constantly 0 or rising from or falling to 0, so that somewhere $H_i = 1$ and $H_j = 0$; therefore, $\|H_i - H_j\|_0 = 1$.

3. Nowhere. The upper horizontals for each group (graphs number $2^k + 1$ to 2^{k+1}) cover the whole segment from $(0, 1)$ to $(1, 1)$, and the lower horizontals cover $(0, 0)$ to $(1, 0)$ more than once. Therefore, at each $x = a$, there exist an infinity of i for which $H_i(a) = 1$, along with even more j with $H_j(a) = 0$. The sequence $(H_i(a))$ diverges.

4. (a) $g_1(x) \equiv 0$ and $g_2(x) \equiv 1$. If $g(a) = 0$ and $g(b) = 1$, then (intermediate value theorem) $g = \frac{1}{2}$ someplace. Hence if g is always 0 or 1, then it cannot take on both values.

   (b) If h is continuous at $x = \frac{1}{2}$, then $\lim_{i\to\infty} h(\frac{1}{2} - \frac{1}{i}) = h(\frac{1}{2}) = \lim_{i\to\infty} h(\frac{1}{2} + \frac{1}{i})$. The values of h cannot then be 0 to the left of $x = \frac{1}{2}$ and 1 to the right.

5. (a) See the solution to Exercise 7b in Section 2.3.

   (b) Assume $(\mathbf{x}_{j(i)}) \to \mathbf{x}$, and let $\varepsilon > 0$. By convergence, there is I such that $i \ge I$ implies $\|\mathbf{x}_{j(i)} - \mathbf{x}\| < \varepsilon/2$; by Cauchyness, there is J such that $i, j \ge J$ implies $\|\mathbf{x}_i - \mathbf{x}_j\| < \varepsilon/2$. Then $i \ge I + J$, which makes $j(i)$ also $\ge I + J$, forces $\|\mathbf{x}_i - \mathbf{x}\| \le \|\mathbf{x}_i - \mathbf{x}_{j(i)}\| + \|\mathbf{x}_{j(i)} - \mathbf{x}\| < \varepsilon$. We conclude that $(\mathbf{x}_i) \to \mathbf{x}$.

   (c) Assume that $(\mathbf{x}_i)$ is Cauchy. By (a), it is bounded. By BWT, it has a subsequence converging to a vector. By (b), $(\mathbf{x}_i)$ itself must converge to that vector.

Chapter 3

Section 3.1

1. $\Rightarrow$ By definition, $\mathbf{f} := (f_1, \ldots, f_m)$ is of first degree if $f_i(\mathbf{x}) = a_{i1}x_1 + \cdots + a_{in}x_n + b_i$ for each i. Then

$$\mathbf{f}(\mathbf{x}) - \mathbf{f}(\mathbf{y}) = \begin{bmatrix} f_1(\mathbf{x}) - f_1(\mathbf{y}) \\ \cdots \\ f_m(\mathbf{x}) - f_m(\mathbf{y}) \end{bmatrix} = \begin{bmatrix} a_{11} & \cdots & a_{1n} \\ & \ddots & \\ a_{m1} & \cdots & a_{mn} \end{bmatrix} \begin{bmatrix} x_1 - y_1 \\ \cdots \\ x_n - y_n \end{bmatrix}.$$

$\Leftarrow$ If $\mathbf{f}(\mathbf{x}) - \mathbf{f}(\mathbf{y}) = \mathbf{A}[\mathbf{x} - \mathbf{y}]$ for all vectors, then $\mathbf{f}(\mathbf{x}) - \mathbf{A}\mathbf{x} = \mathbf{f}(\mathbf{y}) - \mathbf{A}\mathbf{y} =$ constant $\mathbf{b}$, and we have $\mathbf{f}(\mathbf{x}) = \mathbf{A}\mathbf{x} + \mathbf{b}$. (Compare Example 2.) This says that each row $f_i(\mathbf{x})$ is of the form $a_{i1}x_1 + \cdots + a_{in}x_n + b_i$.

2. (a) $\mathbf{g}_1(x, y) := \left(1/[x^2 + y^2], 1/[x^2 + y^2], 1/[(x-1)^2 + (y-1)^2]\right)$ is undefined at precisely (0, 0) and (1, 1).

   (b) $\mathbf{g}_2(x, y) := (x, y, 1/[y - ax - b])$ is undefined along $y = ax + b$.

3. No. You would need $\mathbf{f}(\mathbf{x}) = (p_1(\mathbf{x})/q_1(\mathbf{x}), \ldots, p_m(\mathbf{x})/q_m(\mathbf{x}))$ such that at each place in the square, one of the denominators is zero. Then $q(\mathbf{x}) := q_1(\mathbf{x}) \cdots q_m(\mathbf{x})$ would be zero throughout the square. Organize q as the hint says: $q(x, y) = r_k(x)y^k + \cdots + r_1(x)y + r_0(x)$. For a fixed $t \in [-1, 1]$, $q(t, y)$ would be a polynomial in y with infinitely many zeros. Hence each coefficient $r_j(t)$ would be zero. This being true for an infinity of t, each polynomial $r_j(x)$ would have to be identically zero. Hence $q(x, y)$ would be identically zero, and $\mathbf{f}$ would have empty domain.

4. $ae - bd \neq 0$. The hypothesis says that

$$\mathbf{h}(x, y) = \begin{pmatrix} a & b \\ d & e \end{pmatrix} \begin{pmatrix} x \\ y \end{pmatrix} + \begin{pmatrix} c \\ f \end{pmatrix}.$$

   Clearly, c and f are irrelevant; $\mathbf{h}$ is one-to-one iff the matrix is one-to-one, which occurs iff it has nonzero determinant.

5. Yes. The question becomes whether the system represented by $\mathbf{F}(x, y, z) = (13, 14, 15)$ has solutions. Reduce the augmented matrix

$$\left[\begin{array}{ccc|c} 1 & 2 & 3 & 9 \\ 5 & 6 & 7 & 6 \\ 9 & 10 & 11 & 3 \end{array}\right] \quad \text{to} \quad \left[\begin{array}{ccc|c} 1 & 2 & 3 & 9 \\ 0 & 1 & 2 & 39/4 \\ 0 & 0 & 0 & 0 \end{array}\right].$$

   The system is consistent.

6. Let $\mathbf{i} = (j, k, \ldots, p)$, $\mathbf{j} = (l, m, \ldots, q)$.

   (a) $(\alpha\mathbf{x}^{\mathbf{i}})(\beta\mathbf{x}^{\mathbf{j}}) := (\alpha {x_1}^j {x_2}^k \ldots {x_n}^p)(\beta {x_1}^l {x_2}^m \ldots {x_n}^q) = (\alpha\beta){x_1}^{j+l}{x_2}^{k+m} \ldots {x_n}^{p+q} = (\alpha\beta)\mathbf{x}^{\mathbf{i}+\mathbf{j}}$.

   (b) A polynomial is a sum of terms, and the term $\alpha {x_1}^j {x_2}^k \ldots {x_n}^p$ is $\alpha\mathbf{x}^{(j,k,\ldots,p)}$.

   (c) $g(x, y, z) = x^2y^3z^4 + 5x^6y^7z^8$ has degree 21. In general, $\alpha\mathbf{x}^{\mathbf{i}} + \cdots + \beta\mathbf{x}^{\mathbf{j}}$ has degree $\max\{j + k + \cdots + p, \ldots, l + m + \cdots + q\}$.

Section 3.2

1. (a) Yes. $f(\mathbf{x}_i) = \|\mathbf{x}_i\|_2^2$ converges to ∞ as $\mathbf{x}_i \to \infty$.

   (b) Yes, similarly. $f(\mathbf{x}_i) = \|\mathbf{x}_i\|_2^2$ converges to 0 as $\mathbf{x}_i \to \mathbf{O}$.

2. (a) Yes. If $(x_i, y_i) \to \infty$, then $\|(x_i, y_i)\|_2 > 2$ for $i \geq$ some I. Thereafter, $\max\{{x_i}^2, {y_i}^2\}$ must exceed 1, so that $g(x_i, y_i) := {x_i}^4 + {y_i}^4 \geq \max\{{x_i}^4, {y_i}^4\} > \max\{{x_i}^2, {y_i}^2\} \geq ({x_i}^2 + {y_i}^2)/2 \to \infty$.

(b) Yes. If $(x_i, y_i) \to 0$, then eventually $\|(x_i, y_i)\|_2 < 1$. Thereafter, ${x_i}^2$ and ${y_i}^2$ are both ≤ 1, so that $g(x_i, y_i) := {x_i}^4 + {y_i}^4 \le {x_i}^2 + {y_i}^2 \to 0$.

(c) Yes, because $g(x_i, y_i) \to 0^+$ (part (b)) forces $1/g(x_i, y_i) \to \infty$.

3. (a) No, because $G(i, i) = i^2 \to \infty$, while $G(i, 0) = 0 \to 0$.

(b) Yes; $0 \le |x_i y_i| \le \left({x_i}^2 + {y_i}^2\right)/2$ (property of reals) $= \|(x_i, y_i)\|_2^2/2 \to 0$.

(c) Yes, under our definition. By (b), $(x_i, y_i) \to \mathbf{O}$ implies $G(x_i, y_i) \to 0$. The last means that $|1/G(x_i, y_i)| \to \infty$, which we accept as $1/G \to \infty$, even though the actual values can go to either ∞ or $-\infty$.

(d) Yes. $\sqrt{G}$ is undefined in quadrants II and IV, but if $\mathbf{x}_i \to \mathbf{O}$ in the domain of $\sqrt{G}$, then $\sqrt{G}(\mathbf{x}_i) \to 0$ by part (b).

4. (a) At $\mathbf{O}$, limit is ∞, because $\sqrt{x_i y_i} \to 0^+$, as in Exercise 3d. At ∞, no limit; $f_1(x, y) \to 5$ along the curve $y = 1/x$, $\to \infty$ along $y = 1/x^3$.

(b) At $\mathbf{O}$, limit is 1: $x_i y_i \to 0$, so $f_2(x_i, y_i)$ acts like $\sin s/s \to 1$. At ∞, no limit: $f_2(x, y) = \sin 1$ along $y = 1/x$, but $f_2(x, y) \to 0$ along $y = x$.

(c) At $\mathbf{O}$, limit is $(0, 0)$: $\|\mathbf{f}_3(x_i, y_i)\|_2^2 = \left({x_i}^2 - {y_i}^2\right)^2 + (2x_i y_i)^2 = ({x_i}^2 + {y_i}^2)^2 \to 0$, so $\mathbf{f}_3(x_i, y_i) \to \mathbf{O}$. The same equation shows that the limit is ∞ at ∞.

(d) No limit at either place. Near $\mathbf{O}$, $\mathbf{f}_4(x, y) \to (0, 0)$ along $x = 0$, but $\|\mathbf{f}_4(x, y)\|_2^2 = 2x^2/2x^2$ along $y = x$. Near ∞, $\mathbf{f}_4(x, y) = (0, 1)$ along $y = x$, but tends to ∞ along $y = 0$.

5. Let $\mathbf{c}$ be a closure point of S. By definition, some sequence $(\mathbf{x}_i)$ from S has limit $\mathbf{c}$. Given any neighborhood N of $\mathbf{c}$, by Theorem 2.5, there exists I such that $\mathbf{x}_I, \mathbf{x}_{I+1}, \ldots$ are all in N. Thus, $\mathbf{x}_I$ is a point from S in N.

6. (a) $\Rightarrow$ Assume that $\mathbf{d}$ is an accumulation point of S. The neighborhood $N(\mathbf{d}, 1)$ has some $\mathbf{x}_1 \ne \mathbf{d}$ from S, the neighborhood $N(\mathbf{d}, 1/2)$ has some $\mathbf{x}_2 \ne \mathbf{d}$ from $S, \ldots$. We thereby define $(\mathbf{x}_i)$ from S with $\mathbf{x}_i \ne \mathbf{d}$ and $\|\mathbf{d} - \mathbf{x}_i\| < 1/i$, so that $(\mathbf{x}_i) \to \mathbf{d}$.

$\Leftarrow$ Assume $(\mathbf{x}_i) \to \mathbf{d}$ and $\mathbf{d} \ne \mathbf{x}_i \in S$. Let N be a neighborhood of $\mathbf{d}$. By Theorem 2.5, $x_I \in N$ for some I. Thus, N has elements of S different from $\mathbf{d}$. Hence $\mathbf{d}$ is an accumulation point of S.

(b) $\Rightarrow$ Assume that $\mathbf{d}$ is an accumulation point of S. The neighborhood $N\left(\mathbf{d}, \frac{1}{2}\right)$ has some $\mathbf{x}_1 \ne \mathbf{d}$ from S, the neighborhood $N(\mathbf{d}, \|\mathbf{d} - \mathbf{x}_1\|/2)$ has some $\mathbf{x}_2 \ne \mathbf{d}$ from $S, \ldots$. For this sequence, each term is less than half as far from $\mathbf{d}$ as the previous. Hence the $(\mathbf{x}_i)$ are different from each other as well as from $\mathbf{d}$, plus $(\mathbf{x}_i) \to \mathbf{d}$.

Now let N be a neighborhood of $\mathbf{d}$. As in Exercise 5, we find $\mathbf{x}_I$, $\mathbf{x}_{I+1}, \ldots$, an infinity of members of S in N.

$\Leftarrow$ Trivial; the hypothesis is stronger than the definition of accumulation point.

7. (a) Along $y = ax$, $H(x, y) = a^2x^2/x = a^2x$. Since $(x, y) \to (0, 0)$ forces $x \to 0$ (limit of coordinates), we have $H \to 0$ along the line.

   (b) Along $y = \sqrt{x}$, H is constantly 1. Hence $H \to 1$ on the curve; H has no limit at $(0, 0)$.

8. (a) No. If $x_i := (-1)^i/i$, then $x_i \in \mathbf{R}^{\#}$ and $x_i \to 0$, but $K(x_i) = (-1)^i$ does not have a limit.

   (b) Yes. For every member t of $\mathbf{R}^+$, $|t| = t$, so $K^+(t) = 1$. Thus, if $t_i \in \mathbf{R}^+$ and $t_i \to 0$, then $K^+(t_i) = 1$ has a limit.

   Understand the difference from (a) to (b): Restricting the domain eliminates some testing sequences, exposing the restricted function to fewer tests, and giving it increased chance to qualify. (See the similar comment in Example 1 of Section 3.4.)

Section 3.3

1. We need the result $\lim_{s\to 0} s\,(\ln s) = 0$ (from L'Hospital's rule, or the more elementary fact that $t/2^t \to 0$ as $t \to \infty$). It implies that for any $\varepsilon > 0$, there exists $\Delta > 0$ for which $0 < s < \Delta \Rightarrow |s(\ln s)| < \varepsilon$. By Theorems 3.7 and 3.3(b), we see that $xy \to 0$ as $(x, y) \to (0, 0)$. Hence (Theorem 3.4) given Δ, there is δ such that $0 < \|(x, y)\|_2 < \delta$ implies $|xy| < \Delta$. We have shown that ε leads to δ with $\|(x, y)\|_2 < \delta \Rightarrow |xy| < \Delta \Rightarrow |xy \ln(xy)| < \varepsilon$. By Theorem 3.4, $\lim_{(x,y)\to(0,0)} xy \ln(xy) = 0$.

2. Compare (1): $(x, y) \to (\pi, 0) \Rightarrow x \to \pi, y \to 0$ (Theorem 3.7) $\Rightarrow xy \to 0$ (limit of product), $\sin x/[x - \pi] \to -1$ and $[\cos xy - 1]/x^2y^2 \to -.5$ (L'Hospital) $\Rightarrow f(x, y) \to (-1, -.5)$.

3. Let $L := \lim_{\mathbf{x}\to\mathbf{b}} f(\mathbf{x}) < 0$. Set $\varepsilon := -L/2$. By Theorem 3.4, there is a neighborhood $N(\mathbf{b}, \delta)$ in which $\mathbf{b} \neq \mathbf{x} \in D$ implies $|f(\mathbf{x}) - L| < \varepsilon$. For such $\mathbf{x}$, $L - \varepsilon < f(\mathbf{x}) < L + \varepsilon = L/2 = -\varepsilon$.

4. (a) By various theorems, $\lim \mathbf{f}(\mathbf{x}) = (\lim 4 - [x-1]^2, \lim 5 - [y-2]^2 - [z-3]^2) = (4, 5)$; by Theorems 3.3, 3.7, and L'Hospital, $\lim g(\mathbf{y}) = 6 + \lim([u-4]\ln[u-4]) + \lim([v-5]\ln[v-5]) = 6$; but $g(\mathbf{f}(\mathbf{x})) = 6 - (x-1)^2 \ln(-[x-1]^2) + \cdots$ is not defined anywhere, because the two logarithms' arguments are ≤ 0.

   (b) g^+ matches g in the vicinity of $(4, 5)$, so irrespective of the value of g^+ at $(4, 5)$, $\lim g^+ = \lim g = 6$. The composite $g^+(\mathbf{f})$ is undefined, except at the place where $\mathbf{f} = (4, 5)$, which is $(1, 2, 3)$. Hence $(1, 2, 3)$ is an isolated point of the domain of $g^+(\mathbf{f})$, and $\lim g^+(\mathbf{f}) = g^+(\mathbf{f}(1, 2, 3)) = 7$.

(c) $\mathbf{F}$ is polynomial, so its limit is $\mathbf{F}(1, 2, 3) = (4, 5)$; $\lim G = 6$, its value in the vicinity of $(4, 5)$; $G(\mathbf{F})$ is always defined, because G and $\mathbf{F}$ are. But $G(\mathbf{F})$ has no limit: along the line $x = 1$, $\mathbf{F}$ has value $(4, 5)$, so $G(\mathbf{F}) = 8$; but along the line $x = 1+t, y = 2+t, z = 3+t$, we have $\mathbf{F} = (4 + t^3, 5 + t^3)$, $G(\mathbf{F}) = 6$ for $t \neq 0$.

(d) $\mathbf{F}^{\#}$ has a limit equal to its constant value; $G(\mathbf{F}^{\#}(\mathbf{x}))$ is always defined because both functions are; and $\lim G(\mathbf{F}^{\#}(\mathbf{x}))$ is the constant value 8.

5. Assume that $\mathbf{b}$ is an accumulation point of D; as usual, the isolated-point case is trivial.

   $\Rightarrow$ Suppose $\mathbf{f}$ has finite limit $\mathbf{L}$ at $\mathbf{b}$. Then (Theorem 3.4) for any $\varepsilon > 0$, there is $\delta > 0$ such that $\|\mathbf{x} - \mathbf{b}\|_2 < \delta \Rightarrow \|\mathbf{f}(\mathbf{x}) - \mathbf{L}\|_2 < \varepsilon/2$, $\|\mathbf{y} - \mathbf{b}\|_2 < \delta \Rightarrow \|\mathbf{f}(\mathbf{y}) - \mathbf{L}\|_2 < \varepsilon/2$ (for $\mathbf{x}, \mathbf{y} \neq \mathbf{b}$ in D). For such $\mathbf{x}$ and $\mathbf{y}$, $\|\mathbf{f}(\mathbf{x}) - \mathbf{f}(\mathbf{y})\|_2 \leq \|\mathbf{f}(\mathbf{x}) - \mathbf{L}\|_2 + \|\mathbf{L} - \mathbf{f}(\mathbf{y})\|_2 < \varepsilon$.

   $\Leftarrow$ Suppose $(\mathbf{x}_i)$ is a sequence from D converging to and not reaching $\mathbf{b}$. Let $\varepsilon > 0$. By hypothesis, there exists δ such that $\|\mathbf{f}(\mathbf{x}) - \mathbf{f}(\mathbf{y})\|_2 < \varepsilon$ for $\mathbf{x}$ and $\mathbf{y}$ from D with $0 < \|\mathbf{x} - \mathbf{b}\|_2 < \delta$, $0 < \|\mathbf{y} - \mathbf{b}\|_2 < \delta$. Because $(\mathbf{x}_i)$ must be Cauchy, there exists $I(\varepsilon)$ such that $i, j \geq I(\varepsilon)$ implies $\|\mathbf{x}_i - \mathbf{x}_j\|_2 < \delta$. Therefore, $i, j \geq I(\varepsilon) \Rightarrow \|\mathbf{f}(\mathbf{x}_i) - \mathbf{f}(\mathbf{x}_j)\|_2 < \varepsilon$. This shows that $(\mathbf{f}(\mathbf{x}_i))$ is a Cauchy sequence. Hence this value-sequence converges, proving that $\mathbf{f}$ has a limit at $\mathbf{b}$.

   (Why was it essential to assume that $\mathbf{f}$ maps into $\mathbf{R}^m$?)

6. Assume $\lim \mathbf{f}(\mathbf{x}) = \mathbf{u}$ and $\lim \mathbf{g}(\mathbf{x}) = \mathbf{v}$. If $(\mathbf{x}_i) \to \mathbf{b}$, then $(\alpha\mathbf{f}(\mathbf{x}_i) + \beta\mathbf{g}(\mathbf{x}_i))$ converges to $\alpha\mathbf{u} + \beta\mathbf{v}$ (Theorem 2.3(b)). Hence $\alpha\mathbf{f} + \beta\mathbf{g}$ has a limit, and the limit is $\alpha\mathbf{u} + \beta\mathbf{v}$.

7. See Example 2 in Section 3.4.

Section 3.4

1. If $\|\mathbf{a}\| < 1$ (or $\|\mathbf{a}\| > 1$) and $(\mathbf{x}_i) \to \mathbf{a}$, then (Theorem 2.4(b)) $\lim \|\mathbf{x}_i\| = \|\mathbf{a}\| < 1$ (respectively, > 1). Hence (property of limits) beginning with some I, $\|\mathbf{x}_i\| < (\|\mathbf{a}\| + 1)/2$ (resp. $>$). Thus, $i \geq I \Rightarrow \|\mathbf{x}_i\| < 1$ (resp. $>$) $\Rightarrow \mathbf{f}(\mathbf{x}_i) = 1$ (resp. $= 0$) $= f(\mathbf{a})$. We have shown that $(\mathbf{x}_i) \to \mathbf{a}$ implies $f(\mathbf{x}_i) \to f(\mathbf{a})$; f is continuous at $\mathbf{a}$.

   If instead $\|\mathbf{b}\| = 1$, then $\mathbf{y}_i := (1+1/i)\mathbf{b}$ converges to $\mathbf{b}$ with $\|\mathbf{y}_i\| = 1+1/i$, so $f(\mathbf{y}_i) = 0$. Hence $f(\mathbf{y}_i)$ does not approach $1 = f(\mathbf{b})$; f is discontinuous at $\mathbf{b}$.

2. From $|\,\|\mathbf{y} - \mathbf{b}\| - \|\mathbf{x} - \mathbf{b}\|\,| \leq \|\mathbf{y} - \mathbf{x}\|$ (Theorem 1.8), we see that $\|\mathbf{y} - \mathbf{b}\| \to \|\mathbf{x} - \mathbf{b}\|$ as $\mathbf{y} \to \mathbf{x}$. Thus, the limit equals the value.

3. The domain of f is $\{\mathbf{x}\colon \|\mathbf{x}\| \leq 1\} \cup \{\mathbf{O}\}$. On this domain, $\|\mathbf{x}\|$ is continuous (Exercise 2), $\|\mathbf{x}\| - 1$ is a continuous linear combination, and their product

is in the domain of the continuous real-variable function $h(s) := \sqrt{s}$. By Theorem 3.12, f is continuous. In general, if $g(\mathbf{x}) = G(\|\mathbf{x}\|)$, where $G(t)$ is continuous for $t \geq 0$, then g is continuous on its domain.

4. The key is Cauchy's inequality: $|\mathbf{y}\bullet\mathbf{b}-\mathbf{x}\bullet\mathbf{b}| = |(\mathbf{y}-\mathbf{x})\bullet\mathbf{b}| \leq \|\mathbf{y}-\mathbf{x}\|_2\|\mathbf{b}\|_2$, so that $\mathbf{y} \to \mathbf{x}$ forces $\mathbf{y} \bullet \mathbf{b} \to \mathbf{x} \bullet \mathbf{b}$. Since only properties of inner product spaces are needed, the answer to (b) is yes.

5. (a) $\Rightarrow$ (b) Because $\mathbf{b} \in D$, $\mathbf{b}$ is trivially a closure point of D, so it is legal to discuss limits at $\mathbf{b}$. Assume that $\mathbf{f}$ is continuous at $\mathbf{b}$. This means that for any sequence $(\mathbf{x}_i)$ from D converging to $\mathbf{b}$, $\mathbf{f}(\mathbf{x}_i) \to \mathbf{f}(\mathbf{b})$. By definition of function limit (Section 3.2), $\mathbf{f}$ has a limit at $\mathbf{b}$, and the limit is $\mathbf{f}(\mathbf{b})$.

   (b) $\Rightarrow$ (c) Assume $\lim_{\mathbf{x}\to\mathbf{b}} \mathbf{f}(\mathbf{x}) = \mathbf{f}(\mathbf{b})$.

   If $\mathbf{b}$ is an isolated point of D, then there is some $N(\mathbf{b}, \delta)$ in which $\mathbf{b}$ is the only point of D. Hence $\mathbf{x} \in D$ with $\|\mathbf{x} - \mathbf{b}\| < \delta \Rightarrow \mathbf{x} = \mathbf{b} \Rightarrow \|\mathbf{f}(\mathbf{x}) - \mathbf{f}(\mathbf{b})\| = 0 < \varepsilon$, no matter what ε is.

   Suppose instead that $\mathbf{b}$ is an accumulation point of D. Theorem 3.4 says that given $\varepsilon > 0$, there exists δ such that $\|\mathbf{f}(\mathbf{x}) - \mathbf{f}(\mathbf{b})\| < \varepsilon$ for all $\mathbf{x} \in D$ with $0 < \|\mathbf{x} - \mathbf{b}\| < \delta$. Clearly, we may drop the condition $0 < \|\mathbf{x} - \mathbf{b}\|$.

   (c) $\Rightarrow$ (d) Part (d) is merely a translation into the language of neighborhoods of what part (c) says in terms of inequalities.

   (d) $\Rightarrow$ (a) Assume part (d). Suppose $(\mathbf{x}_i) \to \mathbf{b}$ and $N(\mathbf{f}(\mathbf{b}), \varepsilon)$ is a neighborhood of $\mathbf{f}(\mathbf{b})$. By assumption, there is δ such that any $\mathbf{x} \in N(\mathbf{b}, \delta) \cap D$ has $\mathbf{f}(\mathbf{x}) \in N(\mathbf{f}(\mathbf{b}), \varepsilon)$. By convergence, there exists I such that $i \geq I \Rightarrow \mathbf{x}_i \in N(\mathbf{b}, \delta)$ (Theorem 2.5). Hence $i \geq I \Rightarrow \mathbf{f}(\mathbf{x}_i) \in N(\mathbf{f}(\mathbf{b}), \varepsilon)$. This shows (Theorem 2.5 again) that $\mathbf{f}(\mathbf{x}_i) \to \mathbf{f}(\mathbf{b})$; $\mathbf{f}$ is continuous at $\mathbf{b}$.

6. (a) Assume that $\mathbf{f}$ is continuous at $\mathbf{b}$. Then $\lim_{\mathbf{x}\to\mathbf{b}} \mathbf{f}(\mathbf{x}) = \mathbf{f}(\mathbf{b})$ (Theorem 3.9). By Theorem 3.5(a), there are m and δ with $\|\mathbf{f}(\mathbf{x})\| < m$ for $\mathbf{x} \neq \mathbf{b}$ in $N(\mathbf{b}, \delta)$. Then $M := m + \|\mathbf{f}(\mathbf{b})\| + 1$ has $\|\mathbf{f}(\mathbf{x})\| < M$ for every $\mathbf{x}$ in the neighborhood.

   (b) The limit of $\mathbf{f}(\mathbf{x}) - \mathbf{y}$ is $\mathbf{f}(\mathbf{b}) - \mathbf{y}$ (linear combination) $\neq \mathbf{O}$. Theorem 3.5(b) guarantees some ε_1 and neighborhood N in which $\|\mathbf{f}(\mathbf{x}) - \mathbf{y}\| > \varepsilon_1$, except maybe at $\mathbf{b}$. Then $\|\mathbf{f}(\mathbf{x}) - \mathbf{y}\| > \varepsilon := \min\{\varepsilon_1, \|\mathbf{f}(\mathbf{b}) - \mathbf{y}\|/2\}$ throughout N.

   (c) Apply Theorem 3.9(c) to $\varepsilon := f(\mathbf{b})/2$ (analogously if $f(\mathbf{b})$ is negative). Then in some neighborhood of $\mathbf{b}$,
   $|f(\mathbf{x}) - f(\mathbf{b})| < \varepsilon$, so $f(\mathbf{b})/2 = f(\mathbf{b}) - \varepsilon < f(\mathbf{x}) < f(\mathbf{b}) + \varepsilon$.

7. (a) By properties of $\mathbf{R}$, there is an integer j with $j \leq 10s < j+1$, and $j = 10s$ is ruled out by the irrationality of s. From $j/10 < s < (j+1)/10$, we see that $(j-1)/10$ and $(j \pm 2)/10$, $(j \pm 3)/10$, ... are more than 1/10 away from s. Hence one or both of $j/10$, $(j+1)/10$ is closest to s, at distance $\delta_1 := \min\{s - j/10, (j+1)/10 - s\}$.

(b) and (c) Same arguments as (a), beginning from $j \le 100s < j+1$, etc., or $j \le Js < j+1$.

8. (a) If $\mathbf{x} = (x, y)$ is on the unit circle, then $\|\mathbf{x} - (3,0)\|_2 = ([x-3]^2 + y^2)^{1/2} > 3 - x$ unless $y = 0$, and $3 - x > 2$ unless $x = 1$. Hence $\mathbf{x} := (1, 0)$ gives the minimum $\|\mathbf{x} - (3,0)\|_2 = 2$, and this is necessarily the infimum.

(b) As in (a), $\|\mathbf{x} - (3,0)\|_2 \ge 2$, but no point of the neighborhood gives equality. However, $\mathbf{x}_i := (1 - 1/i, 0)$ is in $N(\mathbf{O}, 1)$, and $\|\mathbf{x}_i - (3,0)\|_2 = 2 + 1/i$. Hence $\inf \|\mathbf{x} - (3,0)\|_2 = 2$.

(c) No and yes. First, let $\mathbf{b} := (1, 0)$. From $\|\mathbf{b} - (1 - 1/i, 0)\|_2 = 1/i$, we judge that $d(\mathbf{b}, N(\mathbf{O}, 1)) = 0$. Hence $d(\mathbf{b}, N) = 0$ does not imply $\mathbf{b} \in N$. Second, let $\mathbf{c} \in S$. Then trivially $d(\mathbf{c}, S) := \inf\{\|\mathbf{c} - \mathbf{x}\| : \mathbf{x}$ is in $S\} = \|\mathbf{c} - \mathbf{c}\| = 0$.

(d) Assume $S = \{\mathbf{x}_1, \ldots, \mathbf{x}_k\}$. Then $\inf\{\|\mathbf{b} - \mathbf{x}\| : \mathbf{x} \in S\} = \min\{\|\mathbf{b} - \mathbf{x}_1\|, \ldots, \|\mathbf{b} - \mathbf{x}_k\|\} = \|\mathbf{b} - \mathbf{x}_i\|$ for some i. This is 0 iff $\mathbf{b} = \mathbf{x}_i \in S$.

(e) In part (d), let $\mathbf{b} := \sqrt{2}$. The distance from $\mathbf{b}$ to $T := \{i/j \in \mathbf{Q}: 0 < j < 10^{100}\}$ is the same as the distance to the subset $S := \{i/j \in T: 0 < j < 10^{100}$ and $j \le i \le 2j\}$ (the ones between 1 and 2), because the rest of T is more than 0.4 away from $\mathbf{b}$. Since S is clearly finite and $\mathbf{b} \in S$, part (d) tells us that $\Delta := d(\mathbf{b}, S) > 0$. Therefore, $i/j \in T \Rightarrow |i/j - \mathbf{b}| \ge d(\mathbf{b}, S) > .9\Delta$.

Section 3.5

1. Because $\|\mathbf{G}(\mathbf{x}_i) - \mathbf{G}(\mathbf{b})\|_2^2 = (G_1(\mathbf{x}_i) - G_1(\mathbf{b}))^2 + \cdots + (G_m(\mathbf{x}_i) - G_m(\mathbf{b}))^2 \le (G_j(\mathbf{x}_i) - G_j(\mathbf{b}))^2$ for each j, we see that $\mathbf{G}(\mathbf{x}_i) \to \mathbf{G}(\mathbf{b})$ as $i \to \infty$ iff $G_j(\mathbf{x}_i) \to G_j(\mathbf{b})$ for each j. It follows that $\mathbf{G}$ is continuous iff every G_j is continuous. (Compare the argument in Theorem 3.13.)

2. We have seen that additive is subtractive: $\Phi(\mathbf{x} - \mathbf{y}) + \Phi(\mathbf{y}) = \Phi(\mathbf{x})$ implies $\Phi(\mathbf{x} - \mathbf{y}) = \Phi(\mathbf{x}) - \Phi(\mathbf{y})$. If Φ is also of bounded magnification, then $\|\Phi(\mathbf{x}) - \Phi(\mathbf{y})\| = \|\Phi(\mathbf{x} - \mathbf{y})\| \le M\|\mathbf{x} - \mathbf{y}\|$, and $\mathbf{y} \to \mathbf{x}$ forces $\Phi(\mathbf{y}) \to \Phi(\mathbf{x})$; Φ is continuous.

3. $\mathbf{K}$ is a linear combination of functions that are separately continuous: $\mathbf{K}_1(f) := -3f + 1$ is of first degree; $\mathbf{H}(f) := f^2$ was proved continuous in $C_0[0,1]$ just below Theorem 3.14; and $\mathbf{K}_2(f) := f^3$ is continuous because $\|f_i{}^3 - f^3\|_0 := \sup|f_i - f|\,|f_i{}^2 + f_i f + f^2| \le \|f_i - f\|_0(\|f_i\|_0^2 + \|f_i\|_0\|f\|_0 + \|f\|_0^2) \to 0$ as $f_i \to f$.

4. A homogeneous map has to map $\mathbf{O}$ to $\mathbf{O}$: $\mathbf{U}(2\mathbf{O}) = 2\mathbf{U}(\mathbf{O})$ forces $\mathbf{U}(\mathbf{O}) = \mathbf{O}$. If $\mathbf{U}$ has unbounded magnification, then to each i there corresponds $\mathbf{x}_i$ with $\|\mathbf{U}(\mathbf{x}_i)\| > i\|\mathbf{x}_i\|$. Then $\mathbf{y}_i := \mathbf{x}_i/(i\|\mathbf{x}_i\|)$ (Why is $\|\mathbf{x}_i\| \ne 0$?) has $\mathbf{y}_i \to \mathbf{O}$, but $\|\mathbf{U}(\mathbf{y}_i)\| = \|\mathbf{U}(\mathbf{x}_i/[i\|\mathbf{x}_i\|])\| = \|\mathbf{U}(\mathbf{x}_i)/[i\|\mathbf{x}_i\|]\| > 1$, so that $(\mathbf{U}(\mathbf{y}_i))$ does not converge to $\mathbf{U}(\mathbf{O})$.

5. In any vector space, let $\mathbf{b}$ be a fixed nonzero vector. Then $\mathbf{f}(\mathbf{x}) := \mathbf{x} + \mathbf{b}$ is of first degree, but not additive.

6. We assume that $\mathbf{L}$ is linear. (a) $\Rightarrow$ (b) is trivial.

   (b) $\Rightarrow$ (c) Assume that $\mathbf{L}$ is continuous at $\mathbf{b}$. Then $\mathbf{x}_i \to \mathbf{O} \Rightarrow \mathbf{x}_i + \mathbf{b} \to \mathbf{b} \Rightarrow \mathbf{L}(\mathbf{x}_i) + \mathbf{L}(\mathbf{b}) = \mathbf{L}(\mathbf{x}_i + \mathbf{b}) \to \mathbf{L}(\mathbf{b}) \Rightarrow \mathbf{L}(\mathbf{x}_i) \to \mathbf{O} = \mathbf{L}(\mathbf{O})$. That makes $\mathbf{L}$ continuous at $\mathbf{O}$.

   (c) $\Rightarrow$ (a) Assume that $\mathbf{L}$ is continuous at $\mathbf{O}$. For any $\mathbf{c}$, $\mathbf{x}_i \to \mathbf{c} \Rightarrow \mathbf{x}_i - \mathbf{c} \to \mathbf{O} \Rightarrow \mathbf{L}(\mathbf{x}_i) - \mathbf{L}(\mathbf{c}) = \mathbf{L}(\mathbf{x}_i - \mathbf{c}) \to \mathbf{L}(\mathbf{O})$. By Exercise 4, $\mathbf{L}(\mathbf{O}) = \mathbf{O}$. Thus, $\mathbf{x}_i \to \mathbf{c} \Rightarrow \mathbf{L}(\mathbf{x}_i) - \mathbf{L}(\mathbf{c}) \to \mathbf{O} \Rightarrow \mathbf{L}(\mathbf{x}_i) \to \mathbf{L}(\mathbf{c})$.

7. (a) Yes. In fact, T is contractive: Let $\max f(x) = f(a)$, $\max g(x) = g(b)$; then $T(f) - T(g) = f(a) - g(b) \le f(a) - g(a) \le \|f - g\|_0$ and $T(f) - T(g) \ge f(b) - g(b) \ge -|f(b) - g(b)| \ge -\|f - g\|_0$. Hence $|T(f) - T(g)| \le \|f - g\|_0$.

   (b) No. For the functions h_k discussed just above Theorem 3.15, $h_k \to \mathbf{O}$, but $T(h_k) = h_k\left(\frac{1}{2}\right) = 1$.

8. No to (c), yes to the others. $\mathbf{U}$ is clearly linear, so we look at its magnification. We have $|\mathbf{U}(f(x))| = |f(x)|\,|g_0(x)|$ for every x. Taking sups, we find that $\|\mathbf{U}(f)\|_0 \le \|g_0\|_0 \|f\|_0$, so that (a) $\mathbf{U}$ has bounded magnification on C_0. Doing integrals instead, we obtain either

$$\begin{aligned}\|\mathbf{U}(f)\|_2 &:= \left(\int_0^1 [f(x)g_0(x)]^2\,dx\right)^{1/2}\\ &\le \left(\|g_0\|_0^2 \int_0^1 f(x)^2\,dx\right)^{1/2} = \|g_0\|_0\|f\|_2,\end{aligned}$$

   meaning that (b) $\mathbf{U}$ has bounded magnification on C_2; or

$$\|\mathbf{U}(f)\|_2 \le \left(\|f\|_0^2 \int_0^1 g(x)^2\,dx\right)^{1/2} = \|g_0\|_2\|f\|_0,$$

   which says that (d) $\mathbf{U}$ is bounded as a mapping from C_0 to C_2. To get a counterexample for (c), look at the functions h_k defined above Theorem 3.15. They have $\|h_k\|_2 \leqq \sqrt{3}/2^{k/2}$ and $\|h_k\|_0 = 1$. Setting $g_0 := 1$, we obtain $\|\mathbf{U}(h_k)\|_0 \geqq 2^{k/2}/\sqrt{3}\|h_k\|_2$, and $\mathbf{U}$ is unbounded from C_2 to C_0.

Chapter 4

Section 4.1

1. (a) Yes. f is a composite of continuous functions, so it is continuous on the box. By Theorem 4.3, it attains a maximum.

   (b) No. g is not even bounded above: Along the line $y = x$, the value $g(1/i, 1/i)$ is $2\exp(i^2)/i^2$, which tends to ∞ as $i \to \infty$.

(c) Yes. h is undefined on the two axes, but its limit is 0 at each place there: Near $(a, 0)$, $|xy\sin(1/xy)| \leq |xy| \leq (|a|+1)|y| \to 0$ as $(x, y) \to (a, 0)$; and similarly near $(0, b)$. Hence h can be extended to a continuous h^+, defined throughout the box. This h^+ must have a maximum $h^+(\mathbf{b})$ in the box. Point $\mathbf{b}$ is not on the axes, because $h^+ = 0$ on the axes, whereas h^+ has such positive values as $h^+(1, 1) = \sin 1$. Therefore, $\mathbf{b}$ is one of the places where h is defined. Then clearly, $h(\mathbf{b})$ is maximal.

2. (a) The biggest $\left(x^2+y^2\right)^{1/2}$—such a maximum must occur—comes when $|x|$ and $|y|$ are maximal. The box is defined by $-3 \leq x \leq 1, 3 \leq y \leq 4$. Therefore, $|x| < 3$, except where $x = -3$, and $|y| < 4$, except where $y = 4$. The farthest place is $(-3, 4)$.

   (b) Generalizing (a) in $\mathbf{R}^2$: In $[\mathbf{a}, \mathbf{b}]$, we have $a_1 \leq x \leq b_1$. If $0 \leq a_1$, then $|x| < b_1$, except where $x = b_1$; if $b_1 \leq 0$, then $|x| < -a_1$, except where $x = -a_1$; and if $a_1 < 0 < b_1$, then $|x| < \max\{-a_1, b_1\}$, except where x is the one of a_1, b_1 farther from 0. Similarly for y. Thus, the outlying place has coordinates (a_1 or b_1, a_2 or b_2). Those name the four "corners."

   The argument extends to the 2^n corners in $\mathbf{R}^n$.

3. On $[\mathbf{O}, \infty)$ (the first octant plus its edges):

   (a) $x^2 + y^2 + z^2$ is unbounded;

   (b) $\tan^{-1}(xy)\sin z$ is between $-\pi/2$ and $\pi/2$, but cannot reach either;

   (c) x^2 is nonuniformly continuous. Set $\varepsilon := 1$. No matter what δ you name, $\mathbf{u} := (1/\delta, 0, 0)$ and $\mathbf{v} := (1/\delta + \delta/2, 0, 0)$ have $\|\mathbf{u} - \mathbf{v}\|_2 < \delta$, but $(1/\delta + \delta/2)^2 - (1/\delta)^2 = 1 + \delta^2/4 > \varepsilon$.

4. (a) For $i \geq 2$, $\mathbf{x}_i := \mathbf{a} + (\mathbf{b} - \mathbf{a})/i$ is in $(\mathbf{a}, \mathbf{b})$, but $\mathbf{x}_i \to \mathbf{a}$, which is not.

   (b) $f(\mathbf{x}) := 1/\|\mathbf{x} - \mathbf{a}\|_2$ is a continuous composite defined in $(\mathbf{a}, \mathbf{b})$ and tending to ∞ as $\mathbf{x} \to \mathbf{a}$.

   (c) $g(\mathbf{x}) := \|\mathbf{x} - \mathbf{a}\|_2$ can approach 0 but not reach it; and, because the diagonal is the longest segment in the closed box, g can get near to, but not as large as, $\|\mathbf{b} - \mathbf{a}\|_2$.

   (d) The function in (b) is nonuniformly continuous, because for the sequence $(\mathbf{x}_i)$ in (a), the distances $\|\mathbf{x}_i - \mathbf{x}_j\|$ become small, but the value differences $f(\mathbf{x}_i) - f(\mathbf{x}_j)$ get large.

5. (a) trivial.

   (b) If each $a_j \leq b_j$ and $b_j \leq a_j$, then $a_j = b_j$ for every j, and $\mathbf{a} = \mathbf{b}$.

   (c) If each $a_j \leq b_j$ and $b_j \leq c_j$, then $a_j \leq c_j$ for every j, and $\mathbf{a} \leq \mathbf{c}$.

6. "$\mathbf{a}$ is not to the left of $\mathbf{b}$" denies that every $a_j \leq b_j$. Hence it means that some a_k exceeds the corresponding b_k. Since $a_k > b_k$, no $\mathbf{x}$ can have $a_k \leq x_k \leq b_k$. This last is required for $\mathbf{a} \leq \mathbf{x} \leq \mathbf{b}$.

Section 4.2

1. (a) Closed and unbounded. Closed because distinct members of $\mathbf{Z}^2$ are $\sqrt{2}$ or more apart. Hence any Cauchy sequence from $\mathbf{Z}^2$ must eventually stabilize: $(\mathbf{x}_i) \to \mathbf{x}$ forces $\mathbf{x} = \mathbf{x}_I = \mathbf{x}_{I+1} = \cdots \in \mathbf{Z}^2$. Unbounded because $(i, i) \to \infty$; apply Theorem 4.6(b).

   (b) Not closed, because $(1/i, 1/i) \to \mathbf{O} \notin 1/\mathbf{Z}^2$. Bounded, because $\|(1/x, 1/y)\|_2 \leq \sqrt{1+1}$.

   (c) Closed, because $(x_i, 0) \to (x, y)$ forces $y = 0$ (Theorem 2.8). Unbounded, because $(i, 0) \to \infty$.

   (d) This part is deceptive, because the graph "approaches" an asymptote. It is nevertheless closed: If $(x_i, \exp(x_i)) \to (x, y)$, then by Theorem 2.8 $x_i \to x$ and $\exp(x_i) \to y$; since $f(s) := e^s$ is continuous, the last forces $y = \exp(x)$, which says that (x, y) is on the graph. The graph is unbounded, because $\left(i, e^i\right) \to \infty$.

   (e) Not closed, because $(1/i, 1) \to (0, 1)$, which is not on the graph. Unbounded, because $(i, 1) \to \infty$.

2. (a) This set $T + T$ is bounded iff T is. If T is bounded, then $|x + y| \leq |x| + |y| \leq 2 \sup |x|$, making $T + T$ bounded. If T is unbounded, then some (x_i) from T tends to ∞, and so does $(x_i + x_1)$ from $T + T$.

   (b) Same argument, with $|x - y|$ and sequence $(x_i - x_1)$.

   (c) Same argument, with xy and sequence $(x_i x_I)$, where x_I is any nonzero term in (x_i). Such terms must exist when $(x_i) \to \infty$. Note that we are unworried about the sign of the product.

   (d) New argument is essential: T/T is bounded iff T is bounded and does not have 0 as accumulation point.

   $\Rightarrow$ If either T is unbounded (some $(x_i) \to \infty$) or 0 is an accumulation point (some $(y_i) \to 0$ with $y_i \neq 0$), then $x_i/x_I \to \infty$ for appropriate I or $y_1/y_i \to \infty$.

   $\Leftarrow$ If there are ε and M such that $\varepsilon < |y| < M$ for those $y \neq 0$ in T, then $|x/y| < M/\varepsilon$ for $x/y \in T/T$.

3. $\mathbf{x}_i \in (-\infty, \mathbf{b}]$ means that $\pi_1(\mathbf{x}_i) \leq b_1, \ldots, \pi_n(\mathbf{x}_i) \leq b_n$. If $(\mathbf{x}_i) \to \mathbf{c}$, then $\pi_k(\mathbf{x}_i) \to c_k$ for each k (Theorem 2.8), forcing $c_k \leq b_k$, or $\mathbf{c} \in (-\infty, \mathbf{b}]$. Hence $(-\infty, \mathbf{b}]$ is closed. Similarly for $[\mathbf{a}, \infty)$.

4. (a) Suppose $(\mathbf{x}_i)$ is a sequence from $S := \{\mathbf{a}_1, \ldots, \mathbf{a}_k\}$. Necessarily some member, say $\mathbf{a}_1$, is repeated infinitely often. Thus some subsequence $(\mathbf{x}_{j(i)})$ has $\mathbf{x}_{j(i)} = \mathbf{a}_1$, and converges to $\mathbf{a}_1$. If now $(\mathbf{x}_i) \to \mathbf{x}$, then $\mathbf{x} = \mathbf{a}_1 \in S$ (Theorem 2.4(a)); S is closed. The converse is false, since $\mathbf{R}^n$ is closed and infinite.

   (b) If $(\mathbf{x}_i)$ comes from $B(\mathbf{b}, \delta)$ and $(\mathbf{x}_i) \to \mathbf{x}$, then $\|\mathbf{x} - \mathbf{b}\| = \lim \|\mathbf{x}_i - \mathbf{b}\|$ (Theorem 2.4(b)) $\leq \delta$, so that $\mathbf{x} \in B(\mathbf{b}, \delta)$. The ball is closed.

(c) Similar argument with $\lim \|\mathbf{x}_i - \mathbf{b}\| = \delta$.

(d) Suppose that S and T are closed. If $(\mathbf{x}_i)$ comes from $S \cap T$ and $(\mathbf{x}_i) \to \mathbf{x}$, then $\mathbf{x}_i \in S$ forces $\mathbf{x} \in S$ and $\mathbf{x}_i \in T$ forces $\mathbf{x} \in T$, so $\mathbf{x} \in S \cap T$; the intersection is closed. If instead $(\mathbf{x}_i)$ comes from $S \cup T$, then it cannot be that each of S and T contributes finitely many terms; that is, a subsequence $(\mathbf{x}_{j(i)})$ must come from just one of S and T. Since $(\mathbf{x}_{j(i)})$ must converge to $\mathbf{x}$ also, we have $\mathbf{x}$ in the same one, and $\mathbf{x} \in S \cup T$. The union is closed.

5. (a) We have seen (Exercise 2 in Section 3.4) that $f(\mathbf{x}) := \|\mathbf{x} - \mathbf{b}\|_2$ is a continuous function on S. By Theorem 4.8, it has a minimum value $f(\mathbf{c})$. Then $d(\mathbf{b}, S) := \inf\{\|\mathbf{x} - \mathbf{b}\|_2 \colon \mathbf{x} \in S\} = \|\mathbf{c} - \mathbf{b}\|_2$.

(b) Yes, there must be a maximum $f(\mathbf{d}) = \|\mathbf{d} - \mathbf{b}\|_2$.

(c) $\left(\frac{3}{5}, \frac{4}{5}\right)$.

Using vectors: Suppose $\mathbf{x}$ is on the circle. Let θ be the angle between $\mathbf{x}$ and $(3, 4)$. By the law of cosines, $\|\mathbf{x}-(3,4)\|_2^2 = \|\mathbf{x}\|_2^2 + \|(3,4)\|_2^2 - 2\|\mathbf{x}\|_2\|(3,4)\|_2 \cos\theta = 26 - 10\cos\theta$. Least value occurs when $\theta = 0$. Necessarily (Theorem 1.4(d)) $\mathbf{x} = t(3, 4)$ for a positive t, which then has to be $\frac{1}{5}$.

Using calculus instead: $d^2 := (3 - \cos u)^2 + (4 - \sin u)^2$ is minimal when $\tan u = \frac{4}{3}$, which describes the closest point $\left(\frac{3}{5}, \frac{4}{5}\right)$ and the farthest $\left(-\frac{3}{5}, -\frac{4}{5}\right)$.

(d) Your sketch will make the answer obvious; here is an analytical proof. For any $\mathbf{x}$ in the ball, let θ be the angle between $\mathbf{x} - \mathbf{a}$ and $\mathbf{b} - \mathbf{a}$. By the law of cosines (compare (c)), $\|\mathbf{x}-\mathbf{b}\|_2^2 = \|\mathbf{x}-\mathbf{a}\|_2^2 + \|\mathbf{b}-\mathbf{a}\|_2^2 - 2\|\mathbf{x}-\mathbf{a}\|_2\|\mathbf{b}-\mathbf{a}\|_2\cos\theta = (\|\mathbf{b}-\mathbf{a}\|_2 - \|\mathbf{x}-\mathbf{a}\|_2)^2 + 2\|\mathbf{x}-\mathbf{a}\|_2\|\mathbf{b}-\mathbf{a}\|_2(1-\cos\theta)$. Clearly, $\theta = 0$ is necessary for a minimum, so we need $\mathbf{x} - \mathbf{a} = t(\mathbf{b} - \mathbf{a})$, or $\mathbf{x} := \mathbf{a} + t(\mathbf{b} - \mathbf{a})$.

If $\mathbf{b}$ is outside the ball, then $\|\mathbf{b} - \mathbf{a}\|_2 > \delta \geq \|\mathbf{x} - \mathbf{a}\|_2$ for every $\mathbf{x}$, so the minimum comes with $\|\mathbf{x} - \mathbf{a}\|_2 = \delta$, forcing $t = \delta/\|\mathbf{b} - \mathbf{a}\|_2$. That describes the place $\mathbf{a} + \delta(\mathbf{b} - \mathbf{a})/\|\mathbf{b} - \mathbf{a}\|_2$ where the segment $\mathbf{ab}$ crosses the sphere.

If $\mathbf{b}$ is in the ball, then obviously we should make $\|\mathbf{x}-\mathbf{a}\|_2 = \|\mathbf{b}-\mathbf{a}\|_2$. That gives $t = 1$ and $\mathbf{x} = \mathbf{b}$, which makes sense: $\mathbf{b}$ is closest to $\mathbf{b}$.

(e) Pick a fixed $\mathbf{c} \in S$. Let $T := B(\mathbf{b}, \|\mathbf{b} - \mathbf{c}\|_2) \cap S$, which includes $\mathbf{c}$. Clearly, every point of S outside T is more than $\|\mathbf{b}-\mathbf{c}\|_2$ from $\mathbf{b}$. Hence $d(\mathbf{b}, S) := \inf\{\|\mathbf{x} - \mathbf{b}\|_2 \colon \mathbf{x} \in S\} = \inf\{\|\mathbf{y} - \mathbf{b}\|_2 \colon \mathbf{y} \in T\} = d(\mathbf{b}, T)$. Since T is closed (Exercises 4b, d) and bounded (contained in a ball), part (a) applies. Thus, there exists $\mathbf{d} \in T$ with $d(\mathbf{b}, S) = \|\mathbf{b} - \mathbf{d}\|_2$.

There need not be a farthest point: in $\mathbf{R}^2$, no point of the x-axis is farthest from $(0, 1)$.

6. $\{\mathbf{O}\}$ is closed (Exercise 4a), but its set of accumulation points is $\emptyset$.

7. (a) $\Rightarrow$ (b) Suppose every $\mathbf{x} \in S$ satisfies $\|\mathbf{x}\| \le M$. Then clearly, every term of $(\mathbf{x}_i)$ from S satisfies the same inequality.

   (b) $\Rightarrow$ (c) Assume that no neighborhood of the origin contains S. Then for each i, there exists $\mathbf{x}_i \in S$ outside of $N(\mathbf{O}, i)$. Since $\|\mathbf{x}_i\| > i$, the sequence $(\mathbf{x}_i)$ is unbounded.

   (c) $\Rightarrow$ (d) trivial.

   (d) $\Rightarrow$ (a) Assume $S \subseteq N(\mathbf{b}, r)$. Since $N(\mathbf{b}, r) \subseteq N(\mathbf{O}, \|\mathbf{b}\| + r)$ (Theorem 2.2(c)), we have $S \subseteq N(\mathbf{O}, \|\mathbf{b}\| + r)$. Consequently, every $\mathbf{x} \in S$ satisfies $\|\mathbf{x}\| < \|\mathbf{b}\| + r$.

   (a) $\Rightarrow$ (e) Assume $\|\mathbf{x}\|_2^2 := x_1^2 + \cdots + x_n^2 \le M^2$ for all $\mathbf{x} \in S$. Then $-M \le x_k \le M$ for each k, and $\mathbf{x} \in [(-M, \ldots, -M), (M, \ldots, M)]$. Hence S is a subset of a box.

   (e) $\Rightarrow$ (f) Let $S \subseteq [\mathbf{a}, \mathbf{b}]$. Then $a_k \le x_k = \pi_k(\mathbf{x}) \le b_k$ for each k and every $\mathbf{x} \in S$. Hence $\pi_k(S) := \{\pi_k(\mathbf{x})\}$ is bounded.

   (f) $\Rightarrow$ (a) Assume that $\pi_k(S)$ is bounded, say $|\pi_k(\mathbf{x})| \le M_k$ for all $\mathbf{x} \in S$. Then $\mathbf{x} \in S \Rightarrow x_1^2 + \cdots + x_n^2 \le M_1^2 + \cdots + M_n^2$, and S is bounded.

8. part (b) Let $M := \sup\{f(\mathbf{x}) : \mathbf{x} \in S\}$, which is finite by part (a). By definition of sup, there is a sequence $(f(\mathbf{x}_i)) \to M$. Because S is bounded, BWT applies, and there is a subsequence $(\mathbf{x}_{j(i)})$ converging to some $\mathbf{b}$. S is closed, so $\mathbf{b} \in S$. Since $(f(\mathbf{x}_{j(i)}))$ must converge to the parent limit, we have $M = \lim f(\mathbf{x}_{j(i)})$, which is $f(\mathbf{b})$ by continuity. Hence the sup of f is a value $f(\mathbf{b})$, making $f(\mathbf{b})$ the maximum. Similarly for min.

   (c) Suppose there were an $\varepsilon > 0$ for which no closeness δ guaranteed function differences less than ε. Thus, for each i, there would be $\mathbf{x}_i$ and $\mathbf{y}_i$ with $\|\mathbf{x}_i - \mathbf{y}_i\|_2 < 1/i$, yet having $\|f(\mathbf{x}_i) - f(\mathbf{y}_i)\|_2 \ge \varepsilon$. From $(\mathbf{x}_i)$, find a subsequence $(\mathbf{x}_{j(i)}) \to \mathbf{b}$. Necessarily $\mathbf{b} \in S$ and $(\mathbf{y}_{j(i)}) \to \mathbf{b}$. By continuity, $f(\mathbf{x}_{j(i)}) \to f(\mathbf{b})$ and $f(\mathbf{y}_{j(i)}) \to f(\mathbf{b})$. But then $\|f(\mathbf{x}_{j(i)}) - f(\mathbf{y}_{j(i)})\|_2 \to 0$, contradicting the function-difference inequality.

9. (a) $\Rightarrow$ (b) Assume (a), and let $(\mathbf{x}_i)$ be a bounded sequence. If $\{\mathbf{x}_1, \mathbf{x}_2, \ldots\}$ is finite, then at least one $\mathbf{x}_k$ is repeated infinitely many times in the sequence. That $\mathbf{x}_k$ qualifies as a sublimit. If instead $\{\mathbf{x}_1, \mathbf{x}_2, \ldots\}$ is an infinite set, then it fits the hypothesis of (a), so it has an accumulation point $\mathbf{b}$. $N(\mathbf{b}, 1)$ must have infinitely many members of the set (Exercise 6b in Section 3.2). Call one of them $\mathbf{x}_{j(1)}$, and write $r_1 := \|\mathbf{b} - \mathbf{x}_{j(1)}\|/2$. $N(\mathbf{b}, r_1)$ must have infinitely many members; among them, there must be $\mathbf{x}_{j(2)}$ with $j(2) > j(1)$. Set $r_2 := \|\mathbf{b} - \mathbf{x}_{j(2)}\|/2$, and continue the selection process. Then $(\mathbf{x}_{j(i)}) \to \mathbf{b}$, and $\mathbf{b}$ is a sublimit.

(b) $\Rightarrow$ (a) Assume (b), and suppose S is infinite and bounded. By the stated definition, it is possible to find a sequence $(\mathbf{x}_i)$ from S of unequal vectors. By Theorem 4.6(b), $(\mathbf{x}_i)$ is bounded. By hypothesis, it must have a sublimit $\mathbf{c} = \lim \mathbf{x}_{k(i)}$. For each neighborhood $N(\mathbf{c}, \delta)$, there is I such that the terms $\mathbf{x}_{k(I)}, \mathbf{x}_{k(I+1)}, \ldots$ are in $N(\mathbf{c}, \delta)$. These terms being unequal, they constitute an infinity of members of S in $N(\mathbf{c}, \delta)$. We conclude that $\mathbf{c}$ is an accumulation point of S.

Section 4.3

1. Each function in Example 1 is a composite of xy (polynomial in $\mathbf{R}^2$), absolute value, exponential, and inverse tangent (all continuous in $\mathbf{R}$); that makes them continuous.

   The range of xy is $(-\infty, \infty)$. Therefore, in (a), $\tan^{-1} xy$ ranges over $(-\pi/2, \pi/2)$, without extremes, and $|\tan^{-1} xy|$ ranges over $[0, \pi/2)$. Similarly in (b), e^{xy} has range $(0, \infty)$, never hitting 0, and $e^{|xy|}$ has range $[1, \infty)$, value $= 1$ at $\mathbf{O}$.

2. Same words as in the solution to Exercise 8c in Section 4.2, with the conclusions in the third and fourth sentences justified by sequential compactness.

3. (a) Continuity is automatic at isolated points.

   (b) You have to have arbitrarily close points. Let $T := \{(1,0), (\frac{1}{2}, 0), (\frac{1}{3}, 0), \ldots\} \subseteq \mathbf{R}^2$. $(1/i, 0)$ is isolated, because no other point of T is within $1/i - 1/(i+1)$ of it. $f(x, y) := 1/x$ is continuous on T, but not uniformly: Set $\varepsilon := 1$, and see that no matter what δ is, there are points with $\|(1/i, 0) - (1/j, 0)\|_2 < \delta$ but $|f(1/i, 0) - f(1/j, 0)| \not< \varepsilon$.

4. Since $\|\mathbf{x} - \mathbf{b}\|$ is a continuous function of $\mathbf{x} \in S$, it must achieve greatest and least values (Theorem 4.11). Those values occur at the farthest and closest points.

5. Fix $\mathbf{x}$ and $\mathbf{y}$. For $\mathbf{z} \in S$, $\|\mathbf{x} - \mathbf{z}\| \le \|\mathbf{x} - \mathbf{y}\| + \|\mathbf{y} - \mathbf{z}\|$. Hence $d(\mathbf{x}, S) := \inf\{\|\mathbf{x} - \mathbf{z}: \mathbf{z} \in S\} \le \|\mathbf{x} - \mathbf{y}\| + \inf\{\|\mathbf{y} - \mathbf{z}\|: \mathbf{z} \in S\} = \|\mathbf{x} - \mathbf{y}\| + d(\mathbf{y}, S)$. Therefore, $d(\mathbf{x}, S) - d(\mathbf{y}, S) \le \|\mathbf{x} - \mathbf{y}\|$, and the absolute value inequality follows by symmetry.

6. By definition of inf, there are sequences $(\mathbf{x}_i)$ from S and $(\mathbf{y}_i)$ from T such that $\|\mathbf{x}_i - \mathbf{y}_i\| \to d(S, T)$. Let S and T be sequentially compact. Then there is a subsequence $(\mathbf{x}_{j(i)}) \to \mathbf{a} \in S$ and corresponding $(\mathbf{y}_{k(j(i))}) \to \mathbf{b} \in T$. From $\|\mathbf{x}_{k(j(i))} - \mathbf{y}_{k(j(i))}\| \to \|\mathbf{a} - \mathbf{b}\|$, we judge that $d(S, T) = \|\mathbf{a} - \mathbf{b}\|$ (multiple uses of Theorem 2.4).

   Alternative proof: By Exercise 5, $d(\mathbf{x}, T)$ is contractive on S, and therefore continuous. Hence there is a minimum value $d(\mathbf{a}, T)$.

By Exercise 4, there is $\mathbf{b} \in T$ with $d(\mathbf{a}, T) = \|\mathbf{a} - \mathbf{b}\|$. Now take any $\mathbf{x} \in S$, $\mathbf{y} \in T$. Then $\|\mathbf{x} - \mathbf{y}\| \geq d(\mathbf{x}, T) \geq d(\mathbf{a}, T) = \|\mathbf{a} - \mathbf{b}\|$. Of necessity, $d(S, T) := \inf \|\mathbf{x} - \mathbf{y}\| = \|\mathbf{a} - \mathbf{b}\|$.

7. (a) If $\mathbf{x}, \mathbf{y} \in B(\mathbf{b}, r)$, then $\|\mathbf{x} - \mathbf{y}\| \leq \|\mathbf{x} - \mathbf{b}\| + \|\mathbf{b} - \mathbf{y}\| \leq 2r$. Since the points $\mathbf{b} \pm r\mathbf{u}$, for any unit vector $\mathbf{u}$, are in the ball $2r$ apart, we conclude that $\sup \|\mathbf{x} - \mathbf{y}\| = 2r$.

   (b) The diameter d is finite, because S has to be bounded (Theorems 4.11 and 4.10). Let $\|\mathbf{x}_i - \mathbf{y}_i\| \to d$. As in Exercise 6, take subsequences $(\mathbf{x}_{j(i)}) \to \mathbf{a}$, $(\mathbf{y}_{k(j(i))}) \to \mathbf{b}$. Then $\|\mathbf{x}_{k(j(i))} - \mathbf{y}_{k(j(i))}\|$ tends to both d and $\|\mathbf{a} - \mathbf{b}\|$, so these last two must be equal.

8. One example fits all: $S :=$ quadrant I in $\mathbf{R}^2$, $\mathbf{b} := \mathbf{O}$, $T := \{\mathbf{O}\}$.

9. (a) Since $d(\mathbf{x}, T)$ is contractive (Exercise 5) on S, it has a minimum value $d(\mathbf{a}, T)$. That makes $\mathbf{a}$ as close to T as a member of S can get. Also, $\mathbf{x} \in S, \mathbf{y} \in T \Rightarrow \|\mathbf{x} - \mathbf{y}\| \geq d(\mathbf{x}, T) \geq d(\mathbf{a}, T)$, giving $d(S, T) \geq d(\mathbf{a}, T)$; while $d(\mathbf{a}, T) := \inf \|\mathbf{a} - \mathbf{y}\| \geq \inf \|\mathbf{x} - \mathbf{y}\|$ (inf of a smaller set is larger) $= d(S, T)$.

   (b) In (a), we found $\mathbf{a}$ with $d(S, T) = d(\mathbf{a}, T)$. Were this distance 0, there would be $(\mathbf{y}_i)$ from T such that $\|\mathbf{a} - \mathbf{y}_i\| \to 0$; that would make $\mathbf{a}$ a closure point of T, forcing $\mathbf{a} \in T$, violating $S \cap T = \emptyset$. Hence $d(\mathbf{a}, T) > 0$.

   (c) $S := x$-axis and $T := \{(x, y) \colon y = e^x\}$ are both closed (Exercise 1c,d in Section 4.2), but have points in arbitrarily close proximity: $\left(-i, e^{-i}\right)$ on the graph is $1/e^i$ from $(-i, 0)$ on the x-axis.

   (d) $S := \{\mathbf{O}\}$ and $T :=$ quadrant I in $\mathbf{R}^2$.

   (e) By (a), there is $\mathbf{a}$ with $d(S, T) = d(\mathbf{a}, T)$. By definition, there is a sequence $(\mathbf{y}_i)$ in T with $\|\mathbf{a} - \mathbf{y}_i\| \to d(\mathbf{a}, T)$. $(\mathbf{y}_i)$ has to be bounded, because $\|\mathbf{y}_i\|_2 \leq \|\mathbf{a}\|_2 + \|\mathbf{y}_i - \mathbf{a}\|_2$, and the sequence on the right converges. Since we are in $\mathbf{R}^n$, some subsequence $(\mathbf{y}_{j(i)}) \to \mathbf{b} \in T$, and $d(S, T) = d(\mathbf{a}, T) = \|\mathbf{b} - \mathbf{a}\|_2$. This says that $\|\mathbf{b} - \mathbf{a}\|_2 = \inf \|\mathbf{x} - \mathbf{y}\|_2 \leq \inf \|\mathbf{x} - \mathbf{b}\|_2$, and so $\mathbf{b}$ is closest to S.

Section 4.4

1. (a) Open: If $\mathbf{c} := (a, b)$ is in quadrant I, so that a and b are positive, let $r := \min\{a, b\}$; then $N(\mathbf{c}, r) \subseteq$ quadrant I, because $(x, y) \in N(\mathbf{c}, r) \Rightarrow |x - a|, |y - b|$ both $\leq (x - a)^2 + (y - b)^2 < r^2 \Rightarrow x > 0, y > 0$. Not closed: $(1/i, 1/i) \to \mathbf{O}$, which is not in quadrant I.

   (b) Not open: $(1, 1)$ is on the graph, but $N((1, 1), r)$ has the point $(1, 1 + r/2)$, which is not. Closed: Suppose $\mathbf{x}_i := (x_i, 1/x_i)$ is a sequence from the graph converging to (a, b); then $x_i \to a$ and $1/x_i \to b$; a cannot be 0, because then $|1/x_i|$ would approach ∞; by continuity

$1/x_i \to 1/a$; necessarily $b = 1/a$, which says that (a, b) is on the graph.

(c) Not open: $(1, 0)$ (rectangular or polar) is on the graph, but $(1 + r, 0)$ is not for $0 < r < e^{2\pi} - 1$. Not closed: $\mathbf{x}_i :=$ polar $\left(e^{-i}, -i\right) \to \mathbf{O}$, which is not on the graph.

(d) Not open: $(1/2\pi, 0)$ is on, but $(1/2\pi, r/2)$ is not. Not closed: $(1/i\pi, 0) \to (0, 0)$, and no point with $x = 0$ is on the graph.

2. (Compare the solution to Exercise 1a. Also, note that $(\mathbf{a}, \infty)$ is not the complement of $(-\infty, \mathbf{a}]$.)

   Assume $\mathbf{c} \in (\mathbf{a}, \infty)$, so that $\mathbf{a} < \mathbf{c}$. Let r be the least of $c_1 - a_1, \ldots, c_n - a_n$. Then $\mathbf{x} \in N(\mathbf{c}, r) \Rightarrow$ each $|x_k - c_k| < r \Rightarrow$ each $x_k > c_k - r \geq a_k \Rightarrow \mathbf{x} \in (\mathbf{a}, \infty)$. Hence $N(\mathbf{c}, r) \subseteq (\mathbf{a}, \infty)$, and the latter is open. Similarly with $\mathbf{c} \in (-\infty, \mathbf{b})$.

3. (Compare the solution to Exercise 6 in Section 4.1.) $\mathbf{a} < \mathbf{b}$ means that each $a_k < b_k$; to deny it is to say that some $a_j \geq b_j$. In that case, no $\mathbf{x}$ can satisfy $a_j < x_j < b_j$, so no $\mathbf{x}$ has $\mathbf{a} < \mathbf{x} < \mathbf{b}$.

4. For every $\mathbf{x} \in V$, $N(\mathbf{x}, 1) \subseteq V$ trivially, so V is open. At the other extreme, "$\mathbf{x} \in \emptyset \Rightarrow N(\mathbf{x}, 1) \subseteq \emptyset$" is true vacuously.

5. Only the empty set. If O is open and $\mathbf{b} \in O$, then some $N(\mathbf{b}, \delta) \subseteq O$, and O has all the vectors $\mathbf{b} + (\delta/i)\mathbf{b}/\|\mathbf{b}\|$, $i \geq 2$.

6. (a) Assume $T \subseteq W$. In V, $\mathbf{x} \in \mathbf{f}^{-1}(T^*) \Leftrightarrow \mathbf{f}(\mathbf{x}) \in T^* \Leftrightarrow \mathbf{f}(\mathbf{x}) \notin T \Leftrightarrow \mathbf{x} \notin \mathbf{f}^{-1}(T) \Leftrightarrow \mathbf{x} \in \mathbf{f}^{-1}(T)^*$. Hence $\mathbf{f}^{-1}(T^*) = \mathbf{f}^{-1}(T)^*$.

   (b) $\mathbf{f}^{-1}(\text{open}) = \text{open}$ comes from Theorem 4.18 (whose proof, to be sure, depends on Exercise 9 below). Suppose instead T is closed. Then T^* is open (Theorem 4.14, whose proof is Exercise 7). By Theorem 4.18, $\mathbf{f}^{-1}(T^*)$ is open. By (a), $\mathbf{f}^{-1}(T^*) = \mathbf{f}^{-1}(T)^*$. Hence $\mathbf{f}^{-1}(T)$ is closed.

7. $\Rightarrow$ Assume that O is open. Suppose $(\mathbf{x}_i)$ is a sequence from O^* converging to $\mathbf{b}$. If $\mathbf{b}$ were in O, there would exist $N(\mathbf{b}, \delta) \subseteq O$, so that (Theorem 2.5) the $\mathbf{x}_i$ would eventually get into $N(\mathbf{b}, \delta)$ and thereby into O. Hence $\mathbf{b}$ must be in O^*. We have shown that O^* is closed.

   $\Leftarrow$ Assume that O^* is closed. Let $\mathbf{b} \in O$. By Theorem 4.7, $\mathbf{b}$ is not a closure point of O^*. By definition, there exists $N(\mathbf{b}, \delta)$ containing no points from O^*. Necessarily $N(\mathbf{b}, \delta) \subseteq O$, proving that O is open.

8. Let $\mathbf{b} \in D$, $\varepsilon > 0$. The set $O := N(\mathbf{f}(\mathbf{b}), \varepsilon)$ is open, so by hypothesis there exists an open P such that $D \cap P$ is $\mathbf{f}^{-1}(O)$. Since $\mathbf{b} \in \mathbf{f}^{-1}(O) \subseteq P$, there is $N(\mathbf{b}, \delta) \subseteq P$. If now $\mathbf{x} \in N(\mathbf{b}, \delta) \cap D$, then $\mathbf{f}(\mathbf{x}) \in O$. By Theorem 3.9(d), $\mathbf{f}$ is continuous at $\mathbf{b}$.

9. (a) If $\mathbf{x} \in O \cap P$, both open, then there are $N(\mathbf{x}, r) \subseteq O$ and $N(\mathbf{x}, s) \subseteq P$. Let $t := \min\{r, s\}$. Clearly, $N(\mathbf{x}, t)$ is contained in each of O and P, so this neighborhood of $\mathbf{x}$ is a subset of $O \cap P$. Hence $O \cap P$ is open.

   (b) Same argument with $t := \min\{r_1, \ldots, r_k\}$.

10. By one of DeMorgan's laws, $(C_1 \cup \cdots \cup C_k)^* = C_1^* \cap \cdots \cap C_k^*$. If each C_j is closed, then each C_j^* is open (Theorem 4.14 = Exercise 7), so their intersection is open (Exercise 9b). From $(C_1 \cup \cdots \cup C_k)^*$ being open, we conclude that the union is closed.

Section 4.5

1. (a) Not connected, therefore not arc-connected. $(0, 0)$ is in the set, but not $(0, 1)$, even though $y = \sin x/x \to 1$ as $x \to 0$. Accordingly, $O_1 := N(\mathbf{O}, \pi/6)$ and $O_2 := \{\mathbf{x}\colon \|\mathbf{x}\|_2 > \pi/6\}$ disconnect the set.

   (b) Arc-connected, therefore connected. You would think it impossible to get from $(1, 0)$ (polar $\theta = 0, r = 1$) to the origin (polar $\theta = -\infty, r = 0$) along the spiral, since the distance appears to be infinite. Without deciding whether that appearance is reality, we merely remark that you can cover infinite distance in finite time by walking fast. In this case, set $\theta = \tan t$, $r = e^{\tan t}$ (rectangular $x = r\cos\theta = e^{\tan t}\cos(\tan t)$, $y = r\sin\theta = e^{\tan t}\sin(\tan t)$) for $-\pi/2 < t < \pi/2$. Clearly, the mapping $t \mapsto (x, y)$ is continuous, stays on the spiral, covers the whole spiral, and has $x \to 0$, $y \to 0$ as $t \to -\pi/2$. Therefore, we can extend it by setting $x := 0$, $y := 0$ when $t := -\pi/2$. The resulting extension maps $[-\pi/2, \pi/2)$ continuously onto the set, so that we can use it to define an arc from any point to another.

2. (a) $\mathbf{Q}^n$ does not have the IVP, so it is not connected (Theorem 4.21). For example, $\|\mathbf{x}\|_2^2$ is continuous on $\mathbf{Q}^n$, has values $\|\mathbf{O}\|_2^2 = 0$ and $\|(2, 0, \ldots, 0)\|_2^2 = 4$, but never reaches $\sqrt{2}$.

   (b) In $\mathbf{R}^1$, the irrationals are disconnected by $O_1 := \{x > 0\}$ and $O_2 := \{x < 0\}$. In $\mathbf{R}^n$, the not-all-rationals (complement of $\mathbf{Q}^n$) are *arc-connected.* To illustrate, you travel from $(\sqrt{2}, y, z)$ to $(x, \sqrt{3}, u)$ along the segments from $(\sqrt{2}, y, z)$ to $(\sqrt{2}, \sqrt{3}, z)$ to $(x, \sqrt{3}, z)$ to $(x, \sqrt{3}, u)$, staying in $(\mathbf{Q}^n)^*$; and from $(\sqrt{2}, y, z)$ to $(\sqrt{5}, v, w)$ along the segments from $(\sqrt{2}, y, z)$ to $(\sqrt{2}, \sqrt{3}, z)$ to $(\sqrt{5}, \sqrt{3}, z)$ to $(\sqrt{5}, v, z)$ to $(\sqrt{5}, v, w)$.

3. Projections are continuous (Theorem 3.8, among other places). By Theorem 4.23, the projected image of a connected set is connected. The converse is false: In $\mathbf{R}^2$, the union of the two lines $y = x$ and $y = x + 1$ is a disconnected set whose projections are connected.

4. $\emptyset$ is trivially connected, because you cannot nontrivially partition it. V is connected for many reasons; for example, it is convex: If $\mathbf{a}, \mathbf{b} \in V$, then the segment $\mathbf{ab} \subseteq V$.

5. $\Rightarrow$ Let O_1 and O_2 disconnect S. Write $C_1 := O_1^*$, $C_2 := O_2^*$; these are closed, by Theorem 4.14. Since $O_1 \cap S$ and $O_2 \cap S$ partition S, we have $\mathbf{x} \in C_1 \cap S \Leftrightarrow \mathbf{x} \in S$ and $\mathbf{x} \notin O_1 \Leftrightarrow \mathbf{x} \in O_2 \cap S$. Thus, $C_1 \cap S = O_2 \cap S$. Similarly $C_2 \cap S = O_1 \cap S$. Hence $\{C_1 \cap S, C_2 \cap S\}$ is a partition of S.

   $\Leftarrow$ Same argument with C's interchanged with O's.

6. Suppose S is neither $\emptyset$ nor V. Then S and S^* are both nonempty, so $\{S, S^*\}$ is a partition of V. Since V is connected (Exercise 4), S and S^* cannot both be open. Hence either S is not open, or S^* is not open, in which case S is not closed.

7. Let S be arc-connected. According to Theorem 4.19, we need to show that S has the IVP. Suppose f is continuous on S, $\mathbf{a}$ and $\mathbf{b} \in S$, and $f(\mathbf{a}) < y < f(\mathbf{b})$. By the arc-connectedness, there exists a continuous function $\mathbf{g}$ mapping some interval $[r, s]$ to S with $\mathbf{g}(r) = \mathbf{a}$, $\mathbf{g}(s) = \mathbf{b}$. Then $f(\mathbf{g}(t))$ is continuous on $[r, s]$ and has $f(\mathbf{g}(r)) < y < f(\mathbf{g}(s))$. By the IVT for real functions, there is a place $u \in [r, s]$ where $f(\mathbf{g}(u)) = y$. Set $\mathbf{c} := \mathbf{g}(u)$, and we have $\mathbf{c} \in S$ with $f(\mathbf{c}) = y$. Hence S has the IVP.

8. (a) $\mathbf{f}(t) := \mathbf{b}$, $0 \le t \le 1$, defines an arc from $\mathbf{b}$ to $\mathbf{b}$ within S.

   (b) If S is open and $\mathbf{b} \in S$, then there is $N(\mathbf{b}, \delta) \subseteq S$. If now $\mathbf{x} \in N(\mathbf{b}, \delta)$, then $\mathbf{g}(t) := (1-t)\mathbf{x} + t\mathbf{b}$, $0 \le t \le 1$, defines an arc from $\mathbf{x}$ to $\mathbf{b}$ within $N(\mathbf{b}, \delta)$ (Theorem 2.1(b)), and therefore within S.

   (c) See the proof of Theorem 5.5(b).

9. (a) Assume that S is star-shaped and $\mathbf{x}, \mathbf{y} \in S$. By hypothesis, the segments $\mathbf{bx}$ and $\mathbf{by}$ are subsets of S. As in Exercise 8b, these are arcs from $\mathbf{b}$ to $\mathbf{x}$ and from $\mathbf{b}$ to $\mathbf{y}$. As in 8c, $\mathbf{xb}$ is an arc from $\mathbf{x}$ to $\mathbf{b}$. Hence we have an arc from $\mathbf{x}$ to $\mathbf{b}$ and one from $\mathbf{b}$ to $\mathbf{y}$. See the argument in Theorem 5.5(b) to establish an arc from $\mathbf{x}$ to $\mathbf{y}$.

   (b) No; the unit circle in $\mathbf{R}^2$ is itself an arc, but contains no line segments.

   (c) No; draw a "star." Alternatively, look at the union of the two axes in $\mathbf{R}^2$.

10. (a) Let $\mathbf{x} := \alpha_1\mathbf{x}_1 + \cdots + \alpha_k\mathbf{x}_k$ and $\mathbf{y} := \beta_1\mathbf{y}_1 + \cdots + \beta_j\mathbf{y}_j \in \mathrm{CH}(S)$. The segment $\mathbf{xy}$ has vectors $(1-\alpha)\mathbf{x} + \alpha\mathbf{y} = (1-\alpha)\alpha_1\mathbf{x}_1 + \cdots + (1-\alpha)\alpha_k\mathbf{x}_k + \alpha\beta_1\mathbf{y}_1 + \cdots + \alpha\beta_j\mathbf{y}_j$. In this last combination, the vectors are from S and the coefficients are nonnegative and sum to $(1-\alpha)(\alpha_1 + \cdots + \alpha_k) + \alpha(\beta_1 + \cdots + \beta_j) = 1$. Hence $(1-\alpha)\mathbf{x} + \alpha\mathbf{y} \in \mathrm{CH}(S)$, proving that $\mathrm{CH}(S)$ is convex.

(b) If $\mathbf{x} \in S$, then $1\mathbf{x} \in \mathrm{CH}(S)$. Hence $S \subseteq \mathrm{CH}(S)$; the latter is one convex (part (a)) superset of S. Assume that T is any convex superset of S. Then T has the convex combinations $1\mathbf{x}_1$ of a single vector from S. Suppose T includes all combinations of $k-1$ or fewer vectors. Then

$$\mathbf{x} := \alpha_1\mathbf{x}_1 + \cdots + \alpha_k\mathbf{x}_k = (1-\alpha_k)\frac{\alpha_1\mathbf{x}_1 + \cdots + \alpha_{k-1}\mathbf{x}_{k-1}}{(1-\alpha_k)} + \alpha_k\mathbf{x}_k$$

is on the segment from one vector in T to another, unless $1-\alpha_k = 0$, in which case $\mathbf{x} = \mathbf{x}_k \in S$. Either way $\mathbf{x} \in T$, since T is convex. By induction, T possesses all the members of $\mathrm{CH}(S)$. Hence $\mathrm{CH}(S) \subseteq T$, and the hull is the smallest convex superset of S.

(c) $\Leftarrow$ If $S = \mathrm{CH}(S)$, then S is convex by part (a).

$\Rightarrow$ If S is convex, then necessarily S is the smallest convex superset of S; by (b), $S = \mathrm{CH}(S)$.

(d) The combination $1\mathbf{x}_1$ consists of both endpoints of a degenerate segment. The combination $\alpha_1\mathbf{x}_1 + \alpha_2\mathbf{x}_2$ lies on a segment with one endpoint in S and the other also in S. The combination $\alpha_1\mathbf{x}_1 + \alpha_2\mathbf{x}_2 + \alpha_3\mathbf{x}_3 = (1-\alpha_3)(\alpha_1[1-\alpha_3]^{-1}\mathbf{x}_1 + \alpha_2[1-\alpha_3]^{-1}\mathbf{x}_2) + \alpha_3\mathbf{x}_3$ lies on a segment with one endpoint $\mathbf{x}_3$ in S and the other $(\alpha_1[1-\alpha_3]^{-1}\mathbf{x}_1 + \alpha_2[1-\alpha_3]^{-1}\mathbf{x}_2)$ on a segment with one endpoint $\mathbf{x}_2$ in S and the other $\mathbf{x}_1$ also in S. ...

(e) No, no, yes. The convex hull of a neighborhood or ball is the neighborhood or ball, by part (c) and Example 3(a). A hull has to be connected; being convex, it is arc-connected.

Section 4.6

1. (a) Positive-definiteness: Since $\|\mathbf{x}\|_1 \geq$ each $|x_j|$, $\|\mathbf{x}\|_1 = 0$ forces $x_j = 0$ for all j.

Radial homogeneity: $\|\alpha\mathbf{x}\|_1 := |\alpha x_1| + \cdots + |\alpha x_n| = |\alpha|(|x_1| + \cdots + |x_n|) = |\alpha|\,\|\mathbf{x}\|_1$.

Subadditivity: $\|\mathbf{x}+\mathbf{y}\|_1 := |x_1+y_1| + \cdots + |x_n+y_n| \leq |x_1| + |y_1| + \cdots + |x_n| + |y_n| = \|\mathbf{x}\|_1 + \|\mathbf{y}\|_1$.

(b) Clearly, $|x_j| \leq \left(x_1^2 + \cdots + x_n^2\right)^{1/2} \leq \left(\|\mathbf{x}\|_1^2\right)^{1/2}$. Hence $\|\mathbf{x}\|_1 \leq n\|\mathbf{x}\|_2 \leq n\|\mathbf{x}\|_1$. Set $m := 1$, $M := n$.

2. (a) Recall (Exercise 1 in Section 1.6) that in $C_2[0,1]$, these sines and cosines are orthogonal and have length $1/\sqrt{2}$. Hence $\|f\|_2 = \left(\alpha^2/2 + \beta^2/2\right)^{1/2}$. We can find the extremes of f by calculus or trigonometry: $f(x) = \left(\alpha^2+\beta^2\right)^{1/2}\sin(2\pi x + \theta)$, where $\theta := \pm\cos^{-1}\left(\alpha/\left[\alpha^2+\beta^2\right]^{1/2}\right)$ (same sign as β), yielding $\sup|f| = \left(\alpha^2+\beta^2\right)^{1/2}$. Accordingly, $\|f\|_0 = \sqrt{2}\|f\|_2$.

(b) If (f_i) converges uniformly to f, then $f_i \to f$ in $C_0[0, 1]$. Since V, being two-dimensional, must be a closed subset of $C_0[0, 1]$ (Theorem 4.30), we must have $f \in V$.

3. The basic principle here is that the cubic polynomials form a 4-dimensional subspace.

(a) By hypothesis, $\|f_i\|_2^2 \to 28$. Necessarily $(\|f_i\|_2)$ is a bounded sequence of reals, so (Theorem 4.29(a)) $(\|f_i\|_0)$ has to be bounded. If, say, $\|f_i\|_0 < M$, then $|f_i(x)| < M$ for each i and every x.

(b) We just saw that (f_i) is a bounded sequence in a 4-dimensional subspace of $C_0[0, 1]$. Hence some subsequence converges (Theorem 4.27(c)) in $C_0[0, 1]$; that is, some subsequence converges uniformly.

4. From the inequality, we conclude that $(\mathbf{x}_i) \to \mathbf{x}$ relative to $\| \ \|_a \Leftrightarrow \|\mathbf{x}_i - \mathbf{x}\|_a \to 0$ (definition) $\Leftrightarrow \|\mathbf{x}_i - \mathbf{x}\|_b \to 0 \Leftrightarrow (\mathbf{x}_i) \to \mathbf{x}$ relative to $\| \ \|_b$.

5. We use the characterization in Theorem 4.25.

Reflexivity: $1\|\mathbf{x}\|_a \leq \|\mathbf{x}\|_a \leq 1\|\mathbf{x}\|_a$, so $\| \ \|_a$ is equivalent to itself.

Symmetry: If $m\|\mathbf{x}\|_a \leq \|\mathbf{x}\|_b \leq M\|\mathbf{x}\|_a$, then $M^{-1}\|\mathbf{x}\|_b \leq \|\mathbf{x}\|_a \leq m^{-1}\|\mathbf{x}\|_b$.

Transitivity: If $m\|\mathbf{x}\|_a \leq \|\mathbf{x}\|_b \leq M\|\mathbf{x}\|_a$ and $k\|\mathbf{x}\|_b \leq \|\mathbf{x}\|_c \leq K\|\mathbf{x}\|_b$, then $mk\|\mathbf{x}\|_a \leq \|\mathbf{x}\|_c \leq MK\|\mathbf{x}\|_a$.

6. Write $\|\mathbf{x}\|$ for the norm in V and $\|\mathbf{x}\|_B$ for the "Pythagorean norm," as defined in the proof of Theorem 4.26. By Theorems 4.25 and 4.26, there are m and M with $m\|\mathbf{x}\| \leq \|\mathbf{x}\|_B \leq M\|\mathbf{x}\|$ for all $\mathbf{x}$.

(a) $\Rightarrow$ If $(\mathbf{x}_i)$ converges to $\mathbf{y} := \alpha_1\mathbf{v}_1 + \cdots + \alpha_n\mathbf{v}_n$, then for each k, $|\alpha_k - \Pi_k(\mathbf{x}_i)|^2 \leq (\alpha_1 - \Pi_1(\mathbf{x}_i))^2 + \cdots + (\alpha_n - \Pi_n(\mathbf{x}_i))^2 = \|\mathbf{y} - \mathbf{x}_i\|_B^2 \leq M^2\|\mathbf{y} - \mathbf{x}_i\|_2^2 \to 0$. This says that each sequence $(\Pi_k(\mathbf{x}_i))$ converges to $\alpha_k = \Pi_k(\mathbf{y})$.

$\Leftarrow$ If, conversely, each coordinate sequence $(\Pi_k(\mathbf{x}_i))$ converges to β_k, then $\mathbf{z} := \beta_1\mathbf{v}_1 + \cdots + \beta_n\mathbf{v}_n$ has $\|\mathbf{z} - \mathbf{x}_i\| = \big\|(\beta_1 - \Pi_1(\mathbf{x}_i))\mathbf{v}_1 + \cdots + (\beta_n - \Pi_n(\mathbf{x}_i))\mathbf{v}_n\big\| \leq \big|\beta_1 - \Pi_1(\mathbf{x}_i)\big|\,\|\mathbf{v}_1\| + \cdots + \big|\beta_n - \Pi_n(\mathbf{x}_i)\big|\,\|\mathbf{v}_n\| \to 0$, and $(\mathbf{x}_i)$ converges to $\mathbf{z}$.

(c) From $\big|\Pi_k(\mathbf{x}_i)\big| \leq \|\mathbf{x}_i\|_B \leq M\|\mathbf{x}_i\|$, we conclude that if $(\mathbf{x}_i)$ is bounded, then so is each sequence $(\Pi_k(\mathbf{x}_i))$. As in the proof of BWT (Theorem 2.10), we extract a subsequence $(\mathbf{x}_{j(i)})$ for which all the coordinate sequences $\big(\Pi_k(\mathbf{x}_{j(i)})\big)$ converge, and their limits are the coordinates of a finite sublimit for $(\mathbf{x}_i)$.

7. Write $\| \ \|_a$ and $\| \ \|_b$ for two norms in V. We now know there are m and M with $m\|\mathbf{x}\|_a \le \|\mathbf{x}\|_b \le M\|\mathbf{x}\|_a$.

   (a) If $(\mathbf{x}_i) \to \mathbf{x}$ relative to $\| \ \|_a$, meaning that $\|\mathbf{x} - \mathbf{x}_i\|_a \to 0$, then $\|\mathbf{x} - \mathbf{x}_i\|_b \le M\|\mathbf{x} - \mathbf{x}_i\|_a \to 0$, and $(\mathbf{x}_i) \to \mathbf{x}$ relative to $\| \ \|_b$. (Conversely, by symmetry.)

   (b) Case 1: $\mathbf{b}$ is isolated relative to $\| \ \|_a$. Thus, no member of D has $\|\mathbf{x}-\mathbf{b}\|_a < \delta$. Then no member of D has $\|\mathbf{x}-\mathbf{b}\|_b < m\delta$. Consequently, $\mathbf{b}$ is isolated relative to $\| \ \|_b$, and under both norms, the limit is $\mathbf{f}(\mathbf{b})$.

   Case 2: $\mathbf{b}$ is an accumulation point of D relative to $\| \ \|_a$. First, there is a sequence $(\mathbf{x}_i)$ from D with $0 < \|\mathbf{b} - \mathbf{x}_i\|_a \to 0$, and the limit of $\mathbf{f}$ is $\mathbf{c} := \lim \mathbf{f}(\mathbf{x}_i)$. By part (a), $0 < \|\mathbf{b}-\mathbf{x}_i\|_b \to 0$; that is, some sequence from D converges to $\mathbf{b}$, without reaching $\mathbf{b}$, relative to $\| \ \|_b$. Next, let $(\mathbf{y}_i)$ be another such sequence. Then also $\|\mathbf{b} - \mathbf{y}_i\|_a \to 0$. Because $\mathbf{f}$ has an a-limit, we have $\mathbf{f}(\mathbf{y}_i) \to \mathbf{c}$. We have established that for every $(\mathbf{y}_i)$, the value sequence $\mathbf{f}(\mathbf{y}_i)$ has limit $\mathbf{c}$. It follows (Section 3.2) that $\mathbf{f}$ has a b-limit matching its a-limit $\mathbf{c}$.

   (c) The conclusion says that $\mathbf{x}_i \in D$ and $\|\mathbf{b} - \mathbf{x}_i\|_b \to 0$ imply $\|\mathbf{f}(\mathbf{b}) - \mathbf{f}(\mathbf{x}_i)\|_W \to 0$. Its proof is immediate, because $\|\mathbf{b} - \mathbf{x}_i\|_b \to 0$ forces $\|\mathbf{b} - \mathbf{x}_i\|_a \to 0$.

8. (a) Assume that S is a-closed, meaning that S possesses the a-limit of any sequence from S. Thus, $\mathbf{x}_i \in S$ and $\|\mathbf{x} - \mathbf{x}_i\|_a \to 0 \Rightarrow \mathbf{x} \in S$. Since $\|\mathbf{x}-\mathbf{x}_i\|_b \to 0$ forces $\|\mathbf{x}-\mathbf{x}_i\|_a \to 0$ (Theorems 4.26 and 4.24), we see that S possesses the b-limit of any sequence from S, and S is b-closed. The conclusion about boundedness follows from $\|\mathbf{x}\|_b \le M\|\mathbf{x}\|_a$.

   (b) $\Rightarrow$ If $S \subseteq V$ has the EVP, then it is closed and bounded by Theorem 4.10.

   $\Leftarrow$ Assume that S is closed and bounded. Let $(\mathbf{x}_i)$ be a sequence from S. It has to be bounded, so by Theorem 4.26(c), there is a convergent subsequence. Because S is closed, the limit of the subsequence is in S. Hence S is sequentially compact. By Theorem 4.11, S has the EVP.

   (c) Suppose S is closed and bounded and $\mathbf{f}$ is continuous there. By (b), S is sequentially compact. By Exercise 2 in Section 4.3, $\mathbf{f}$ is uniformly continuous.

9. Let $\mathbf{L}\colon V \to W$, with norms $\| \ \|_V$ and $\| \ \|_W$. By theorems from linear algebra, the range $\mathbf{L}(V)$ is a finite-dimensional subspace of W. Let $B := \{\mathbf{v}_1, \ldots, \mathbf{v}_n\}$ and $C := \{\mathbf{w}_1, \ldots, \mathbf{w}_m\}$ be bases for V and $\mathbf{L}(V)$. There are scalars α_{jk} with $\mathbf{L}(\mathbf{v}_k) = \alpha_{1k}\mathbf{w}_1 + \cdots + \alpha_{mk}\mathbf{w}_m$ for each k. For any $\mathbf{x} =$

$\beta_1\mathbf{v}_1 + \cdots + \beta_n\mathbf{v}_n$, we then have

$$\begin{aligned}
\|\mathbf{L}(\mathbf{x})\|_W^2 &\le M^2\|\mathbf{L}(\mathbf{x})\|_C^2 \qquad \text{(by equivalence of norms in } \mathbf{L}(V))\\
&= M^2\|\beta_1(\alpha_{11}\mathbf{w}_1 + \cdots + \alpha_{m1}\mathbf{w}_m)\\
&\qquad + \cdots + \beta_n(\alpha_{1n}\mathbf{w}_1 + \cdots + \alpha_{mn}\mathbf{w}_m)\|_C^2\\
&:= M^2\left([\beta_1\alpha_{11} + \cdots + \beta_n\alpha_{1n}]^2 + \cdots + [\beta_1\alpha_{m1} + \cdots + \beta_n\alpha_{mn}]^2\right)\\
&\le M^2([\beta_1^2 + \cdots + \beta_n^2][\alpha_{11}^2 + \cdots + \alpha_{1n}^2]\\
&\qquad + \cdots + [\beta_1^2 + \cdots + \beta_n^2][\alpha_{11}^2 + \cdots + \alpha_{1n}^2])
\end{aligned}$$

by Cauchy's inequality. Writing K^2 for the sum of all the α_{jk}^2, we have

$$\|\mathbf{L}(\mathbf{x})\|_W^2 \le M^2[\beta_1^2 + \cdots + \beta_n^2]K^2 = M^2K^2\|\mathbf{x}\|_B^2 \le M^2K^2m^2\|\mathbf{x}\|_V,$$

the last by equivalence of norms in V. This says that $\mathbf{L}$ is a bounded linear map from V to W. By Theorem 3.16, $\mathbf{L}$ is continuous.

10. (a) $\Rightarrow$ (b) If the norms are equivalent, then there are M and m with $m\|\mathbf{x}\|_a \le \|\mathbf{x}\|_b \le M\|\mathbf{x}\|_a$, so that $\|\mathbf{x}\|_a/\|\mathbf{x}\|_b \le 1/m$ and $\|\mathbf{x}\|_b/\|\mathbf{x}\|_a \le M$ are both bounded.

(b) $\Rightarrow$ (c) If $\|\mathbf{x}\|_a/\|\mathbf{x}\|_b \le m$ and $\|\mathbf{x}\|_b/\|\mathbf{x}\|_a \le M$, then $1/M \le \|\mathbf{x}\|_a/\|\mathbf{x}\|_b \le m$ says that the fraction is bounded and bounded away from zero.

(c) $\Rightarrow$ (d) Assume $0 < \delta \le \|\mathbf{x}\|_a/\|\mathbf{x}\|_b \le \varepsilon$. Then $|\ \|\mathbf{x}\|_a| \le \varepsilon\|\mathbf{x}\|_b$ says that the a-norm is a function of bounded magnification relative to the b-norm; and analogously $|\ \|\mathbf{x}\|_b| \le (1/\delta)\|\mathbf{x}\|_a$.

(d) $\Rightarrow$ (e) Assume that the a-norm is a function of bounded magnification relative to the b-norm, so that $\|\mathbf{x}\|_a \le M\|\mathbf{x}\|_b$. If $(\mathbf{x}_i) \to \mathbf{x}$ in the b-norm, then $|\ \|\mathbf{x}\|_a - \|\mathbf{x}_i\|_a| \le \|\mathbf{x}-\mathbf{x}_i\|_a$ (difference of the norms) $\le M\|\mathbf{x}-\mathbf{x}_i\|_b \to 0$. This says that $\|\mathbf{x}\|_a$ is a continuous function under the b-norm. Similarly for the opposite direction.

(e) $\Rightarrow$ (a) Assume that each norm is continuous relative to the other. Since $\|\mathbf{x}\|_a$ is b-continuous at the origin, there exists δ such that $\|\mathbf{x} - \mathbf{O}\|_b < \delta \Rightarrow |\ \|\mathbf{x}\|_a - \|\mathbf{O}\|_a| < 1$. Therefore, for every $\mathbf{x} \ne \mathbf{O}$, since the vector $\mathbf{y} := \delta\mathbf{x}/(2\|\mathbf{x}\|_b)$ satisfies $\|\mathbf{y}\|_b < \delta$, we know that $\|\mathbf{y}\|_a < 1$, or $\|\mathbf{x}\|_a < (2/\delta)\|\mathbf{x}\|_b$. Clearly, also $\|\mathbf{O}\|_a \le (2/\delta)\|\mathbf{O}\|_b$. Similarly, we find ε such that $\|\mathbf{x}\|_b \le (2/\varepsilon)\|\mathbf{x}\|_a$, and the norms are equivalent (Theorem 4.25).

Chapter 5

Section 5.1

1. (a) This is just a consequence of continuity. Since the x-value $\pi_1(\mathbf{f}(a))$ is $10/\pi$, there is (Theorem 3.10(c)) a neighborhood $N(a, \delta)$ such that $\pi_1(\mathbf{f}(t)) > 5/\pi$ for $t \in N(a, \delta) \cap [a, b] = [a, a + \delta)$. All these points $\mathbf{f}(t)$ are therefore on the curved part.

(b) Considering (a), let $\varepsilon := a+\sup\{\delta\colon \pi_1(\mathbf{f}(t)) > 0 \text{ for all } t \in [a, a+\delta)\}$. We show that this ε is the first point of the arc on the straight part. First, if $a \le s < \varepsilon$, then by definition of sup there exists δ such that $s < a + \delta \le \varepsilon$ and $\mathbf{f}(t)$ is on the curved part for $t \in [a, a + \delta)$; in particular, $\mathbf{f}(s)$ is on the curved part. Second, $\pi_1(\mathbf{f}(\varepsilon))$ cannot be positive. If it were, then we would have $\varepsilon < b$, plus there would be a neighborhood $(\varepsilon - \Delta, \varepsilon + \Delta)$ in which $\pi_1(\mathbf{f}(t))$ would stay positive, contradicting the definition of ε. Hence $\pi_1(\mathbf{f}(\varepsilon)) = 0$; $\mathbf{f}(\varepsilon)$ is on the y-axis.

Alternatively: Put $\varepsilon := \inf\{t\colon \pi_1(\mathbf{f}(t)) = 0\}$. The indicated set is the inverse image of $\{0\}$ under $\pi_1(\mathbf{f})$, so it is closed. Therefore, its inf is its least element. That makes $\pi_1(\mathbf{f}(\varepsilon)) = 0$ and $\pi_1(\mathbf{f}(t)) > 0$ for $t < \varepsilon$, answering (a) along with (b).

(c) Let $(s, \sin 10/s)$, $0 < s \le 10/\pi$, be a point of the curved part up to $\mathbf{f}(a)$. Since the x-value $\pi_1(\mathbf{f}(t))$ is a continuous function of t, with values $10/\pi$ when $t = a$ and 0 when $t = \varepsilon$, it must achieve the intermediate value s at some $t = c$. Since $\pi_1(\mathbf{f}(c)) = s$ and $\mathbf{f}(c)$ is on the curve, necessarily $\mathbf{f}(c) = (s, \sin 10/s)$.

(d) Observe that the order has to be $a < t_5 < t_9 < \cdots < \varepsilon$. That is, $\pi_1(\mathbf{f}(t_9)) = 20/9\pi < 20/2\pi = \pi_1(\mathbf{f}(a))$, so the intermediate value $20/5\pi$ must have been achieved between $t = a$ and $t = t_9$; that proves $a < t_5 < t_9$, and similarly for the other t_{1+4i}. Thus, (t_{1+4i}) is an increasing sequence, converging to some $d \le \varepsilon$. Since $\pi_1(\mathbf{f}(d)) = \lim \pi_1(\mathbf{f}(t_{1+4i})) = \lim 20/(\pi + 4\pi i) = 0$, and ε is the first place where $x = 0$, we have $d \ge \varepsilon$. We see that $(t_{1+4i}) \to \varepsilon$.

(e) Same argument as (d).

(f) We now have $\pi_2(\mathbf{f}(\varepsilon)) = \lim_{i\to\infty} \pi_2(\mathbf{f}(t_{1+4i})) = 1$ and $\pi_2(\mathbf{f}(\varepsilon)) = \lim_{i\to\infty} \pi_2(\mathbf{f}(t_{3+4i})) = -1$.

2. The interval $(\infty, -\infty)$ is empty, and by convention $\inf \emptyset = \infty$, $\sup \emptyset = -\infty$.

3. (a) See the proof of Theorem 5.5(b).

   (b) See the solution to Exercise 8b in Section 4.5.

4. (a) Exercise 8a in Section 4.5 shows that JOINS is reflexive, and Exercise 8c shows that it is symmetric. Exercise 3a in this section shows it to be transitive.

   (b) Let $J_{\mathbf{a}}$ be an equivalence class. Any two $\mathbf{x}, \mathbf{y} \in J_{\mathbf{a}}$ are related to $\mathbf{a}$, so there are arcs within O from $\mathbf{a}$ to $\mathbf{x}$ and from $\mathbf{a}$ to $\mathbf{y}$. Clearly, any point along these arcs is joined to $\mathbf{a}$ by part of that arc, so these points are in $J_{\mathbf{a}}$. Then we put the arcs together (Exercises 8c in Section 4.5 and 3 in Section 5.1) to give us one arc within $J_{\mathbf{a}}$ from $\mathbf{x}$ to $\mathbf{y}$. We conclude that $J_{\mathbf{a}}$ is arc-connected.

(c) Assume $J_{\mathbf{a}} \cup \{\mathbf{b}\} \subseteq T$ with $\mathbf{b} \notin J_{\mathbf{a}}$. By definition of equivalence class, $\mathbf{b}$ is not related to $\mathbf{a}$. That is, no arc within S joins $\mathbf{a}$ to $\mathbf{b}$. Hence T is not arc-connected, and $J_{\mathbf{a}}$ is maximal.

Section 5.2

1. If $t \in (0, 1]$, then $0 < t \leq 1 < 1 + 1/i$ for each i, so t is in the intersection. Conversely, if t is in the intersection, then $0 < t < 1 + 1/i$ for each i; passing to the limit in the second inequality, we have $t \leq 1$, and $t \in (0, 1]$.

2. (a) Assume $x < 0$. Then $(1 - \alpha)(-1, 0) + \alpha(x, y) = (\alpha - 1 + \alpha x, \alpha y)$ has (strictly) negative x-coordinate for every $0 \leq \alpha \leq 1$.

   (b) Assume $x > 0$, and suppose $\mathbf{g} : [r, s] \to \mathbf{R}^2$ is a continuous function with $\mathbf{g}(r) = (-1, 0), \mathbf{g}(s) = (x, y)$. Then $\pi_1(\mathbf{g})$ is continuous on $[r, s]$, with $\pi_1(\mathbf{g}(r)) = -1, \pi_1(\mathbf{g}(s)) = x > 0$. By the intermediate value theorem, there exists $c \in (r, s)$ with $\pi_1(\mathbf{g}(c)) = 0$. Then $\mathbf{g}(c)$ is a point of the arc on the y-axis.

3. For $x > 0$, $f = 0$ at $x = 1, \frac{1}{2}, \frac{1}{3}, \ldots$. In each of the intervening intervals, f must be of one sign. The intervals of positivity are $(1, \infty)$, $\left(\frac{1}{3}, \frac{1}{2}\right)$, $\left(\frac{1}{5}, \frac{1}{4}\right), \ldots$. Since f is even, we add the corresponding ones on the $x < 0$ side.

4. (a) The sum of the lengths is $\frac{1}{3} + \frac{2}{9} + \frac{4}{27} + \cdots = \left(\frac{1}{3}\right) / \left(1 - \frac{2}{3}\right) = 1$.

   (b) Let $C := [0, 1] - O$. Since $C = [0, 1] \cap O^*$, C is closed. C is also infinite, since the components of O do not abut (do not share endpoints), and those endpoints are in C. On the other hand, (a) says that in some sense C occupies 0% of the space in $[0, 1]$. So C is weird, and deserves more study.

5. (a) and (b) In the figure, the shaded portion shows $Q_2 \cup Q_3 \cup Q_4$ near the unit square. It makes clear that each Q_i crosses the "main canal" Q_2. Accordingly, O is arc-connected, and has just one component.

   (c) By DeMorgan's law, $S - O = S \cap Q_2^* \cap Q_3^* \cap \cdots$. For each power-of-two group (2, then 3, 4, then 5, 6, 7, 8), its intersection cuts the square(s) from the group before into four times as many, each one-third as big. In symbols, $S \cap Q_2^*$ has four squares of size $\frac{1}{3} \times \frac{1}{3}$, $S \cap Q_2^* \cap Q_3^* \cap Q_4^*$ has sixteen $\frac{1}{9} \times \frac{1}{9}$, and so on. Hence for each i, $S - O$ is subset of a union of squares with total area $4^i/9^i$; O occupies 100% of the unit area of S.

6. (a) In the figure, the shaded portion is $P_1 \cup \cdots \cup P_{1+8+64}$. The picture suggests that the gasket is a square with an infinity of square holes.

   (b) Yes. Since the P_i are disjoint convex open sets, they must be the components of O.

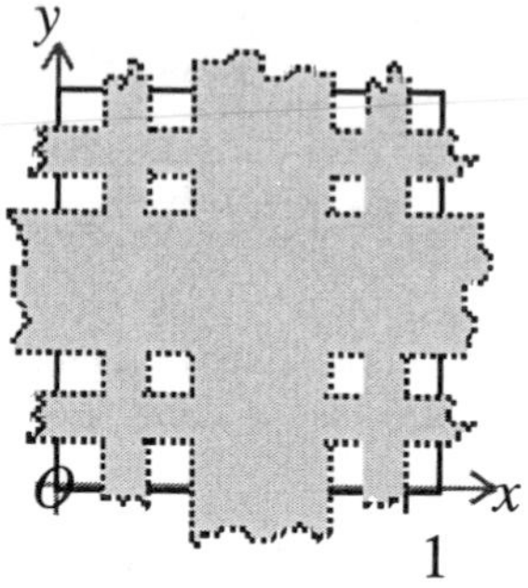

Exercise 5

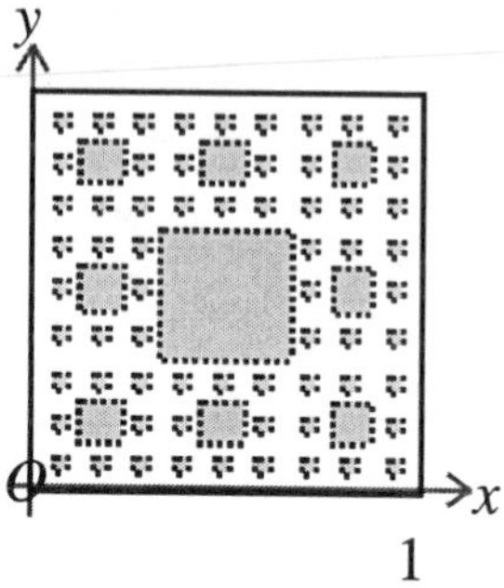

Exercise 6

(c) P_1 has area $\left(\frac{1}{3}\right)^2$, P_2 through P_{1+8} all have area $\left(\frac{1}{9}\right)^2$, and so on. Thus: 1 square of area $\left(\frac{1}{3}\right)^2$, 8 squares of area $\left(1/3^2\right)^2$, 64 squares of area $\left(1/3^3\right)^2, \ldots$; total area $1/3^2 + 8/3^4 + 64/3^6 + \cdots = \left(1/3^2\right)/\left(1 - 8/3^2\right) = 1$. The holes have all the area.

(d) No. All these constructions behave alike: If some neighborhood $N(\mathbf{b}, \delta) \subseteq S$ were free of points of O, then the area occupied by O would be $1 - \pi\delta^2$ or less.

7. (a) This is trivial: The way we defined "component" (elements joined to **a**) matches the definition of "equivalence class" (elements related to **a**).

(b) A set harboring an equivalence relation is always the disjoint union of the equivalence classes, and by (a) the classes are the components.

(c) See the solution to Exercise 4b in Section 5.1.

8. (a) $T := \{\mathbf{O}\}$ has only one equivalence class, and it is not open.

(b) Remember S from Example 1 in Section 5.1 and Exercise 1 in that section. Clearly, you can join any point of the straight part to (0, 0), and you can join any two points in the curved part. We decided that you cannot join from the curved part to (0, 0). Hence the MACS (equivalence classes) are the curved part and the straight part. But neither is alone a maximal connected subset, because S is connected.

(c) Let $U := \{(x, y)\colon x \text{ is irrational}\}$. Then each vertical line contained in U is an equivalence class under JOINS, and the lines are uncountably numerous.

Section 5.3

1. (a) Let Σ be a class of closed sets and $S := \bigcap_{C \in \Sigma} C$ their intersection. Then each complement C^* is open (Theorem 4.14), the union $T :=$

$\bigcup_{C \in \Sigma} C^*$ of the complements must be open (Theorem 5.3(a)), and $T = S^*$ (DeMorgan). Thus, S has open complement, so S is closed.

(b) Same argument with $\Sigma := \{C_1, \ldots, C_k\}$ and Theorem 5.3(b).

2. (a) $B(\mathbf{O}, 1) \cup B\left(\mathbf{O}, \frac{1}{2}\right) \cup B\left(\mathbf{O}, \frac{1}{3}\right) \cup \cdots = B(\mathbf{O}, 1)$, which is closed (Exercise 4b in Section 4.2) but not open (any neighborhood $N((1, 0), \delta)$ of $(1, 0)$ has the point $(1 + \delta/2, 0)$ from outside $B(\mathbf{O}, 1)$).

(b) $B\left(\mathbf{O}, \frac{2}{3}\right) \cup B\left(\mathbf{O}, \frac{3}{4}\right) \cup B\left(\mathbf{O}, \frac{4}{5}\right) \cup \ldots = N(\mathbf{O}, 1)$ (Explain!), which is open but not closed.

(c) The rings $\left\{(x, y)\colon 1/(i + 1) \le x^2 + y^2 \le 1/i\right\}$, $i = 1, 2, \ldots$, add up to $\left\{(x, y)\colon 0 < x^2 + y^2 \le 1\right\}$ (Explain!), which is neither open nor closed (Explain, too!).

(d) $B(\mathbf{O}, 1) \cup B(\mathbf{O}, 2) \cup B(\mathbf{O}, 3) \cup \cdots = \mathbf{R}^2$. By Exercise 6 in Section 4.5, the only open and closed candidates are $\mathbf{R}^2$ and $\emptyset$, and it is impossible to express $\emptyset$ as the union of unequal sets.

3. (a) T is not open, because every neighborhood $N(\mathbf{O}, \delta)$ has points $(0, \delta/2)$ from outside T. It is not closed, because $\mathbf{x}_i := ([1 + 1/i]^i, 0) \to (e, 0) \notin T$.

(b) Every set is the union of its MACS (Exercise 7b in Section 5.2). If the MACS are closed and finitely numerous, then the set is the finite union of closed sets, and is therefore closed.

(c) See the solution to Exercise 8c in Section 5.2. (Why is each line there, like $\left\{\left(\sqrt{2}, y\right)\colon y \in \mathbf{R}\right\}$, closed?)

(d) No. MACS are disjoint, so a collection of open MACS is a collection of disjoint open sets. The argument in Theorem 5.7 applies: Every such collection has to be countable.

4. By Example 2(b), if $b < c$ are in C, then there exists $z \notin C$ with $b < z < c$. Then $\{C \cap (-\infty, z), C \cap (z, \infty)\}$ is a disconnection of C.

5. See it geometrically and analytically. (Consult Example 2.)

For the geometric look, let $c \in C$. Since no two component subintervals from O abut, c cannot be the endpoint of subintervals on both sides. Thus, on one side—say the right—every interval $(c, c + \varepsilon)$ has a point $x \in C$. The interval (c, x) has a point y from O, because C contains no intervals (Example 2(b)). The component of O to which y belongs must be a subset of (c, x), because otherwise (if it stretched left of c or right of x) either c or x would be in it. That gives evidence for (c), which implies the others.

Analytically, let $c \in C$ have the ternamal $\#a_1a_2\ldots$, with each $a_i = 0$ or 2. Then $x_i := \#a_1a_2\ldots a_i111\ldots$, which is not in C, is less than 3^{-i} from c. Hence $x_i \in N(c, \varepsilon)$ for big enough i, proving (a). The same neighborhood has $y_i := \#a_1a_2\ldots a_i1000\ldots$ and $z_i := \#a_1a_2\ldots a_i1222\ldots$. Those are in C, because the 1's are removable. The interval (y_i, z_i) is necessarily a component of O, because anything in it, having an irremovable 1 in ternamal place $i+1$, is in O. That proves (b). It also proves (c), because no two y_i match.

6. (a) Map the endpoints x_i, y_i of component P_i (defined in Example 3 in Section 5.2) into $2i-1, 2i$, respectively. The mapping is well-defined, because no component of O shares an endpoint with any of its predecessors, and is clearly one-to-one.

   (b) Expansions have two properties: Each real $r \in [0, 1]$ has a "binimal" $r = \&b_1b_2\ldots := b_1/2 + b_2/4 + b_3/8 + \cdots$, where every b_i is 0 or 1; and the binimal is unique if we require that the sequence (b_i) not end in $0, 0, \cdots$. With that principle in mind, we see that $f(r) := (2b_1)/3 + (2b_2)/9 + (2b_3)/27 + \cdots = \#(2b_1)(2b_2)\ldots$ maps r into a ternamal with only 0's and 2's. Since two such ternamals are equal only if produced by identical sequences $(2b_i)$, f maps $[0, 1]$ into C one-to-one.

   (c) If C were countable, there would be a one-to-one $g\colon C \to \mathbf{N}$. But then $g(f)$ (using f from part (b)) would map $[0, 1]$ one-to-one into $\mathbf{N}$. That would contradict the uncountability of $[0, 1]$.

Section 5.4

1. (a) The y-axis has empty interior, because the neighborhood $N((0, b), \delta)$ of point $(0, b)$ has the point $(\delta/2, b)$ not on the axis. Hence all its points are boundary points. There are no other boundary points, because if $a \neq 0$, then $N((a, b), |a|/2)$ has no points of the axis. Hence bd(y-axis) $=$ y-axis.

   (b) Quadrant I is an open set (Exercise 1 in Section 4.4), so it is its own interior. Its boundary is the union of the nonnegative axes: Near $(0, b)$, with $b \geq 0$, we have $(\varepsilon/2, b)$ from quadrant I and $(-\varepsilon/2, b)$ from outside it; similarly near $(a, 0)$, with $a \geq 0$; and if either $c < 0$ or $d < 0$, then the smaller of $N((c, d), |c|)$ and $N((c, d), |d|)$ is devoid of quadrant I points.

   (c) We have seen (Example 2 in Section 3.4, among others) that every neighborhood has points from $\mathbf{Q}^2$ and from its complement. That tells us that bd($\mathbf{Q}^2$) is all of $\mathbf{R}^2$. Then nothing is left for int($\mathbf{Q}^2$), so this set must be empty.

2. (a) Every ball is closed (Exercise 4b in Section 4.2); by Theorem 5.11(b), a ball equals its closure. This means that its complement is open, so its exterior := interior of complement = complement.

   (b) $\mathrm{cl}(N(\mathbf{b}, \delta)) = B(\mathbf{b}, \delta)$ and $\mathrm{ext}(N(\mathbf{b}, \delta))$ is the rest of $\mathbf{R}^2$ (outside the ball). The ball $B(\mathbf{b}, \delta)$ is a closed superset of $N(\mathbf{b}, \delta)$, so $B(\mathbf{b}, \delta) \subseteq$

cl($N(\mathbf{b}, \delta)$) (Theorem 5.11(d)). By the argument in Example 2(a), the points outside the ball are not closure points of the ball, let alone of $N(\mathbf{b}, \delta)$, so cl($N(\mathbf{b}, \delta)$) $\subseteq B(\mathbf{b}, \delta)$. We conclude both that the closure is the ball and that the complement of the ball, being open, is ext($N(\mathbf{b}, \delta)$).

(c) Closure = line, exterior = rest. Lines are closed sets: If (x_i, y_i) is a sequence from the line given by $ax + by = c$ and $(x_i, y_i) \to (s, t)$, then $as + bt = \lim(ax_i + by_i) = c$, so (s, t) is on the line. Hence the closure is the line, and exterior = interior of (open) complement = complement.

(d) Adapt the argument from the solution to Exercise 1(b): The points on the nonnegative axes $\{(x, 0)\colon x \geq 0\}$ and $\{(0, y)\colon y \geq 0\}$ are boundary points, and the points outside $[\mathbf{O}, \infty)$ are exterior. Consequently cl(quadrant I) $= [\mathbf{O}, \infty)$, and the rest is the exterior.

(e) Since every point is a boundary point of $\mathbf{Q}^2$, the closure is $\mathbf{R}^2$ and the exterior is $\emptyset$.

3. (a) int(int(S)) = int(S). The interior is open (Theorem 5.9(c)), so it is its own interior (Theorem 5.9(b)).

(b) int(ext(S)) = ext(S). The exterior is by definition int(S^*), which by (a) is its own interior.

(c) ext(bd(S)) = int(S) $\cup$ ext(S). The boundary is closed (Theorem 5.10). Hence its complement is open, is therefore its own interior, is therefore ext(bd(S)). As our figures have shown, the complement bd$(S)^*$ is int(S) $\cup$ ext(S).

4. Exercises 4 and 5 are intended to show that the six combinations not covered in Exercise 3 are as we would expect for a neighborhood, but can be strange in general. Let $N := N(\mathbf{b}, \delta)$ and $S := S(\mathbf{b}, \delta)$. Then:

(a) int(bd(N)) $= \emptyset$. The boundary is the sphere: Points in N are interior (N is open), points outside the ball are exterior (Exercise 2b), and $\mathbf{y} \in S$ has nearby points $\mathbf{y} \pm .5\varepsilon(\mathbf{y} - \mathbf{b})/\delta$ from inside and outside N. Those same points are off the sphere, so they tell us that the sphere has no interior points. Hence int(bd(N)) = int(S) $= \emptyset$.

(b) bd(int(N)) = bd(N). The interior of N is N, so its boundary is that of N (which is S by the previous paragraph).

(c) bd(bd(N)) = bd(N). First, bd(N) = S. Second, we have seen that S has no interior, so every point of S is a boundary point. Last, there are no other boundary points of S, because points of N are surrounded by N and points outside the ball are surrounded by outsiders. Hence bd(S) = S.

(d) bd(ext(N)) = bd(N). Points in N are surrounded by N, because N is open. Similarly, points in ext(N) are surrounded by ext(N). That leaves S, whose points have nearby points from ext(N) (Example 2(a) or part (a) here). Hence bd(ext(N)) $= S$.

(e) ext(int(N)) $=$ ext(N). int(N) is N, so its exterior is ext(N) ($=$ complement of ball, by Exercise 2a).

(f) ext(ext(N)) $=$ int(N). We know ext$(N)^* =$ (complement of ball)$^* =$ ball. Therefore, ext(ext(N)) $:=$ int(ext$(N)^*$) $=$ int(ball) $= N =$ int(N).

5. (a) and (c) In Example 2(b), we saw that bd(T) is the annulus consisting of the unit circle, the radius-2 circle, and the points in between. From the arguments in Exercise 4, it is clear that the in-betweens are interior to this set and the two circles form its boundary. Hence (a) int(bd(T)) $= \{\mathbf{x}\colon 1 < \|\mathbf{x}\|_2 < 2\}$, which is not empty; and (c) bd(bd(T)) $= \{\mathbf{x}\colon \|\mathbf{x}\|_2 = 1 \text{ or } 2\}$, which is only a subset of bd(T).

   (b) and (e) Taking from T the part in the boundary annulus, we see that int(T) $= N(\mathbf{O}, 1)$. Hence (b) bd(int(T)) $= S(\mathbf{O}, 1)$, a subset of bd(T); and (e) ext(int(T)) $=$ complement of unit ball, a superset of ext(T).

   (d) and (f) We have identified int(T) and bd(T), so what is left is exterior: ext(T) $= \{\mathbf{x}\colon \|\mathbf{x}\|_2 > 2\}$. Hence (d) bd(ext($T$)) $=$ radius-2 circle, a subset of bd(T); and (f) ext(ext(T)) $= N(\mathbf{O}, 2)$, a superset of int(T).

   (In these last three paragraphs, those extra remarks about "subset" and "superset" are not coincidences; check that they are always true.)

6. For (c): We know that C is closed, so cl(C) $= C$. For the others, in view of Exercise 5 in Section 5.3:

   (a) int(C) is empty.

   (b) bd(C) $= C$. Since there are no interior points, $C \subseteq$ bd(C). Since the definition of boundary implies that bd(C) $\subseteq$ cl(C), we have bd(C) $\subseteq C$.

   (d) Set is empty. Every $c \in C$ has nearby intervals from C^*, an infinity of these intervals ending in Cantor points $\neq c$. Hence no point of C is isolated.

   (e) Set is C. By (d), every $c \in C$ is an accumulation point of C.

7. (c) Suppose $(\mathbf{x}_i)$ is a sequence from cl(S) converging to $\mathbf{x}$. For any $N(\mathbf{x}, \varepsilon)$, there is I such that $\mathbf{x}_I, \mathbf{x}_{I+1}, \ldots$ are in $N(\mathbf{x}, \varepsilon)$. Since $\mathbf{x}_I$ is in the open set $N(\mathbf{x}, \varepsilon)$, there is a neighborhood $N(\mathbf{x}_I, \delta) \subseteq N(\mathbf{x}, \varepsilon)$. Because $\mathbf{x}_I$ is a closure point of S, some $\mathbf{y} \in S$ must be in $N(\mathbf{x}_I, \delta)$ (Exercise 5 of Section 3.2). This $\mathbf{y}$ is a member of S inside $N(\mathbf{x}, \varepsilon)$. We have shown that $\mathbf{x} \in$ cl(S), leading to the conclusion that cl(S) is closed.

   (d) By part (a), $S \subseteq$ cl(S), and by (c), cl(S) is closed. Thus, cl(S) is a closed set containing S.

Suppose T is another such set. If $\mathbf{x} \in \text{cl}(S)$, then by definition there exists $(\mathbf{x}_i)$ from S converging to $\mathbf{x}$. Since $S \subseteq T$, $(\mathbf{x}_i)$ comes from T. The latter being closed, $(\mathbf{x}_i) \to \mathbf{x}$ forces $\mathbf{x} \in T$. We have proved that $\text{cl}(S)$ is the smallest closed superset.

8. The key is Theorem 3.1. Using that result, we have $\mathbf{x} \in \text{cl}(S)^* \Leftrightarrow \mathbf{x} \notin \text{cl}(S)$ $\Leftrightarrow$ there exists $N(\mathbf{x}, \varepsilon)$ devoid of points of $S \Leftrightarrow$ there exists $N(\mathbf{x}, \varepsilon) \subseteq S^*$ $\Leftrightarrow \mathbf{x} \in \text{int}(S^*)$.

9. (a) Apply Theorem 3.1 again: $\mathbf{x} \in \text{cl}(S) \Leftrightarrow$ every neighborhood of $\mathbf{x}$ has points from $S \Leftrightarrow$ (every neighborhood has points from both S and S^*) or (every neighborhood has points from S and some neighborhoods have points from only S). The first clause in parentheses says that $\mathbf{x} \in \text{bd}(S)$, the second says that $\mathbf{x} \in \text{int}(S)$, and their disjunction ("or") amounts to $\mathbf{x} \in \text{bd}(S) \cup \text{int}(S)$.
   (b) By (a), $\text{cl}(S) = \text{bd}(S) \cup \text{int}(S)$. We know that $\text{int}(S)$ is contained in S. Therefore, $\text{cl}(S) \subseteq S$ iff $\text{bd}(S) \subseteq S$. By parts (a) and (b) of Theorem 5.11, $\text{bd}(S) \subseteq S$ iff S is closed.
   (c) By Theorem 5.9(b), S is open iff every point of S is an interior point. Since every set partitions into interior points and boundary points, we conclude that S is open iff no point of S is in $\text{bd}(S)$.

10. (a) Let $(\mathbf{x}_i)$ be a sequence of accumulation points of S converging to $\mathbf{b}$. Any neighborhood $N(\mathbf{b}, \varepsilon)$ possesses some $\mathbf{x}_I, \mathbf{x}_{I+1}, \ldots$. If $\mathbf{x}_I = \mathbf{b}$, then $\mathbf{b}$ is already an accumulation point of S. If not, let $\delta := \min\{\|\mathbf{x}_I - \mathbf{b}\|, \varepsilon - \|\mathbf{x}_I - \mathbf{b}\|\}$. Then (draw the picture) $N(\mathbf{x}_I, \delta) \subseteq N(\mathbf{b}, \varepsilon)$ and does not reach $\mathbf{b}$. Within $N(\mathbf{x}_I, \delta)$, there must exist $\mathbf{y} \neq \mathbf{x}_I$ from S. This $\mathbf{y}$ cannot be $\mathbf{b}$, so we have found $\mathbf{y} \neq \mathbf{b}$ from S in $N(\mathbf{b}, \varepsilon)$. Hence $\mathbf{b}$ is an accumulation point; the set of such points is closed.
   (b) Partition the space into $\text{int}(S)$, $\text{bd}(S)$, and $\text{ext}(S)$. The points of $\text{int}(S)$ are accumulation points of S but not S^*; those in $\text{ext}(S)$ are accumulators of S^* but not of S. In the boundary, there are points of S surrounded by S^*, meaning isolated points of S, which are accumulators of S^* but not S; points of S^* surrounded by S, which are accumulators of S but not S^*; and points with neighbors from both S and S^*, which are accumulation points for both, irrespective of which they belong to.
   (c) No. $S := \left\{\left(\frac{1}{2}, 0\right), \left(\frac{1}{3}, 0\right), \ldots\right\}$ has only isolated points, and is not closed. (Why?)

11. (a) Yes to all. The closure of the interior of a ball is the ball (equal); cl(int(line)) is empty (smaller); cl(int(neighborhood)) = ball (bigger); $S := N(\mathbf{O}, 1) \cup \{(2, 0)\} \subseteq \mathbf{R}^2$ has $\text{int}(S) = N(\mathbf{O}, 1)$, $\text{cl}(\text{int}(S)) = B(\mathbf{O}, 1)$ (not subset of S, not superset).
   (b) Yes again: int(cl(neighborhood)) = neighborhood; int(cl(line)) = empty; $\text{int}(\text{cl}(\mathbf{Q}^2)) = \mathbf{R}^2$; $S := \{\mathbf{x}\colon 0 < \|\mathbf{x}\| \le 1\}$ has $\text{int}(\text{cl}(S)) = \{\mathbf{x}\colon 0 \le \|\mathbf{x}\| < 1\}$.

(c) First question: $\text{ext}(\text{cl}(S)) = \text{ext}(S)$ by Theorem 5.12 and Exercise 3b. Second question: $\text{cl}(\text{ext}(S))$ is $\text{int}(\text{ext}(S)) \cup \text{bd}(\text{ext}(S))$ by Exercise 9a. Of the last two, $\text{int}(\text{ext}(S))$ is predictable, being $\text{ext}(S)$ because $\text{ext}(S)$ is open; the other is not, as we see by Exercise 5d. It has to be a subset of $\text{bd}(S)$, because it cannot have any of $\text{int}(S)$ or $\text{ext}(S)$, and it would include any isolated points of S.

12. (a) If S is bounded, say $S \subseteq N(\mathbf{O}, M)$, then $B(\mathbf{O}, M)$ is a closed superset of S, so that $\text{cl}(S) \subseteq B(\mathbf{O}, M)$ (Theorem 5.11(d)). The closure is therefore bounded.

(b) Assume that S is convex and $\mathbf{x}, \mathbf{y} \in \text{cl}(S)$. By definition of closure, there exist sequences from S with $(\mathbf{x}_i) \to \mathbf{x}$ and $(\mathbf{y}_i) \to \mathbf{y}$. If α and β are nonnegative reals summing to 1, then $(\alpha\mathbf{x}_i + \beta\mathbf{y}_i)$ is a sequence from S (because S is convex) with $\lim(\alpha\mathbf{x}_i + \beta\mathbf{y}_i) = \alpha\mathbf{x} + \beta\mathbf{y}$. This shows that $\alpha\mathbf{x} + \beta\mathbf{y} \in \text{cl}(S)$, so that the closure is convex.

(c) The result is trivial if $S = \emptyset$, so assume that S is connected and nonempty. Let $\{R, T\}$ be a partition of $\text{cl}(S)$. Examine $R \cap S$ and $T \cap S$. If one of them is empty, say $R \cap S$, then $S \subseteq T$; that says that R, which must have closure points of S, has closure points of T. If instead both are nonempty, then they partition S. Because S is connected, Theorem 5.13 tells us that one, say $R \cap S$, has closure points of the other. This says that R has closure points of T. We have shown that if $\{R, T\}$ partitions $\text{cl}(S)$, then one of R, T holds closure points of the other. By Theorem 5.13, $\text{cl}(S)$ is connected.

Section 5.5

1. (a) The hypotheses make S closed and bounded. By the Heine–Borel theorem, a closed bounded subset of $\mathbf{R}^n$ is compact. Its continuous image is compact (Theorem 5.19), and this compact image is closed (Theorem 5.15).

(b) No. Any constant function maps open sets into nonopen sets.

(c) First, $\mathbf{f}^{-1}(T) = B(\mathbf{O}, 1) - \mathbf{f}^{-1}(T^*)$: $\mathbf{x} \in \mathbf{f}^{-1}(T) \Leftrightarrow \mathbf{f}(\mathbf{x}) \in T \Leftrightarrow \mathbf{f}(\mathbf{x}) \notin T^* \Leftrightarrow \mathbf{x}$ is in the part of $B(\mathbf{O}, 1)$ that does not map to T^*. Second, $\mathbf{f}^{-1}(T^*)$ is the intersection of $B(\mathbf{O}, 1)$ with some open set O: T is closed, so T^* is open, so $\mathbf{f}^{-1}(T^*)$ is $B(\mathbf{O}, 1) \cap O$ (Theorem 4.17). Therefore, $\mathbf{f}^{-1}(T) = B(\mathbf{O}, 1) - O = B(\mathbf{O}, 1) \cap O^*$. This last is an intersection of closed sets, so it is closed. Being also a bounded set in $\mathbf{R}^n$, it is compact.

You should see that the compactness of $B(\mathbf{O}, 1)$ is essential. Any constant function will have $\mathbf{f}^{-1}(\text{range}) = \text{domain}$, with "range" compact but not necessarily "domain."

2. (a) By Theorem 5.20, $\mathbf{f}$ can be extended continuously to $\text{cl}(S)$. The closure of a bounded set is bounded (Exercise 12 in Section 5.4), so $\text{cl}(S)$ is

compact. Therefore, $\mathbf{f}(\mathrm{cl}(S))$ is compact (Theorem 5.19), and its subset $\mathbf{f}(S)$ must be bounded.

(b) No. $\mathbf{R}^n$ is closed and $f(\mathbf{x}) := \tan^{-1} \|\mathbf{x}\|_2$ is uniformly continuous on it (by the mean value theorem), but $f(\mathbf{R}^n) = (-\pi/2, \pi/2)$.

(c) Yes to both. If S is closed and bounded in $\mathbf{R}^n$, then it is compact, and so must $\mathbf{f}(S)$ be.

(d) No, yes. See Exercises 1b and 2a, respectively.

(e) Yes. $f(\mathbf{x}) := 1/(1 - \|\mathbf{x}\|_2)$ maps $N(\mathbf{O}, 1)$ continuously onto $[1, \infty)$.

3. Assume that S is compact and $T \subseteq S$ is closed. Let $(\mathbf{x}_i)$ be a sequence from T. The $\mathbf{x}_i$ belong also to S, and S is sequentially compact (Theorem 5.15(b)). Hence there is $\mathbf{x} \in S$ with $\mathbf{x}_i \to \mathbf{x}$. We now have a sequence from T converging to $\mathbf{x}$, forcing $\mathbf{x} \in T$. We have proved that T is sequentially compact, and by the remarks following Theorem 5.16, T is compact.

4. We have here the hypothesis of Theorem 5.18, plus diameter $\to 0$. Consequently, we know that $T_1 \cap T_2 \cap \cdots$ is not empty. The only question is whether it can hold two different members. Suppose $\mathbf{x}, \mathbf{y}$ are in the intersection. Then $\mathbf{x}, \mathbf{y} \in T_i$ for each i, so that $\|\mathbf{x} - \mathbf{y}\| \le \mathrm{diam}(T_i) \to 0$. We conclude that $\mathbf{x} = \mathbf{y}$.

5. (a) Assume that S is compact. Let Ω be a family of closed subsets of S, and suppose that Ω has empty intersection: $\bigcap_{C\in\Omega} C = \emptyset$. By DeMorgan, $\bigcup_{C\in\Omega} C^*$ is the whole space. In particular, S is covered by $\{C^*\}$, and these sets are open. Hence some finite subcollection $C_1^*, \ldots, C_k^*$ suffices to cover S. That is, each $\mathbf{x} \in S$ is in $C_1^* \cup \cdots \cup C_k^*$, so that $\mathbf{x} \in C_1 \cap \cdots \cap C_k$ is impossible. Thus, from the assumption that $\bigcap_{C\in\Omega} C = \emptyset$, we constructed a finite subfamily $\{C_1, \ldots, C_k\}$ having empty intersection. By contraposition, if Ω has the finite intersection property, then it has nonempty intersection.

(b) $N(\mathbf{O}, 1) \subseteq \mathbf{R}^n$ is not compact, but if Ω is a family of closed subsets of $N(\mathbf{O}, 1)$ with the finite intersection property, then the members of Ω are also subsets of $B(\mathbf{O}, 1)$ and must therefore (part (a)) have nonempty intersection. (If this appears to contradict a theorem from topology, bear in mind that "closed subsets of $N(\mathbf{O}, 1)$" means "closed subsets of $\mathbf{R}^n$ that happen to be contained in $N(\mathbf{O}, 1)$.")

6. (a) Let $\mathbf{y}$ be any member of S. Recall that if $\delta_{\mathbf{y}} := \|\mathbf{x}-\mathbf{y}\|/2$, then the neighborhoods $N(\mathbf{x}, \delta_{\mathbf{y}})$ and $N(\mathbf{y}, \delta_{\mathbf{y}})$ are disjoint. The family $\{N(\mathbf{y}, \delta_{\mathbf{y}})\colon \mathbf{y} \in S\}$ is clearly an open cover for S. There is therefore a finite subcover; label it $N(\mathbf{y}_1, \delta_1), \ldots, N(\mathbf{y}_k, \delta_k)$. If we set $\delta :=$ least of $\delta_1, \ldots, \delta_k$, then $O_1 := N(\mathbf{x}, \delta)$ does not intersect any of the $N(\mathbf{y}_j, \delta_j)$. Consequently, $O_2 := N(\mathbf{y}_1, \delta_1) \cup \cdots \cup N(\mathbf{y}_k, \delta_k)$ is an open set with $S \subseteq O_2$, $O_1 \cap O_2 = \emptyset$, and $\mathbf{x} \in O_1$.

(b) Let $\mathbf{z}$ be any member of S. By (a), there exist disjoint open sets $O_1(\mathbf{z})$ and $O_2(\mathbf{z})$ with $\mathbf{z} \in O_1(\mathbf{z})$, $T \subseteq O_2(\mathbf{z})$. The collection $\{O_1(\mathbf{z})\}$ covers S, so there exists a subcover $O_1(\mathbf{z}_1), \ldots, O_1(\mathbf{z}_m)$. Each corresponding $O_2(\mathbf{z}_j)$ contains T, so $T \subseteq O_2 := O_2(\mathbf{z}_1) \cap \cdots \cap O_2(\mathbf{z}_m)$. O_2, being a subset of every $O_2(\mathbf{z}_j)$, is disjoint from every $O_1(\mathbf{z}_j)$. Hence O_2 is disjoint from their union: $O_1 := O_1(\mathbf{z}_1) \cup \cdots \cup O_1(\mathbf{z}_m)$ does not meet O_2, and $S \subseteq O_1$.

(c) In $\mathbf{R}^2$, $\mathbf{x} := \mathbf{O}$, $T := \{\mathbf{O}\}$, $S :=$ quadrant I offer counterexamples for (a) and (b).

7. We can look back to Exercise 9b in Section 4.3. The hypotheses here give us S sequentially compact, T closed and disjoint from S. Hence $\delta := d(S, T)/2 > 0$. For any $\mathbf{x} \in S$ and $\mathbf{y} \in T$, we must have $\|\mathbf{x} - \mathbf{y}\| \geq 2\delta$. Therefore, $N(\mathbf{x}, \delta)$ does not reach $N(\mathbf{y}, \delta)$, and $O_1 := \bigcup_{\mathbf{x} \in S} N(\mathbf{x}, \delta)$ does not intersect $O_2 := \bigcup_{\mathbf{y} \in T} N(\mathbf{y}, \delta)$. (Why are O_1, O_2 open?)

8. The distance function $d(\mathbf{x}, S)$ is a good start, since it is constantly 0 on S and is at least $\varepsilon := d(S, T) > 0$ on T (Exercise 9b in Section 4.3). It is also continuous, in fact, contractive (Exercise 5 in Section 4.3). Set $f(\mathbf{x}) := d(\mathbf{x}, S)/\varepsilon$ if $d(\mathbf{x}, S) < \varepsilon$, $:= 1$ if $d(\mathbf{x}, S) \geq \varepsilon$. Clearly, $f = 0$ on S, $f = 1$ on T. The reason f is continuous is that f is everywhere the smaller of $d(\mathbf{x}, S)/\varepsilon$ and 1. Whenever you have continuous functions g and h, the function $\min\{g(\mathbf{x}), h(\mathbf{x})\}$ is necessarily continuous: $\min\{g(\mathbf{x}), h(\mathbf{x})\} = (g(\mathbf{x}) + h(\mathbf{x}))/2 - |g(\mathbf{x}) - h(\mathbf{x})|/2$, and the expression on the right is a continuous composite.

Section 5.6

1. $\Leftarrow$ Suppose V has finite dimension. We know, from considerable experience, that $B(\mathbf{O}, 1)$ is closed and bounded. By Theorem 4.29(b), $B(\mathbf{O}, 1)$ has the EVP. By Theorem 5.16, it is compact.

   $\Rightarrow$ Suppose V has infinite dimension. By Theorem 5.23, V has a sequence $(\mathbf{u}_i)$ of unit vectors separated by distance ≥ 1. That separation means that $(\mathbf{u}_i)$ has no Cauchy subsequences (same argument used in Theorem 5.24(a)). Therefore $(\mathbf{u}_i)$ has no convergent subsequences. Thus, $B(\mathbf{O}, 1)$ has a sequence with no sublimits; it is not sequentially compact.

2. Compare the solution to Exercise 3 in Section 4.6.

   (a) Let P_2 be the set of all quadratic polynomials over the domain $[0, 1]$. P_2 is a vector space of dimension 3. Hence all norms on P_2 are equivalent. In particular, $\|p(x)\|_1 := |b| + |c| + |d|$ and $\|p(x)\|_0 := \sup_{-1 \leq x \leq 1} |p(x)|$ are equivalent, so there exists M with $\|p\|_1 \leq M\|p\|_0$ independent of p. The subset (not subspace) P is precisely the $\| \; \|_0$-unit ball: $P = \{p\colon \|p\|_0 \leq 1\}$. Hence for any $p \in P$, we have $|b| \leq \|p\|_1 \leq M\|p\|_0 \leq M$.

(b) Set $p_i(x) := (1-x)^i$. Clearly, $0 \le p_i(x) \le 1$ for every i and x, but the second-degree coefficient in $p_i(x)$ is $i(i-1)/2$.

3. This is a standard vector-space result. The expression $\mathbf{p}$ in brackets is the projection of $\mathbf{x}$ onto the subspace spanned by $\mathbf{u}_1, \ldots, \mathbf{u}_k$. (See the remarks between Theorems 5.21 and 5.22.)

   For any j, $(\mathbf{x}-\mathbf{p}) \bullet \mathbf{u}_j = \mathbf{x} \bullet \mathbf{u}_j - \mathbf{p} \bullet \mathbf{u}_j = \mathbf{x} \bullet \mathbf{u}_j - ([\mathbf{x} \bullet \mathbf{u}_1]\mathbf{u}_1 \bullet \mathbf{u}_j + \cdots + [\mathbf{x} \bullet \mathbf{u}_k]\mathbf{u}_k \bullet \mathbf{u}_j)$. In the parentheses, each of $\mathbf{u}_1 \bullet \mathbf{u}_j, \ldots, \mathbf{u}_k \bullet \mathbf{u}_j$ is 0, save for $\mathbf{u}_j \bullet \mathbf{u}_j$, which is 1. Hence $(\mathbf{x}-\mathbf{p}) \bullet \mathbf{u}_j = \mathbf{x} \bullet \mathbf{u}_j - \mathbf{x} \bullet \mathbf{u}_j = 0$. Thus, $\mathbf{x}-\mathbf{p}$ is orthogonal to every $\mathbf{u}_j$. By linearity of the dot product, $\mathbf{x}-\mathbf{p}$ is orthogonal to any combination $\alpha_1\mathbf{u}_1 + \cdots + \alpha_k\mathbf{u}_k$.

4. Refer to Section 1.6, especially Example 1.

   (a) As we saw, the functions $s_j := \sin 2j\pi x$ are mutually orthogonal and of length $\sqrt{2}/2$. Therefore $\{\sqrt{2}s_j\}$ is an orthonormal set. Clearly, $\alpha_j = f \bullet \sqrt{2}s_j$ and $f_k := \alpha_1\sqrt{2}s_1 + \cdots + \alpha_k\sqrt{2}s_k$ is the projection of f onto the span of $\sqrt{2}s_1, \ldots, \sqrt{2}s_k$. By the principles cited between Theorems 5.21 and 5.22, f_k is the combination of $\sqrt{2}s_1, \ldots, \sqrt{2}s_k$ closest to f, making it likewise the closest combination of $s_1, \ldots, s_k$.

   (b) and (c) This is just the Pythagorean theorem. We have $f = (f - f_k) + f_k$. By Exercise 3, $f - f_k$ is orthogonal to f_k. Hence $\|f\|_2^2 = \|f - f_k\|_2^2 + \|f_k\|_2^2 \ge \|f_k\|_2^2$. The inequality is strict, unless f_k actually is f.

5. Write $C := B(\mathbf{u}_2, 1/2) \cup B(\mathbf{u}_3, 1/3) \cup \ldots$. Since each $\mathbf{u}_i$ has unit norm, the biggest possible norm in C is 1.5, so C is bounded. Also, since the centers are 1 or more apart, the closest two of the component balls can get is $\frac{1}{2} - \frac{1}{3} = \frac{1}{6}$. Let $(\mathbf{x}_i)$ be a sequence from C converging to $\mathbf{x}$. Then $(\mathbf{x}_i)$ is Cauchy, so there is I such that $\|\mathbf{x}_i - \mathbf{x}_j\| < \frac{1}{6}$ for any $i, j \ge I$. Of necessity, all of $\mathbf{x}_I, \mathbf{x}_{I+1}, \ldots$ come from a single ball $B(\mathbf{u}_k, 1/k)$. Since this ball is closed and $(\mathbf{x}_i) \to \mathbf{x}$, we conclude that $\mathbf{x} \in B(\mathbf{u}_k, 1/k) \subseteq C$; the latter is closed.

   Now define $f(\mathbf{y}) := 2k\|\mathbf{y}-\mathbf{u}_k\|$ for $\mathbf{y} \in B(\mathbf{u}_k, 1/k), k = 2, 3, \ldots$. Then f is continuous in each ball separately, because distance from $\mathbf{u}_k$ is continuous. That makes f continuous on C, because the test sequences (those with $\mathbf{y}_i \to \mathbf{y}$) must, as described above, come eventually from a single ball. However, f is not uniformly continuous, because the points $[1 + .25/k]\mathbf{u}_k$ and $[1 - .75/k]\mathbf{u}_k$ are "close together" (distance $= 1/k$) in C, but the function difference $f([1 - .75/k]\mathbf{u}_k) - f([1 + .25/k]\mathbf{u}_k) = 1$ is not "small."

6. Let $S := \{\mathbf{O}\}$ and $T := \{(4/3)\mathbf{u}_2, (5/4)\mathbf{u}_3, (6/5)\mathbf{u}_4, \ldots\}$. S obviously fits the bill. T is closed, because its members are far apart: Each member comes from a different one of the balls in Exercise 5, and those balls are at least $\frac{1}{6}$ from one another; therefore, you cannot make Cauchy sequences from T, except by repetition. But the members of T are at distances $\frac{4}{3}, \frac{5}{4}, \ldots$ from S; there is no minimum.

References

We have made references to just four books. Here we list them, together with comments on why they are our favorites.

Morris Kline, *Mathematical Thought from Ancient to Modern Times*, Oxford University Press, New York, 1972.

> Kline's book is a monumental achievement among histories of the sciences and mathematics. Although our focus on continuous functions restricts our interest in history to Europeans during (roughly) 1810–1870, Kline is a wonderful source of information about the development of mathematical ideas worldwide and over thousands of years.

David C. Lay, *Linear Algebra and Its Applications*, 2nd edition, Addison Wesley Longman, Reading MA, 1997.

> Lay is a good source for the material we have assumed from elementary linear algebra.

Kenneth A. Ross, *Elementary Analysis: The Theory of Calculus*, Springer–Verlag, New York, 1980.

> This is the best introduction to advanced calculus we know. The subtitle suggests that Ross's mission is an axiomatization of elementary calculus. The book carries out that mission with an admirable combination of mathematical rigor and attention to pedagogy, the latter absolutely essential in a student's entry to this level of theory.

H. L. Royden, *Real Analysis*, 3rd edition, Macmillan, New York, 1988.

This graduate-level text is best described as "elegant." It may therefore seem like cheating to refer to it in a treatment that claims to be elementary. Royden's first two chapters, however, are introductory, and happen to cover the properties of continuous functions of one real variable.

Index

Forthcoming by Alberto Guzman

Derivatives and Integrals of Multivariable Functions
ISBN: 0-8176-4274-9

This work examines derivatives and integrals of functions of several real variables. Topics from advanced calculus are covered, including: differentiability and its relation to partial derivatives, directional derivatives and the gradient, surfaces, inverse and implicit functions, integrability and properties of integrals, and the theorems of Fubini, Stokes, and Gauss. The order of topics reflects the order of development in calculus: limits, continuity, derivatives, integrals---a sequencing that allows generalizations from and analogies to elementary results, such as the second-derivative test and the Chain Rule.

Derivatives and Integrals of Multivariable Functions has a definition-theorem-proof format, together with a conversational style, including historical comments, an abundance of questions, and discussions about strategy, difficulties, and alternative paths. It is aimed at advanced undergraduate pure mathematics majors whose next course will be real analysis with measure theory. Required background includes theoretical work in linear algebra, one-variable calculus, properties of continuous functions, and related topological material. The last two are used in the context of Euclidean space, but a strong grounding in the corresponding real-line topics will suffice.

Table of Contents: